QUANTUM MECHANICS AND PATH INTEGRALS

Emended Edition

Richard P. Feynman
Albert R. Hibbs

Emended by Daniel F. Styer
Schiffer Professor of Physics
Oberlin College

DOVER PUBLICATIONS, INC.
MINEOLA, NEW YORK

Copyright

Copyright © 1965 by Richard P. Feynman and Albert R. Hibbs
Emended Edition © 2005 by Daniel F. Styer.
All rights reserved

Bibliographical Note

This Dover edition, first published in 2010, is an unabridged, emended republication of the work originally published in 1965 by McGraw-Hill Companies, Inc., New York.

Library of Congress Cataloging-in-Publication Data

Feynman, Richard Phillips.
Quantum mechanics and path integrals / Richard P. Feynman, Albert R. Hibbs, and Daniel F. Styer.— Emended ed.
p. cm.
Originally published: Emended edition. New York : McGraw-Hill, 2005.
Includes bibliographical references and index.
ISBN-13: 978-0-486-47722-0
ISBN-10: 0-486-47722-3
1. Quantum Theory. I. Hibbs, Albert R. II. Styer, Daniel F. III. Title.

QC174.12.F484 2010
530.12—dc22

2010004550

Manufactured in the United States by Courier Corporation
47722307 2014
www.doverpublications.com

Preface

The fundamental physical and mathematical concepts which underlie the path integral approach to quantum mechanics were first developed by R.P. Feynman in the course of his graduate studies at Princeton, although more fully developed ideas, such as those described in this volume, were not worked out until a few years later. These early inquiries were involved with the problem of the infinite self-energy of the electron. In working on that problem, a "least-action" principle using half advanced and half retarded potentials was discovered. The principle could deal successfully with the infinity arising in the application of classical electrodynamics.

The problem then became one of applying this action principle to quantum mechanics in such a way that classical mechanics could arise naturally as a special case of quantum mechanics when $\hbar$ was allowed to go to zero.

Feynman searched for any ideas which might have been previously worked out in connecting quantum-mechanical behavior with such classical ideas as the lagrangian or, in particular, Hamilton's principle function S, the indefinite integral of the lagrangian. During some conversations with a visiting European physicist, Feynman learned of a paper in which Dirac had suggested that the exponential function of $i\epsilon$ times the lagrangian was analogous to a transformation function for the quantum-mechanical wave function in that the wave function at one moment could be related to the wave function at the next moment (a time interval ϵ later) by multiplying with such an exponential function.

The question that then arose was what Dirac had meant by the phrase "analogous to," and Feynman determined to find out whether or not it would be possible to substitute the phrase "equal to." A brief analysis showed that indeed this exponential function could be used in this manner directly.

Further analysis then led to the use of the exponent of the time integral of the lagrangian, S (in this volume referred to as the *action*), as the transformation function for finite time intervals. However, in the application of this function it is necessary to carry out integrals over all space variables at every instant of time.

In preparing an article[1] describing this idea, the idea of "integral over all paths" was developed as a way of both describing and evaluating the required integrations over space coordinates. By this time a number of mathematical devices had been developed for applying the path integral technique and a number of special applications had been worked out, although the primary direction of work at this time was toward quantum electrodynamics. Actually, the path integral did not then provide, nor has it since provided, a truly satisfactory method of avoiding the divergence difficulties of quantum electrodynamics, but it has been found to be most useful in solving other problems in that field. In particular, it provides an expression for quantum-electrodynamic laws in a form that makes their relativistic invariance obvious. In addition, useful applications to other problems of quantum mechanics have been found.

The most dramatic early application of the path integral method to an intractable quantum-mechanical problem followed shortly after the

[1] R.P. Feynman, Space-Time Approach to Non-relativistic Quantum Mechanics, *Rev. Mod. Phys.*, vol. 20, pp. 367–387, 1948.

discovery of the Lamb shift and the subsequent theoretical difficulties in explaining this shift without obviously artificial means of getting rid of divergent integrals. The path integral approach provided one way of handling these awkward infinities in a logical and consistent manner.

The path integral approach was used as a technique for teaching quantum mechanics for a few years at the California Institute of Technology. It was during this period that A.R. Hibbs, a student of Feynman's, began to develop a set of notes suitable for converting a lecture course on the path integral approach to quantum mechanics into a book on the same subject.

Over the succeeding years, as the book itself was elaborated, other subjects were brought into both the lectures of Dr. Feynman and the book; examples are statistical mechanics and the variational principle. At the same time, Dr. Feynman's approach to teaching the subject of quantum mechanics evolved somewhat away from the initial path integral approach. At the present time, it appears that the operator technique is both deeper and more powerful for the solution of more general quantum-mechanical problems. Nevertheless, the path integral approach provides an intuitive appreciation of quantum-mechanical behavior which is extremely valuable in gaining an intuitive appreciation of quantum-mechanical laws. For this reason, in those fields of quantum mechanics where the path integral approach turns out to be particularly useful, most of which are described in this book, the physics student is provided with an excellent grasp of basic quantum-mechanical principles which will permit him to be more effective in solving problems in broader areas of theoretical physics.

R.P. Feynman

A.R. Hibbs

Preface to Emended Edition

In the forty years since the first publication of *Quantum Mechanics and Path Integrals*, the physics and the mathematics introduced here has grown both rich and deep. Nevertheless this founding book — full of the verve and insight of Feynman — remains the best source for learning about the field. Unfortunately, the 1965 edition was flawed by extensive typographical errors as well as numerous infelicities and inconsistencies. This edition corrects more than 879 errors, and many more equations are recast to make them easier to understand and interpret. Notation is made uniform throughout the book, and grammatical errors have been corrected. On the other hand, the book is stamped with the rough and tumble spirit of a creative mind facing a great challenge. The objective throughout has been to retain that spirit by correcting, but not polishing. This edition does not attempt to add new topics to the book or to bring the treatment up to date. However, some comments are added in an appendix of notes. (The existence of a relevant comment is signaled in the text through the symbol°.) Equation numbers are the same here as in the 1965 edition, except that equations (10.63) and (10.64) are swapped.

I thank Edwin Tayor for encouragement and Daniel Keren, Jozef Hanc, and especially Tim Hatamian for bringing errors to my attention. A research status leave from Oberlin College made this project possible.

I can well remember the day thirty years ago when I opened the pages of Feynman-Hibbs, and for the first time saw quantum mechanics as a living piece of nature rather than as a flood of arcane algorithms that, while lovely and mysterious and satisfying, ultimately defy understanding or intuition. It is my hope and my belief that this emended edition will open similar doors for generations to come.

Daniel F. Styer

Contents

1

The Fundamental Concepts of Quantum Mechanics

1-1 PROBABILITY IN QUANTUM MECHANICS[1]

From about the beginning of the twentieth century experimental physics amassed an impressive array of strange phenomena which demonstrated the inadequacy of classical physics. The attempts to discover a theoretical structure for the new phenomena led at first to a confusion in which it appeared that light, and electrons, behaved sometimes like waves and sometimes like particles. This apparent inconsistency was completely resolved in 1926 and 1927 in the theory called quantum mechanics. The new theory asserts that there are experiments for which the exact outcome is fundamentally unpredictable and that in these cases one has to be satisfied with computing probabilities of various outcomes. But far more fundamental was the discovery that in nature the laws of combining probabilities were *not* those of the classical probability theory of Laplace. The quantum-mechanical laws of the physical world approach very closely the laws of Laplace as the size of the objects involved in the experiments increases. Therefore, the laws of probabilities which are conventionally applied are quite satisfactory in analyzing the behavior of the roulette wheel but not the behavior of a single electron or a single photon of light.

A Conceptual Experiment. The concept of probability is not altered in quantum mechanics. When we say the probability of a certain outcome of an experiment is p, we mean the conventional thing, i.e., that if the experiment is repeated many times, one expects that the fraction of those which give the outcome in question is roughly p. We shall not be at all concerned with analyzing or defining this concept in more detail; for no departure from the concept used in classical statistics is required.

What is changed, and changed radically, is the method of calculating probabilities. The effect of this change is greatest when dealing with objects of atomic dimensions. For this reason we shall illustrate the laws of quantum mechanics by describing the results to be expected in some conceptual experiments dealing with a single electron.

Our imaginary experiment is illustrated in Fig. 1-1. At A we have a source of electrons S. The electrons at S all have the same energy

[1]Much of the material appearing in this chapter was originally presented as a lecture by R.P. Feynman and published as "The Concept of Probability in Quantum Mechanics" in the Second Berkeley Symposium on Mathematical Statistics and Probability, University of California Press, Berkeley, Calif., pp. 533–541, 1951.

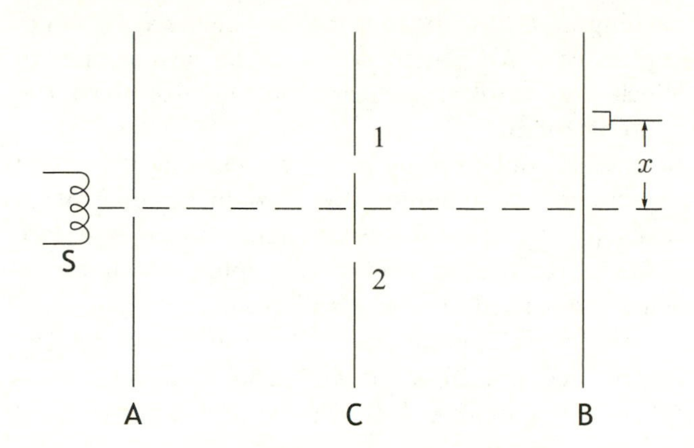

Fig. 1-1 The experimental arrangement. Electrons emitted at A make their way to the detector at screen B, but a screen C with two holes is interposed. The detector registers a count for each electron which arrives; the fraction which arrives when the detector is placed at a distance x from the center of the screen is measured and plotted against x, as in Fig. 1-2.

but come out in all directions to impinge on a screen C. The screen C has two holes, 1 and 2, through which the electrons may pass. Finally, behind the screen C at plane B we have a detector of electrons which may be placed at various distances x from the center of the screen.°

If the detector is extremely sensitive (as a Geiger counter is) it will be discovered that the current arriving at x is not continuous, but corresponds to a rain of particles. If the intensity of the source S is very low, the detector will record pulses representing the arrival of individual particles, separated by gaps in time during which nothing arrives. This is the reason we say electrons are particles. If we had detectors simultaneously all over the screen B, with a very weak source S, only one detector would respond, then after a little time, another would record the arrival of an electron, etc. There would never be a half response of the detector; either an entire electron would arrive or nothing would happen. And two detectors would never respond simultaneously (except for the coincidence that the source emitted two electrons within the resolving time of the detectors — a coincidence whose probability can be decreased by further decrease of the source intensity). In other words, the detector of Fig. 1-1 records the passage of a single corpuscular entity traveling from S to the point x.

This particular experiment has never been done in just this way.° In the following description we are stating what the results would be according to the laws which fit every experiment of this type which has ever been performed. Some experiments which directly illustrate the conclusions we are reaching here have been done, but such experiments are usually more complicated. We prefer, for pedagogical reasons, to select experiments which are simplest in principle and disregard the difficulties of actually doing them.

Incidentally, if one prefers, one could just as well use light instead of electrons in this experiment. The same points would be illustrated. The source S could be a source of monochromatic light and the sensitive detector a photoelectric cell or, better, a photomultiplier which would record pulses, each being the arrival of a single photon.

What we shall measure for various positions x of the detector is the mean number of pulses per second. In other words, we shall determine experimentally the (relative) probability P that the electron passes from S to x, as a function of x.

The graph of this probability as a function of x is the complicated curve illustrated qualitatively in Fig. 1-2a. It has several maxima and minima, and there are locations near the center of the screen at which electrons hardly ever arrive. It is the problem of physics to discover the laws governing the structure of this curve.

We might suppose (since the electrons behave as particles) that

> I. Each electron which passes from S to x must go through either hole 1 or hole 2.

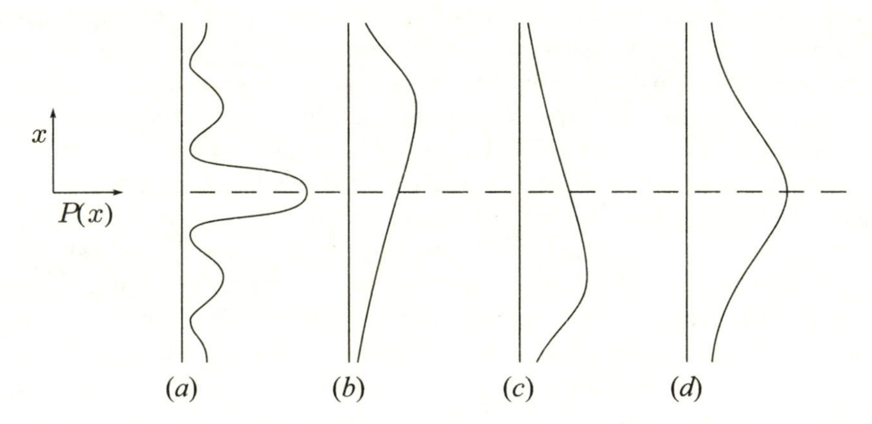

Fig. 1-2 Results of the experiment. Probability of arrival of electrons at x plotted against the position x of the detector. The result of the experiment of Fig. 1-1 is plotted here at (a). If hole 2 is closed, so the electrons can go through just hole 1, the result is (b). For just hole 2 open, the result is (c). If we imagine that each electron goes through one hole or the other, we expect the curve (d) = (b) + (c) when both holes are open. This is considerably different from what we actually get, (a).

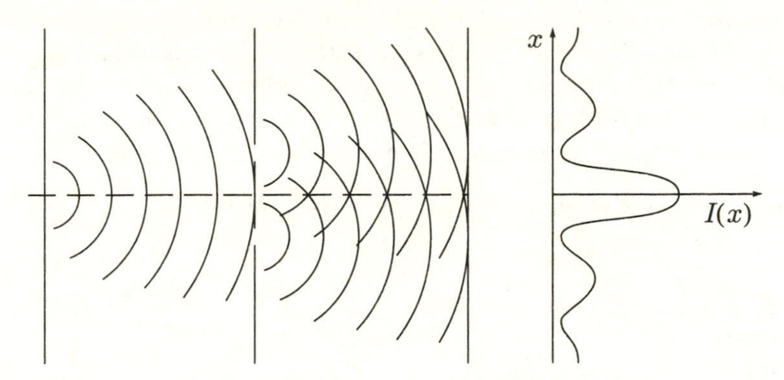

Fig. 1-3 An analogous experiment in wave interference. The complicated curve $P(x)$ in Fig. 1-2*a* is the same as the intensity $I(x)$ of waves which would arrive at x starting from S and coming through the holes. At some points x the wavelets from holes 1 and 2 interfere destructively (e.g., a crest from hole 1 arrives at the same time as a trough from hole 2); at others, constructively. This produces the complicated minima and maxima of the curve $I(x)$.

As a consequence of I we expect that

> II. The chance of arrival at x is the sum of two parts: P_1, the chance of arrival coming through hole 1, plus P_2, the chance of arrival coming through hole 2.

We may find out if this is true by direct experiment. Each of the component probabilities is easy to determine. We simply close hole 2 and measure the chance of arrival at x with only hole 1 open. This gives the chance P_1 of arrival at x for electrons coming through hole 1. The result is given in Fig. 1-2*b*. Similarly, by closing hole 1 we find the chance P_2 of arrival through hole 2 (Fig. 1-2*c*).

The sum of these (Fig. 1-2*d*) clearly is not the same as curve (*a*). Hence experiment tells us definitely that $P \neq P_1 + P_2$, or that assertion II is false.

The Probability Amplitude. The chance of arrival at x with both holes open is *not* the sum of the chance with just hole 1 open plus the chance with just hole 2 open.

Actually, the complicated curve $P(x)$ is familiar, inasmuch as it is exactly the intensity of distribution in the interference pattern to be expected if waves starting from S pass through the two holes and impinge on the screen B (Fig. 1-3). The easiest way to represent wave amplitudes is by complex numbers. We can state the correct law for $P(x)$ mathematically by saying that $P(x)$ is the absolute square of a certain complex quantity (if electron spin is taken into account, it is a hypercomplex quantity) $\phi(x)$ which we call the *probability amplitude* of arrival at x. Furthermore, $\phi(x)$ is the sum of two contributions: $\phi_1(x)$, the amplitude for arrival at x through hole 1, plus $\phi_2(x)$, the amplitude for arrival at x through hole 2. In other words,

III. There are complex numbers ϕ_1 and ϕ_2 such that

$$P = |\phi|^2 \tag{1.1}$$

$$\phi = \phi_1 + \phi_2 \tag{1.2}$$

and

$$P_1 = |\phi_1|^2 \qquad P_2 = |\phi_2|^2 \tag{1.3}$$

In later chapters we shall discuss in detail the actual calculation of ϕ_1 and ϕ_2. Here we say only that ϕ_1, for example, may be calculated as a solution of a wave equation representing waves spreading from the source to hole 1 and from hole 1 to x. This reflects the wave properties of electrons (or in the case of light, photons).

To summarize: We *compute* the intensity (i.e., the absolute square of the amplitude) of waves which would arrive in the apparatus at x and then *interpret* this intensity as the probability that a particle will arrive at x.

Logical Difficulties. What is remarkable is that this dual use of wave and particle ideas does not lead to contradictions. This is so only if great care is taken as to what kind of statements one is permitted to make about the experimental situation.

To discuss this point in more detail, consider first the situation which arises from the observation that our new law III of composition of probabilities implies, in general, that it is not true that $P = P_1 + P_2$. We must conclude that when both holes are open, it is *not true* that the particle goes through one hole or the other. For if it had to go through one or the other, we could classify all the arrivals at x into two disjoint classes, namely, those arriving through hole 1 and those arriving through hole 2; and the frequency P of arrival at x would surely be the sum of the frequency P_1 of particles coming through hole 1 and the frequency P_2 of those coming through hole 2.

To extricate ourselves from the logical difficulties introduced by this startling conclusion, we might try various artifices. We might say, for example, that perhaps the electron travels in a complex trajectory going through hole 1, then back through hole 2 and finally out through hole 1 in some complicated manner. Or perhaps the electron spreads out somehow and passes partly through both holes so as to eventually produce the interference result III. Or perhaps the chance P_1 that the electron passes through hole 1 has not been determined correctly inasmuch as closing hole 2 might have influenced the motion near hole 1. Many such classical mechanisms have been tried to explain the result. When light photons are used (in which case the same law III applies), the two interfering paths 1 and 2 can be made to be many centimeters apart (in space), so that the two alternative trajectories must almost certainly be independent. That the actual situation is more profound than might at first be supposed is shown by the following experiment.

The Effect of Observation. We have concluded on logical grounds that since $P \neq P_1 + P_2$, it is not true that the electron passes through either hole 1 or hole 2. But it is easy to design an experiment to test our conclusion directly. We have merely to have a source of light behind the holes and watch to see through which hole the electron passes (see Fig. 1-4). For electrons scatter light, so that if light is scattered behind hole 1, we may conclude that an electron passed through hole 1; and if it is scattered behind hole 2; the electron has passed through hole 2.

The result of this experiment is to show unequivocally that the electron *does* pass through either hole 1 or hole 2! That is, for every electron which arrives at the screen B (assuming the light is strong enough that we do not miss seeing it) light is scattered either behind hole 1 or behind hole 2, and never (if the source S is very weak) at both places. (A more delicate experiment could even show that the charge passing through the holes passes through either one or the other and is in all cases the complete charge of one electron and not a fraction of it.)

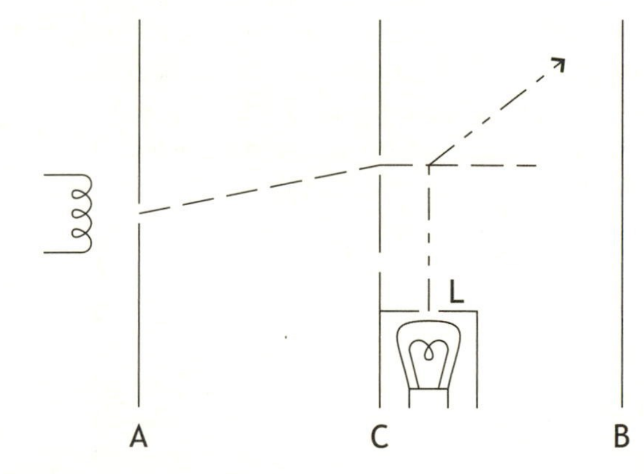

Fig. 1-4 A modification of the experiment of Fig. 1-1. Here we place a lamp L behind the screen C and look for light scattered by the electrons passing through hole 1 or hole 2. With a strong lamp every electron is indeed found to pass by one or the other hole. But now the probability of arrival at x is no longer given by the curve of Fig. 1-2a, but is instead given by Fig. 1-2d.

It now appears that we have come to a paradox. For suppose that we combine the two experiments. We watch to see through which hole the electron passes and at the same time measure the chance that the electron arrives at x. Then for each electron which arrives at x we can say experimentally whether it came through hole 1 or hole 2. First we may verify that P_1 is given by the curve in Fig. 1-2*b*, because if we select, of the electrons which arrive at x, only those which appear to come through hole 1 (by scattering light there), we find they are, indeed, distributed very nearly as in curve (*b*). (This result is obtained whether hole 2 is open or closed, so we have verified that there is no subtle influence of closing hole 2 on the motion near hole 1.) If we select the electrons scattering light at hole 2, we get (very nearly) P_2 of Fig. 1-2*c*. But now each electron appears at either 1 or 2 and we can separate our electrons into disjoint classes. So, if we take both together, we *must* get the distribution $P = P_1 + P_2$ illustrated in Fig. 1-2*d*. And experimentally we do! Somehow now the distribution does *not* show the interference effects III of curve (*a*)!

What has been changed? When we watch the electrons to see through which hole they pass, we obtain the result $P = P_1 + P_2$. When we do not watch, we get a different result,

$$P = |\phi_1 + \phi_2|^2 \neq P_1 + P_2$$

Just by watching the electrons, we have changed the chance that they arrive at x. How is this possible? The answer is that, to watch them, we used light and the light in collision with the electrons may be expected to alter its motion, or, more exactly, to alter its chance of arrival at x.

On the other hand, can we not use weaker light and thus expect a weaker effect? A negligible disturbance certainly cannot be presumed to produce the finite change in distribution from (*a*) to (*d*). But weak light does not mean a weaker disturbance. Light comes in photons of energy $h\nu$, where ν is the frequency, or of momentum h/λ, where λ is the wavelength. Weakening the light just means using fewer photons, so that we may miss seeing an electron. But when we do see one, it means a complete photon was scattered and a finite momentum of order h/λ is given to the electron.

The electrons that we miss seeing are distributed according to the interference law (*a*), while those we do see and which therefore have scattered a photon arrive at x with the probability $P = P_1 + P_2$ in (*d*). The net distribution in this case is therefore the weighed mean of (*a*) and (*d*). In strong light, when nearly all electrons scatter light, it is nearly (*d*); and in weak light, when few scatter, it becomes more like (*a*).

It might still be suggested that since the momentum carried by the light is h/λ, weaker effects could be produced by using light of a longer wavelength λ. But there is a limit to this. If light of too long a wavelength is used, we shall not be able to tell whether it was scattered from behind hole 1 or hole 2; for the source of light of wavelength λ cannot be located in space with precision greater than order λ.

We thus see that any physical agency designed to determine through which hole the electron passes must produce, lest we have a paradox, enough disturbance to alter the distribution from (a) to (d).

It was first noticed by Heisenberg, and stated in his uncertainty principle, that the consistency of the then-new mechanics required a limitation to the subtlety to which experiments could be performed. In our case the principle says that an attempt to design apparatus to determine through which hole the electron passed, and delicate enough so as not to deflect the electron sufficiently to destroy the interference pattern, must fail. It is clear that the consistency of quantum mechanics requires that it must be a general statement involving all the agencies of the physical world which might be used to determine through which hole an electron passes. The world cannot be half quantum-mechanical, half classical. No exception to the uncertainty principle has been discovered.

1-2 THE UNCERTAINTY PRINCIPLE

We shall state the uncertainty principle as follows: Any determination of the alternative taken by a process capable of following more than one alternative destroys the interference between alternatives. Heisenberg's original statement of the uncertainty principle was not given in the form we have used here. We shall interrupt our argument for a few paragraphs to discuss Heisenberg's original statement.

In classical physics a particle can be described as moving along a definite trajectory and having, for example, a precise position and velocity at any particular time. Such a picture would not lead to the odd results that we have seen are characteristic of quantum mechanics. Heisenberg's uncertainty principle gives the limits of accuracy of such classical ideas. For example, the idea that a particle has both a definite position and a definite momentum has its limitations. A real system (i.e., one obeying quantum mechanics) looked upon from a classical view appears to be one in which the position or momentum is not definite, but is uncertain. The uncertainty in position can be reduced by careful measurement, and (by applying different techniques) the uncertainty in momentum can be reduced by careful measurement. But, as Heisenberg stated in his principle, both cannot be accurately known simultaneously; the product of the uncertainties of momentum and position involved in any experiment cannot be smaller than a number with the order of magnitude of $\hbar$. (Here $\hbar = h/2\pi = 1.055 \times 10^{-27}$ erg·sec, where h is Planck's constant.) That such a result is required by physical consistency in the situation we have been discussing can be shown by considering still another way of trying to determine through which hole the electron passes.

Example. Notice that if an electron is deflected in passing through one of the holes, its vertical component of momentum is changed. Furthermore, an electron arriving at the detector at x after passing through hole 1 is deflected by a different amount, and thus suffers a different change in momentum, than an electron arriving at x via hole 2. Suppose that the screen at C is not rigidly supported, but is free to move up and down (Fig. 1-5). Any change in the vertical component of the momentum of an electron upon passing through a hole will be accompanied by an equal and opposite change in the momentum of the screen. This change in momentum can be measured by measuring the velocity of the screen before and after the passage of an electron. Call δp the difference in momentum change between electrons passing through hole 1 and hole 2. Then an unambiguous determination of the hole used by a particular electron requires a momentum determintation of the screen to an accuracy of better than δp.

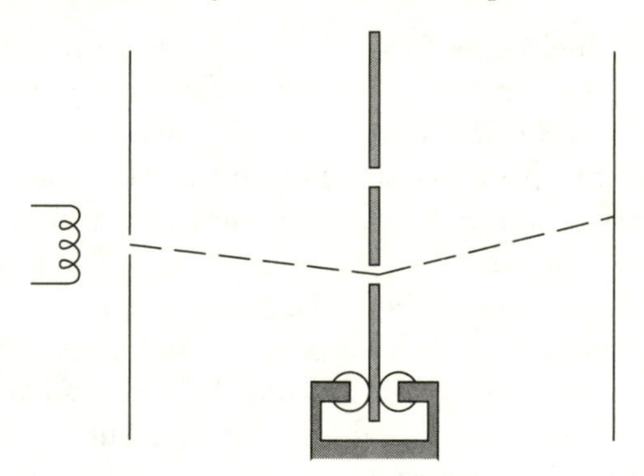

Fig. 1-5 Another modification of the experiment of Fig. 1-1. The screen C is left free to move vertically. If the electron passes hole 2 and arrives at the detector (at $x = 0$, for example), it is deflected upward and the screen C will recoil downward. The hole through which the electron passes can be determined for each passage by starting with the screen at rest and measuring whether it is recoiling up or down afterward. According to Heisenberg's uncertainty principle, however, such precise momentum measurements on screen C are inconsistent with accurate knowledge of its vertical position, so we could not be sure that the center line of the holes is correctly set. Instead of $P(x)$ of Fig. 1-2a, we get this smeared a little in the vertical direction, so it looks like Fig. 1-2d.

If the experiment is set up in such a way that the momentum of screen C can be measured to the required accuracy, then, since we can determine the hole passed through, we must find that the resulting distribution of electrons is that of curve (*d*) of Fig. 1-2. The interference pattern of curve (*a*) must be lost. How can this happen? To understand, note that the construction of a distribution curve in the plane B requires an accurate knowledge of the vertical position of the two holes in screen C. Thus we must measure not only the momentum of screen C but also its position. If the interference pattern of curve (*a*) is to be established, the vertical position of C must be known to an accuracy of better than $d/2$, where d is the spacing between maxima of the curve (*a*). For suppose the vertical position of C is not known to this accuracy; then the vertical position of every point in Fig. 1-2*a* cannot be specified with an accuracy greater than $d/2$ since the zero point of the vertical scale must be lined up with some nominal zero point on C. Then the value of P at any particular height x must be obtained by averaging over all values within a distance $d/2$ of x. Clearly, the interference pattern will be smeared out by this averaging process. The resulting curve will look like Fig. 1-2*d*.

The interference pattern in the original experiment is the sign of wave-like behavior of the electron. The pattern is the same for any wave motion, so we may use the well-known result from the theory of light interference that the relation between the separation a of the holes, the distance l between screen C and plane B, the wavelength λ of the light, and d is

$$\frac{a}{l} = \frac{\lambda}{d} \tag{1.4}$$

as shown in Fig 1.6. In Chap. 3 (at Eq. 3.10) we shall find that the wavelength of the electron wave is intimately connected with the momentum of the electron by the relation

$$p = h/\lambda \tag{1.5}$$

If p is the total momentum of an electron (and we assume all the electrons have the same total momentum), then for $l \gg a$,

$$\frac{\delta p}{p} \approx \frac{a}{l} \tag{1.6}$$

as shown in Fig. 1-7. It follows that

$$d = \frac{h}{\delta p} \tag{1.7}$$

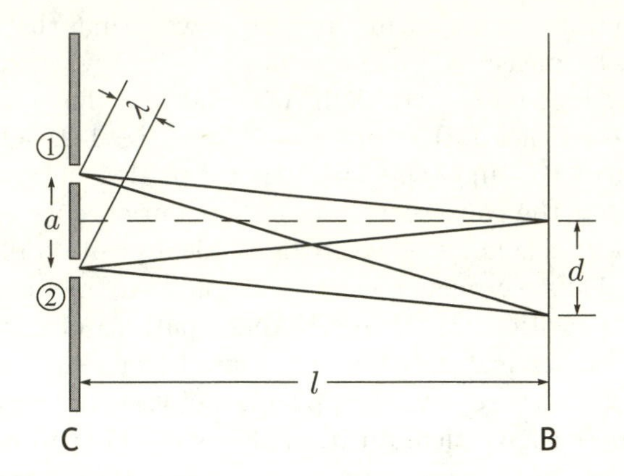

Fig. 1-6 Two beams of light, starting in phase at holes 1 and 2, will interfere constructively when they reach the screen B if they take the same time to travel from C to B. This means that a maximum in the interference pattern for light beams passing through two holes will occur at the center of the screen. As we move down the screen, the next maximum will occur at a distance d, which is far enough from the center that, in traveling to this point, the beam from hole 1 will have traveled exactly one wavelength λ farther than the beam from hole 2.

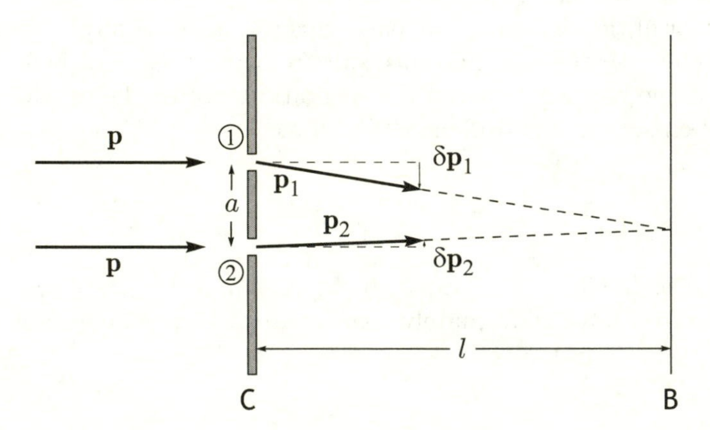

Fig. 1-7 The deflection of an electron in passing through a hole in the screen C involves a change in momentum $\delta\mathbf{p}$. This change amounts to the addition of a small component of momentum in a direction approximately perpendicular to the original momentum vector. The change in energy is completely negligible. For small deflection angles, the total momentum vector keeps the same magnitude (approximately). Then the deflection angle is represented to a very good approximation by $|\delta\mathbf{p}|/|\mathbf{p}|$. If two electrons, one starting from hole 1 with momentum $\mathbf{p}_1$ and the other starting from hole 2 with momentum $\mathbf{p}_2$, reach the same point on the screen B, then the angles through which they were deflected must differ by approximately a/l. Since we cannot say through which hole an electron has come, the uncertainty in the vertical component of momentum which the electron receives on passing through the screen C must be equivalent to this uncertainty in deflection angle. This gives the relation $|\mathbf{p}_1 - \mathbf{p}_2|/|\mathbf{p}| = |\delta\mathbf{p}|/|\mathbf{p}| = a/l$.

Since experimentally we find that the interference pattern has been lost, it must be that the uncertainty δx in the measurement of the position of C is larger than $d/2$. Thus

$$\delta x\ \delta p \geq \frac{h}{2} \tag{1.8}$$

which agrees (in order of magnitude) with the usual statement of the uncertainty principle.

A similar analysis can be applied to the previous measuring device where the scattering of light was used to determine through which hole the electron passed. Such an analysis produces the same lower limit for the uncertainties of measurement.

The uncertainty principle is not "proved" by considering a few such experiments. It is only illustrated. The evidence for it is of two kinds. First, no one has yet found any experimental way to defeat the limitations in measurements which it implies. Second, the laws of quantum mechanics seem to require it if their consistency is to be maintained, and the predictions of these laws have been confirmed again and again with great precision.

1-3 INTERFERING ALTERNATIVES

Two Kinds of Alternatives. From a physical standpoint the two routes are independent alternatives, yet the implications that the probability is the sum $P_1 + P_2$ is false. This means that either the premise or the reasoning which leads to such a conclusion must be false. Since our habits of thought are very strong, many physicists find that it is much more convenient to deny the premise than to deny the reasoning. To avoid the logical inconsistencies into which it is so easy to stumble, they take the following view: When no attempt is made to determine through which hole the electron passes, one cannot say it must pass through one hole or the other. Only in a situation where apparatus is operating to determine through which hole the electron goes is it permissible to say that it passes through one or the other. When you watch, you find that it goes through either one hole or the other hole; but if you are not looking, you cannot say that it goes either one way or the other! Nature demands that we walk a logical tightrope if we wish to describe her.

Contrary to that way of thinking, we shall in this book follow the suggestion made in Sec. 1-1 and deny the reasoning; i.e., we shall not compute probabilities by adding probabilities for all alternatives. In order to make definite the new rules for combining probabilities, it will be convenient to define two meanings for the word "alternative." The first of these meanings carries with it the concept of exclusion. Thus holes 1 and 2 are *exclusive alternatives* if one of them is closed or if some apparatus that can unambiguously determine which hole is used is operating. The other meaning of the word "alternative" carries with it a concept of combination or interference. (The term *interference* has the same meaning here as it has in optics, i.e., either constructive or destructive interference.) Thus we shall say that holes 1 and 2 present *interfering alternatives* to the electron when (1) both holes are open and (2) no attempt is made to determine through which hole the electron passes. When the alternatives are of this interfering type, the laws of probability must be changed to the form given in Eqs. (1.1) and (1.2).

The concept of interfering alternatives is fundamental to all of quantum mechanics. In some situations we may have both kinds of alternatives present. Suppose we ask, in the two-hole experiment, for the probability that the electron arrives at some point, say, within 1 cm of the center of the screen. We mean by this the probability that if there were counters arranged all over the screen (so one or another would go off when the electron arrived), the counter which went off was within 1 cm of $x = 0$. Here the various possibilities are that the electron arrives at some counter via some hole. The holes represent interfering alternatives, but the counters represent exclusive alternatives. Thus we first add $\phi_1 + \phi_2$ for a fixed x, square that, and then sum those resultant probabilities over x from -0.5 to $+0.5$ cm.

It is not hard, with a little experience, to tell which kind of alternative is involved. For example, suppose that information about the alternatives is available (or could be made available without altering the result), but this information is not used. Nevertheless, in this case a sum of probabilities (in the ordinary sense) must be carried out over *exclusive* alternatives. These exclusive alternatives are those which *could* have been separately identified by the information.

Some Illustrations. When alternatives cannot possibly be resolved by any experiment, they always interfere. A striking illustration of this is the scattering of two nuclei at 90°, say, in the center-of-gravity system, as illustrated in Fig. 1-8. Suppose the nucleus starting at A is an alpha particle and the one starting at B is some other nucleus. Ask for the probability that the nucleus starting from A is scattered to position 1 and that from B to 2. The amplitude is, say, $\phi(1, 2; \mathsf{A}, \mathsf{B})$. The probability of this is $p = |\phi(1, 2; \mathsf{A}, \mathsf{B})|^2$. Suppose we do not distinguish what kind of nucleus arrives at 1, that is, whether it is from A or from B. If it is the nucleus from B, the amplitude is $\phi(2, 1; \mathsf{A}, \mathsf{B})$ (which equals $\phi(1, 2; \mathsf{A}, \mathsf{B})$,

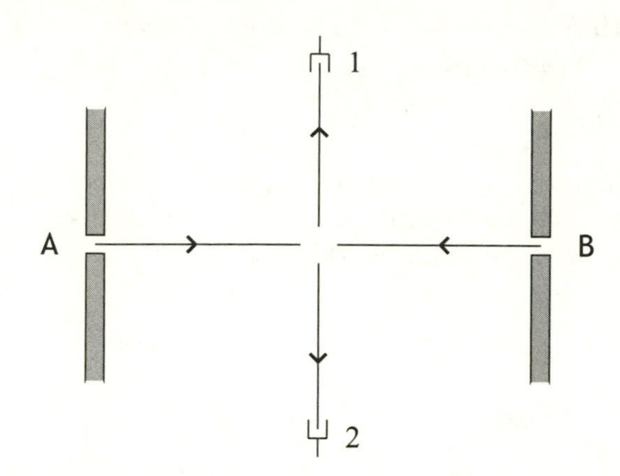

Fig. 1-8 Scattering of one nucleus by another in the center-of-gravity system. The scattering of two identical nuclei shows striking interference effects. There are two interfering alternatives here. The particle which arrives at 1, say, could have started either from A or from B. If the original nuclei were not identical, tests of identity at 1 could determine which alternative had actually been taken, so they are exclusive alternatives and the special interference effects do not arise in this case.

because we have taken a 90° angle). The chance that some nucleus arrives at 1 and the other at 2 is

$$|\phi(1,2;\mathsf{A},\mathsf{B})|^2 + |\phi(2,1;\mathsf{A},\mathsf{B})|^2 = 2p \tag{1.9}$$

We have added the probabilities. The cases "A to 1 and B to 2" and "A to 2 and B to 1" are exclusive alternatives because we could, if we wished, determine the character of the nucleus at 1 without disturbing the previous scattering process.

But what would happen if both A and B released alpha particles? Then no experiment can distinguish which is which, and we cannot know whether the nucleus arriving at 1 started from A or B. We have interfering alternatives, and the probability is

$$|\phi(1,2;\mathsf{A},\mathsf{B}) + \phi(2,1;\mathsf{A},\mathsf{B})|^2 = 4p \tag{1.10}$$

This interesting result is readily verified experimentally.

If electrons scatter electrons, the result is different in two ways. First, the electron has a quality we call *spin*, and a given electron may be in one of the two states called *spin up* and *spin down*. The spin is not changed to first approximation for scattering at low energy. The spin carries a magnetic moment. At low velocities the main forces are electrical, owing to charge, and the magnetic influences make only a small correction, which we neglect. So if the electron from A has spin up and the electron from B has spin down, we could later tell which arrived at 1 by measuring its spin. If up, it is from A; if down, from B. The scattering probability is then

$$|\phi(1,2;\mathsf{A},\mathsf{B})|^2 + |\phi(2,1;\mathsf{A},\mathsf{B})|^2 = 2p \tag{1.11}$$

in this case.

If, however, electrons at both A and B start with spin up, we cannot later tell which is which and we would expect

$$|\phi(1,2;\mathsf{A},\mathsf{B}) + \phi(2,1;\mathsf{A},\mathsf{B})|^2 = 4p \tag{1.12}$$

Actually this is wrong and, remarkably, electrons obey a different rule. The amplitude for an event in which the identity of a pair of electrons is reversed contributes 180° out of phase. That is, the case of both spin up gives

$$|\phi(1,2;\mathsf{A},\mathsf{B}) - \phi(2,1;\mathsf{A},\mathsf{B})|^2 \tag{1.13}$$

In our case of 90° scattering $\phi(1,2;\mathsf{A},\mathsf{B}) = \phi(2,1;\mathsf{A},\mathsf{B})$, so this is zero.

Fermions and Bosons. This rule of the 180° phase shift for alternatives involving exchange in identity of electrons is very odd, and its ultimate reason in nature is still only imperfectly understood. Other particles besides electrons obey it. Such particles are called fermions, and are said to obey Fermi, or antisymmetric, statistics. Electrons, protons, neutrons, neutrinos, and μ mesons are fermions. So are composites of an odd number of these such as a nitrogen atom, which contains seven electrons, seven protons, and seven neutrons. This 180° rule was first stated by Pauli and is the full quantum-mechanical basis of his exclusion principle, which controls the character of the chemists' periodic table.

Particles for which interchange does not alter the phase are called bosons and are said to obey Bose, or symmetrical, statistics. Examples of bosons are photons, π mesons, and composites containing an even number of Fermi particles such as an alpha particle, which is two protons and two neutrons. All particles are either one or the other, bosons or fermions. These interference properties can have profound and mysterious effects. For example, liquid helium made of atoms of atomic mass 4 (bosons) at temperatures of one or two degrees Kelvin can flow without any resistance through small tubes, whereas the liquid made of atoms of mass 3 (fermions) does not have this property.

The concept of identity of particles is far more complete and definite in quantum mechanics than it is in classical mechanics. Classically, two particles which seem identical could be nearly identical, or identical for all practical purposes, in the sense that they may be so closely equal that present experimental techniques cannot detect any difference. However, the door is left open for some future technique to establish the difference. In quantum mechanics, however, the situation is different. We can give a direct test to determine whether or not particles are completely indistinguishable.

If the particles in the experiment diagramed in Fig. 1-8, starting from A and B, were only approximately identical, then improvements in experimental techniques would enable us to determine by close scrutiny of the particle arriving at 1, for example, whether it came from A or B. In this situation the alternatives of the two initial positions must be exclusive, and there must be no interference between the amplitudes describing these alternatives. Now the important point is that this act of scrutiny would take place *after* the scattering had taken place. This means that the observation could not possibly affect the scattering process, and this in turn implies that we would expect no interference between the amplitudes describing the alternatives (that it is either the particle from A or the particle from B which arrives at 1). In this case we must conclude from the uncertainty principle that there is no way, even in principle, to ever distinguish between these possibilities. That is, when a particle arrives at 1, it is completely impossible by any test whatsoever, now or in the future, to determine whether the particle started from A or B. In this more rigorous sense of identity, all electrons are identical, as are all protons, etc.

As a second example we consider the scattering of neutrons from a crystal. When neutrons of wavelength somewhat shorter than the atomic spacing are scattered from the atoms in a crystal, we get very strong interference effects. The neutrons emerge only in certain discrete directions determined by the Bragg law of reflection, just as for X-rays. The interfering alternatives which enter this example are the alternative possibilities that it is one, or another, atom which does the scattering of a particular neutron. (The amplitude to scatter neutrons from any atom is so small that we need not consider alternatives in which a neutron is scattered by two or more atoms.) The waves of amplitude describing the motion of a neutron which start from these atoms interfere constructively only in certain definite directions.

Now there is an interesting complication which enters this apparently simple picture. Neutrons, like electrons, carry a spin, which can be analyzed in two states, spin up and spin down. Suppose the scattering material is composed of an atomic species which has a similar spin property, such as carbon-13. In this case an experiment will reveal two apparently different types of scattering. It is found that besides the scattering in discrete directions, as described in the preceding paragraph, there is a diffuse scattering in all directions. Why should this be?

A clue to the source of these two types of scattering is provided by the following observation. Suppose all the neutrons which enter the experiment are prepared with spin up. If the spin direction of the emerging neutrons is analyzed, it will be found that some are up and some are down; those which still have spin up are scattered only at the discrete Bragg angles, while those whose spin has been changed to down come out scattered diffusely in all directions!

Now in order that a neutron flip its spin from up to down, the law of conservation of angular momentum requires that the spin of the scattering nucleus change from down to up. Therefore, in principle, the particular nucleus which was responsible for scattering that particular neutron could be determined. We could, in principle, note down before the experiment the spin state of all the scattering nuclei in the crystal. Then, after the neutron is scattered, we could reinvestigate the crystal and see which nucleus had changed its spin from down to up. If no crystal nucleus underwent such a change in spin, then neither did the neutron, and we cannot tell from which nucleus the neutron actually scattered. In this case the alternatives interfere and the Bragg law of scattering results.

If, on the other hand, one crystal nucleus is found to have changed spin, then we know that this nucleus did the scattering. There are no interfering alternatives. The spherical waves of amplitude which emerge from this particular nucleus describe the motion of the scattered neutron, and only the waves emerging from this nucleus enter into that description. In this case there is equal likelihood to find the scattered neutron coming out in any direction.

The concept of searching through all the nuclei in a crystal to find which one has changed its spin state is surely a needle-in-the-haystack type of activity, but nature is not concerned with the practical difficulties of experimentation. The important fact is that in principle it is possible without producing any disturbance of the scattered neutron to determine (in the latter case where the spin states change) which crystal nucleus actually did the scattering. The existence of this possibility means that even if we do not actually carry out this determination, we are nevertheless dealing with exclusive (and thus noninterfering) alternatives.

On the other hand, the fact that we get interference between alternatives in the situation where the spin of the neutron was not changed means that it is impossible, even in principle, to ever discover which particular crystal nucleus did the scattering — impossible, at least, without disturbing the situation during or before the scattering.

1-4 SUMMARY OF PROBABILITY CONCEPTS

Alternatives and the Uncertainty Principle. The purpose of this introductory chapter has been to explain the meaning of a probability amplitude and its importance in quantum mechanics and to discuss the rules for manipulation of these amplitudes. Thus we have stated that there is a quantity called a *probability amplitude* associated with every method whereby an event in nature can take place. For example, an electron going from source S (Fig. 1-1) to a detector at x has one amplitude for completing this course while passing through hole 1 of the screen at C and another amplitude for passing through hole 2. Further, we can associate an amplitude with the overall event by adding together the amplitudes of each alternative method. Thus, for example, the overall amplitude for arrival at x is given in Eq. (1.2) as

$$\phi = \phi_1 + \phi_2 \tag{1.14}$$

Next, we interpret the absolute square of the overall amplitude as the probability that the event will happen. For example, the probability that an electron reaches the detector is

$$P = |\phi_1 + \phi_2|^2 \tag{1.15}$$

If we interrupt the course of the event before its conclusion with an observation of the state of the particles involved in the event, we disturb the construction of the overall amplitude. Thus if we observe the system of particles to be in one particular state, we exclude the possibility that it can be in any other state, and the amplitudes associated with the excluded states can no longer be added in as alternatives in computing the overall amplitude. For example, if we determine with the help of some sort of measuring equipment that the electron passes through hole 1, the amplitude for arrival at the detector is just ϕ_1. Further, it does not matter if we actually observe and record the outcome of the measurement or not, so long as the measurement equipment is working. Obviously, we could observe the outcome at any time we wished. The operation of the measuring equipment is sufficient to disturb the system and its probability amplitude.

This latter fact is the basis of the Heisenberg uncertainty principle, which states that there is a natural limit to the subtlety of any experiment or the refinement of any measurement.

The Structure of the Amplitude. The amplitude for an event is the sum of the amplitudes for the various alternative ways that the event can occur. This permits the amplitude to be analyzed in many different ways depending on the different classes into which the alternatives can be divided. The most detailed analysis results from considering that a particle going from a to b, for example, in a given time interval, can be considered to have done this by going in a certain motion (position vs. time) or path in space and time. We shall therefore associate an amplitude with each possible motion. The total amplitude will be the sum of a contribution from each of the paths.

This idea can be made more clear by a further consideration of our experiment with the two holes. Suppose we put a couple of extra screens between the source and the holes. Call these screens E and D. In each of them we drill a few holes which we call E_1, E_2, ... and D_1, D_2, ... (Fig. 1-9). For simplicity, we shall assume the electrons are constrained to move in the xy plane. Then there are several alternative paths which an electron may take in going from the source to either hole in screen C. It could go from the source to E_2, and then D_3, and then the hole 1; or it could go from the source to E_3, then D_1, and finally to the hole 1; etc. Each of these paths has its own amplitude. The complete amplitude is the sum of all of them.

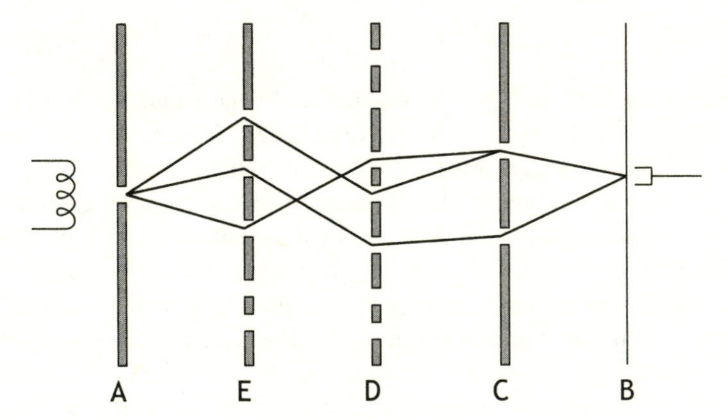

Fig. 1-9 When several holes are drilled in the screens E and D placed between the source at screen A and the final position at screen B, several alternative routes are available for each electron. For each of these routes there is an amplitude. The result of any experiment in which all of the holes are open requires the addition of all these amplitudes, one for each possible path.

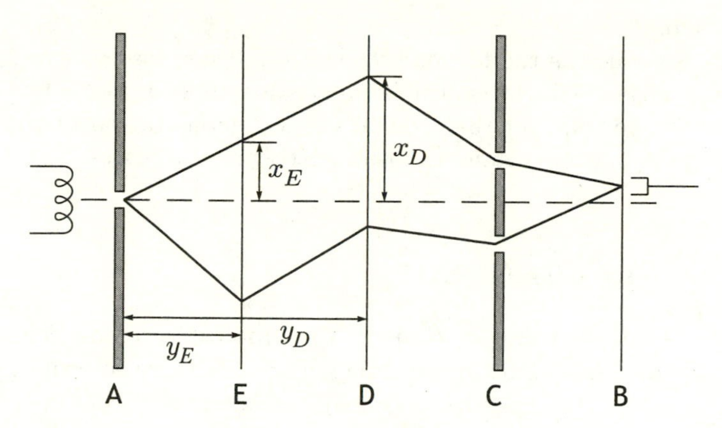

Fig. 1-10 More and more holes are cut in the screens at y_E and y_D. Eventually, the screens are completely riddled with holes, and the electron has a continuous range of positions, up and down along each screen, at which it can pass through the position of the screen. In this case the sum of alternatives becomes a double integral over the continuous variables x_E and x_D describing the alternative heights at which the electron passes the position of the screens at y_E and y_D.

Next, suppose we continue to drill holes in the screens E and D until there is nothing left of the screens. The path of an electron must now be specified by the height x_E at which the electron passes the position y_E at the nonexistent screen E, together with the height x_D, at the position y_D as in Fig. 1-10. To each pair of heights there corresponds an amplitude. The principle of superposition still applies, and we must take the sum (or by now, the integral) of these amplitudes over all possible values of x_E and x_D.

Clearly, the next thing to do is to place more and more screens between the source and hole 1 and in each screen drill so many holes that there is nothing left. Throughout this process we continue to refine the definition of the path of the electron, until finally we arrive at the sensible idea that a path is merely height as a particular function of distance, or $x(y)$. We also continue to apply the principle of superposition, until we arrive at the integral over all paths of the amplitude for each path.

Now we can make a still finer specification of the motion. Not only can we think of the particular path $x(y)$ in space, but we can specify the *time* at which it passes each point in space.° That is, a path will (in our two-dimensional case) be given if the two functions $x(t)$, $y(t)$ are given. Thus we have the idea of an amplitude to take a certain path $x(t)$, $y(t)$. The total amplitude to arrive is the sum or integral of this amplitude over all possible paths. The problem of defining this concept of a sum or integral over all paths in a mathematically more precise way will be taken up in Chap. 2.

Chapter 2 also contains the formula for the amplitude for any given path. Once this is given, the laws of nonrelativistic quantum mechanics are completely stated, and all that remains is a demonstration of the application of these laws in a number of interesting special cases.

1-5 SOME REMAINING THOUGHTS

We shall find that in quantum mechanics, the amplitudes ϕ are solutions of a completely deterministic equation (the Schrödinger equation). Knowledge of ϕ at $t = 0$ implies its knowledge at all subsequent times. The interpretation of $|\phi|^2$ as the probability of an event is an indeterministic interpretation. It implies that the result of an experiment is not exactly predictable. It is very remarkable that this interpretation does not lead to any inconsistencies. That this is true has been amply demonstrated by analyses of many particular situations by Heisenberg, Bohr, Born, von Neumann, and many other physicists. In spite of all these analyses the fact that no inconsistency can arise is not thoroughly obvious. For this reason quantum mechanics appears as a difficult and somewhat mysterious subject to a beginner. The mystery gradually decreases as more examples are tried out, but one never quite loses the feeling that there is something peculiar about the subject.

There are a few interpretational problems on which work may still be done. They are very difficult to state until they are completely worked out. One is to show that the probability interpretation of ϕ is the *only* consistent interpretation of this quantity. We and our measuring instruments are part of nature and so are, in principle, described by an amplitude functions satisfying a deterministic equation. Why can we only predict the probability that a given experiment will lead to a definite result? From what does the uncertainty arise? Almost without doubt it arises from the need to amplify the effects of single atomic events to such a level that they may be readily observed by large systems. The details of this have been analyzed only on the assumption that $|\phi|^2$ is a probability, and the consistency of this assumption has been shown. It would be an interesting problem to show that *no other* consistent interpretation can be made.°

Other problems which may be further analyzed are those dealing with the theory of knowledge. For example, there seems to be a lack of symmetry in time in our knowledge. Our knowledge of the past is qualitatively different from that of the future. In what way is only the probability of a future event accessible to us, whereas the certainty of a past event can often apparently be asserted? These matters again have been analyzed to a great extent. Possibly a little more can be said to clarify the situation, however. Obviously, we are again involved in the consequences of the large size of ourselves and of our measuring equipment. The usual separation of observer and observed which is now needed in analyzing measurements in quantum mechanics should not really be necessary, or at least should be even more thoroughly analyzed. What seems to be needed is the statistical mechanics of amplifying apparatus.°

The analyses of such problems are, of course, in the nature of philosophical questions. They are not necessary for the further development of physics. We know we have a consistent interpretation of ϕ and, almost without doubt, the only consistent one. The problem of today seems to be the discovery of the laws governing the behavior of ϕ for phenomena involving nuclei and mesons. The interpretation of ϕ is interesting. But the much more intriguing question is: What new modifications of our thinking will be required to permit us to analyze phenomena occurring within nuclear dimensions?

1-6 THE PURPOSE OF THIS BOOK

So far, we have given the form the quantum-mechanical laws must take, i.e., that a probability amplitude exists, and we have outlined one possible scheme for calculating this amplitude. There are other ways to formulate this. In a more usual approach to quantum mechanics the amplitude is calculated by solving a kind of wave equation. For particles of low velocity, it is called the Schrödinger equation. A more accurate equation valid for electrons of velocity arbitrarily close to the velocity of light is the Dirac equation. In this case the probability amplitude is a kind of hypercomplex number. We shall not discuss the Dirac equation in this book, nor shall we investigate the effects of spin. Instead, we limit our attention to low-velocity electrons, extending our horizon somewhat in the direction of quantum electrodynamics by investigating photons, particles whose behavior is determined by Maxwell's equations.

In this book we shall give the laws to compute the probability amplitude for nonrelativistic problems in a manner which is somewhat unconventional. In some ways, particularly in developing a conceptual understanding of quantum mechanics, it may be preferred, but in others, e.g., in making computations for the simpler problems and for understanding the literature, it is disadvantageous.

The more conventional view, via the Schrödinger equation, is already presented in many books, but the views to be presented here have appeared only in abbreviated form in papers in the journals.[1] A central aim of this book is to collect this work into one volume where it may be expounded with sufficient clarity and detail to be of use to the interested student.

In order to keep the subject within bounds, we shall not make a complete development of quantum mechanics. Instead, whenever a topic has reached such a point that further elucidation would best be made by conventional arguments appearing in other books, we refer to those books. Because of this incompleteness, this book cannot serve as a complete textbook of quantum mechanics. It can serve as an introduction to the ideas of the subject if used in conjunction with another book that deals with the Schrödinger equation, matrix mechanics, and applications of quantum mechanics.

On the other hand, we shall use the space saved (by our not developing all of quantum mechanics in detail) to consider the application of the mathematical methods used in the formulation of quantum mechanics to other branches of physics (Chaps. 10–12).

It is a problem of the future to discover the exact manner of computing amplitudes for processes involving the apparently more complicated particles, namely, neutrons, protons, and mesons. Of course, one can doubt that, when the unknown laws are discovered, we shall find ourselves computing amplitudes at all. However, the situation today does not seem analogous to that preceding the discovery of quantum mechanics.

In the 1920's there were many indications that the fundamental theorems and concepts of classical mechanics were wrong, i.e., there were many paradoxes. General laws could be proved independently of the detailed forces involved. Some of these laws did not hold. For example, each spectral line showed a degree of freedom for an atom, and at temperature T each degree of freedom should have an energy kT, contributing R to the specific heat. Yet this very high specific heat expected from the enormous number of spectral lines did not appear.

Today, any general law that we have been able to deduce from the principle of superposition of amplitudes, such as the characteristics of angular momentum, seems to work. But the detailed interactions still elude us. This suggests that amplitudes will exist in a future theory, but their method of calculation may be strange to us.

---

[1] R.P. Feynman, Space-Time Approach to Non-relativistic Quantum Mechanics, *Rev. Mod. Phys.*, vol. 20, pp. 367–387, 1948.

2

The Quantum-mechanical Law of Motion

IN this chapter we intend to complete our specification of nonrelativistic quantum mechanics which we began in Chap. 1. There we noted the existence of an amplitude for each trajectory; here we shall give the form of the amplitude for each trajectory. For a while, for simplicity, we shall restrict ourselves to the case of a particle moving in one dimension. Thus the position at any time can be specified by a coordinate x, a function of time t. By the path, then, we mean a function $x(t)$.

If a particle at an initial time t_a starts from point x_a and goes to a final point x_b at time t_b, we shall say simply that the particle goes from a to b and our function $x(t)$ will have the property that $x(t_a) = x_a$ and $x(t_b) = x_b$. In quantum mechanics, then, we shall have an amplitude, often called° a *kernel*, which we may write $K(b, a)$, to get from the point a to the point b. This will be the sum over all of the trajectories that go between the end points a and b of a contribution from each. This is to be contrasted with the situation in classical mechanics in which there is only one specific and particular trajectory which goes from a to b, the so-called *classical trajectory*, which we shall label $\bar{x}(t)$. Before we go on to give the rule for the quantum-mechanical case, let us remind ourselves of the situation in classical mechanics.

2-1 THE CLASSICAL ACTION

One of the most elegant ways of expressing the condition that determines the particular path $\bar{x}(t)$ out of all the possible paths is the *principle of least action.* That is, there exists a certain quantity S which can be computed for each path. The classical path $\bar{x}(t)$ is that for which S is a minimum. Actually, the real condition is that S be merely an extremum. That is to say, the value of S is unchanged in the first order if the path $\bar{x}(t)$ is modified slightly.

The quantity S is given by the expression

$$S = \int_{t_a}^{t_b} L(\dot{x}, x, t)\, dt \tag{2.1}$$

where L is the lagrangian for the system. For a particle of mass m subject to a potential energy $V(x, t)$, which is a function of position and time, the lagrangian is

$$L = \frac{m}{2}\dot{x}^2 - V(x, t) \tag{2.2}$$

The form of the extremum path $\bar{x}(t)$ is determined through the usual procedures of the calculus of variations. Thus, suppose the path is varied away from $\bar{x}(t)$ by an amount $\delta x(t)$; the condition that the end points of $\bar{x}(t)$ are fixed requires

$$\delta x(t_a) = \delta x(t_b) = 0 \tag{2.3}$$

The condition that $\bar{x}(t)$ be an extremum of S means

$$\delta S = S[\bar{x} + \delta x] - S[\bar{x}] = 0 \tag{2.4}$$

to first order in $\delta x(t)$. Using the definition of Eq. (2.1) we may write

$$\begin{aligned} S[x + \delta x] &= \int_{t_a}^{t_b} L(\dot{x} + \delta\dot{x}, x + \delta x, t)\,dt \\ &= \int_{t_a}^{t_b} \left[L(\dot{x}, x, t) + \delta\dot{x}\frac{\partial L}{\partial \dot{x}} + \delta x \frac{\partial L}{\partial x} \right] dt \\ &= S[x] + \int_{t_a}^{t_b} \left[\delta\dot{x}\frac{\partial L}{\partial \dot{x}} + \delta x \frac{\partial L}{\partial x} \right] dt \end{aligned} \tag{2.5}$$

Upon integration by parts, the variation in S becomes

$$\delta S = \left[\delta x \frac{\partial L}{\partial \dot{x}} \right]_{t_a}^{t_b} - \int_{t_a}^{t_b} \delta x \left[\frac{d}{dt}\left(\frac{\partial L}{\partial \dot{x}} \right) - \frac{\partial L}{\partial x} \right] dt \tag{2.6}$$

Since $\delta x(t)$ is zero at the end points, the first term on the right-hand side of the equation is zero. Between the end points $\delta x(t)$ can take on any arbitrary value. Thus the extremum is that curve along which the following condition is always satisfied:

$$\frac{d}{dt}\left(\frac{\partial L}{\partial \dot{x}} \right) - \frac{\partial L}{\partial x} = 0 \tag{2.7}$$

This is, of course, the classical lagrangian equation of motion.

In classical mechanics, the *form* of the action integral $S = \int L\,dt$ is interesting, not just the extreme value S_{cl}. This interest derives from the necessity to know the action along a set of neighboring paths in order to determine the path of least action.

In quantum mechanics both the form of the integral and the value of the extremum are again important. In the following problems we shall evaluate the extremum in a variety of situations.

Problem 2-1 For a free particle $L = (m/2)\dot{x}^2$. Show that the action S_{cl} corresponding to the classical motion of a free particle is

$$S_{cl} = \frac{m}{2}\frac{(x_b - x_a)^2}{t_b - t_a} \tag{2.8}$$

Problem 2-2° For a harmonic oscillator $L = (m/2)(\dot{x}^2 - \omega^2 x^2)$. With T equal to $t_b - t_a$, show that the classical action is

$$S_{cl} = \frac{m\omega}{2\sin\omega T}[(x_b^2 + x_a^2)\cos\omega T - 2x_b x_a] \tag{2.9}$$

Problem 2-3° Find S_{cl} for a particle under a constant force f, that is, $L = (m/2)\dot{x}^2 + fx$.

Problem 2-4 Classically, the momentum is defined as

$$p = \frac{\partial L}{\partial \dot{x}} \tag{2.10}$$

Show that the momentum at a final point is

$$\left(\frac{\partial L}{\partial \dot{x}}\right)_{x=x_b} = +\frac{\partial S_{cl}}{\partial x_b} \tag{2.11}$$

while the momentum at an initial point is

$$\left(\frac{\partial L}{\partial \dot{x}}\right)_{x=x_a} = -\frac{\partial S_{cl}}{\partial x_a}$$

Hint: Consider the effect on Eq. (2.6) of a change in the end points.

Problem 2-5 Classically, the energy is defined as

$$E = \dot{x}p - L \tag{2.12}$$

Show that the energy at a final point is

$$\dot{x}_b \left(\frac{\partial L}{\partial \dot{x}}\right)_{x=x_b} - L(x_b) = -\frac{\partial S_{cl}}{\partial t_b} \tag{2.13}$$

while the energy at an initial point is

$$+\frac{\partial S_{cl}}{\partial t_a}$$

Hint: A change in the time of an end point requires a change in path, since all paths must be classical paths.

2-2 THE QUANTUM-MECHANICAL AMPLITUDE

Now we can give the quantum-mechanical rule. We must say how much each trajectory contributes to the total amplitude to go from a to b. It is not just the particular path of extreme action that contributes;

rather, all the paths contribute. They contribute equal magnitudes to the total amplitude, but contribute at different phases. The phase of the contribution for a given path is the action S for that path in units of the quantum of action $\hbar$. That is, to summarize: The probability $P(b,a)$ to go from a point x_a at time t_a to the point x_b at time t_b is the absolute square $P(b,a) = |K(b,a)|^2$ of an amplitude $K(b,a)$ to go from a to b. This amplitude is the sum of contributions $\phi[x(t)]$ from each path.

$$K(b,a) = \sum_{\text{paths from } a \text{ to } b} \phi[x(t)] \tag{2.14}$$

The contribution of a path has a phase proportional to the action S:

$$\phi[x(t)] = \text{const } e^{(i/\hbar)S[x(t)]} \tag{2.15}$$

The action is that for the corresponding classical system (see Eq. 2.1). The constant will be chosen to normalize K correctly, and it will be taken up later (Sec. 2.4) when we discuss more mathematically just what we mean in Eq. (2.14) by a sum over paths.

2-3 THE CLASSICAL LIMIT

Before we go on to making the mathematics more complete, we shall compare this quantum law with the classical rule. At first sight, from Eq. (2.15) all paths contribute equally, although their phases vary, so it is not clear how, in the classical limit, some particular path becomes most important. The classical approximation, however, corresponds to the case that the dimensions, masses, times, etc., are so large that S is enormous in relation to $\hbar$ ($= 1.05 \times 10^{-27}$ erg·sec). Then the phase of the contribution $S/\hbar$ is some very, very large angle. The real (or imaginary) part of ϕ is the cosine (or sine) of this angle. This is as likely to be plus as minus. Now if we move the path as shown in Fig 2-1 by a small amount δx, *small on the classical scale*, the change in S is likewise small on the classical scale, but not when measured in the tiny units of $\hbar$. These small changes in path will, generally, make enormous changes in phase, and our cosine or sine will oscillate exceedingly rapidly between plus and minus values. The total contribution will then add to zero; for if one path makes a positive contribution, another infinitesimally close (on a classical scale) makes an equal negative contribution, so that no net contribution arises.

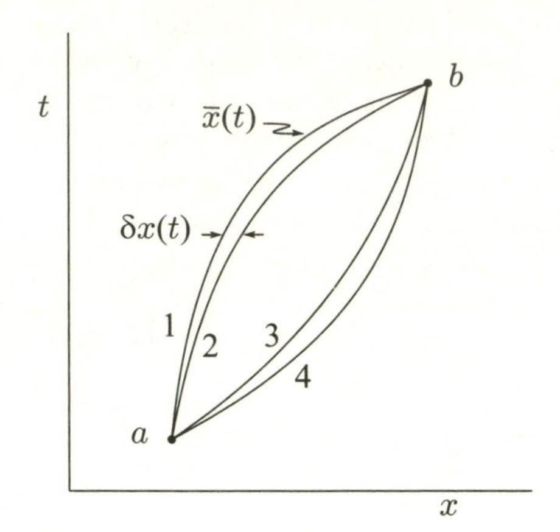

Fig. 2-1 The classical path 1, $\bar{x}(t)$, is that for which a certain integral, the action S, is minimum. If the path is varied by $\delta x(t)$, to path 2, the integral suffers no first-order change. This determines the classical equation of motion.

In quantum mechanics, the amplitude to go from a to b is the sum of amplitudes for each interfering alternative path. The amplitude for a given path, $e^{iS/\hbar}$, has a phase proportional to the action.

If the action is very large compared to $\hbar$, neighboring paths such as 3 and 4 have slightly different actions — slightly different on a classical scale. Such paths will (because of the smallness of $\hbar$) have very different phases. Their contributions will cancel out. Only in the vicinity of the classical path $\bar{x}(t)$, where the action changes little when the path varies, will neighboring paths, such as 1 and 2, contribute in the same phase and constructively interfere. That is why the approximation of classical physics — that only the path $\bar{x}(t)$ need be considered — is valid when the action is very large compared to $\hbar$.

Therefore, no path really needs to be considered if the neighboring path has a different action; for the paths in the neighborhood cancel out the contribution. But for the special path $\bar{x}(t)$, for which S is an extremum, a small change in path produces, in the first order at least, no change in S. All the contributions from the paths in this region are nearly in phase, at phase $S_{cl}/\hbar$, and do not cancel out. Therefore, only for paths in the vicinity of $\bar{x}(t)$ can we get important contributions, and in the classical limit we need only consider this particular trajectory as being of importance. In this way the classical laws of motion arise from the quantum laws.

We may note that trajectories which differ from $\bar{x}(t)$ contribute as long as the action is still within about $\hbar$ of S_{cl}. The classical trajectory is indefinite to this slight extent, and this rule serves as a measure of the limitations of the precision of the classically defined trajectory.

Next consider the dependence of the phase on the position of the end point (x_b, t_b). If we change the end point a little, this phase changes a great deal, and $K(b,a)$ changes very rapidly. If by a "smooth function" we mean one like S_{cl} which changes only when changes in argument which are appreciable on a classical scale are made, we note that $K(b,a)$ is far from smooth, but in this classical approximation our arguments show that it is of the form

$$K(b,a) = \text{"smooth function"} \cdot e^{iS_{cl}/\hbar} \tag{2.16}$$

All these approximate considerations apply to a situation on a scale for which classical physics might be expected to work ($S \gg \hbar$). But at an atomic level, S may be comparable with $\hbar$, and then all trajectories must be added in Eq. (2.14) in detail. No particular trajectory is of overwhelming importance, and of course Eq. (2.16) is not necessarily a good approximation. To deal with such cases, we shall have to find out how to carry out such sums as are implied by Eq. (2.14).

2-4 THE SUM OVER PATHS

Analogy with the Riemann Integral. Although the qualitative idea of a sum of a contribution for each of the paths is clear, a more precise mathematical definition of such a sum must be given. The number of paths is a high order of infinity, and it is not evident what measure is to be given to the space of paths. It is our purpose in this section to give such a mathematical definition. This definition will be found rather cumbersome for actual calculation. In the succeeding chapters we shall describe other and more efficient methods of computing the sum over all paths. As for this section, it is hoped that the mathematical difficulty, or rather inelegance, will not distract the reader from the physical content of the ideas.

We can begin our understanding with a consideration of the ordinary Riemann integral. We could say, very roughly, that the area A under a curve is the sum of all its ordinates. Better, we could say that it is proportional to that sum. But to make the idea precise, we do this: take a subset of all ordinates (e.g., those spaced at equal intervals h). Adding these ordinates, we obtain

$$A \sim \sum_i f(x_i) \tag{2.17}$$

where the summation is carried out over the finite set of points x_i, as shown in Fig. 2.2.

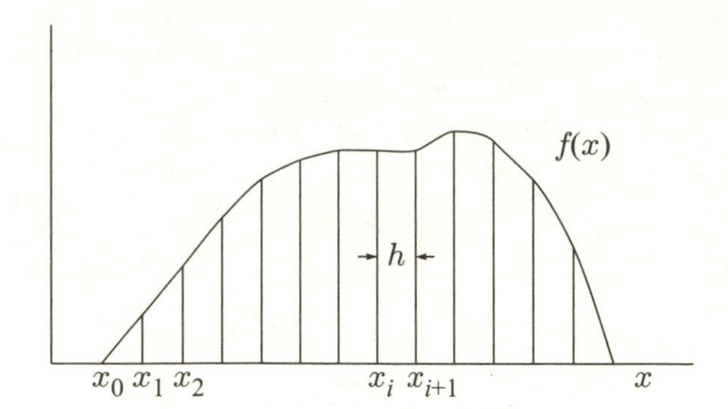

Fig. 2-2 In the definition of the ordinary Riemann integral, a set of ordinates is drawn from the abscissa (the x-axis) to the curve. The ordinates are spaced a distance h apart. The integral (area between the curve and the abscissa) is approximated by h times the sum of the ordinates. This approximation approaches the correct value as h approaches zero.

An analogous definition can be used for path integrals. The measure which goes to zero in the limit process is the time interval ϵ between discrete points on the paths.

The next step is to define A as the limit of this sum as the subset of points, and thus the subset of ordinates, becomes more complete or — because a finite set is never any measurable part of the infinite continuum — we may better say as the subset becomes more representative of the complete set. We can pass to the limit in an orderly manner by taking continually smaller and smaller values of h. In so doing, we would obtain a different sum for each value of h. No limit exists. In order to obtain a limit to this process, we must specify some normalizing factor which should depend on h. Of course, for the Riemann integral, this factor is just h itself. Now the limit exists and we may write the expression

$$A = \lim \left[h \sum_i f(x_i) \right] \tag{2.18}$$

Constructing the Sum. We can follow through an analogous procedure in defining the sum over all paths. First, we choose a subset of all paths. To do this, we divide the independent variable time into steps of width ϵ. This gives us a set of values t_i spaced an interval ϵ apart between the values t_a and t_b. At each time t_i we select some special point x_i. We construct a path by connecting all the points so selected with straight lines. It is possible to define a sum over all paths constructed in this manner by taking a multiple integral over all values of x_i for i

between 1 and $N-1$, where

$$\begin{aligned} N\epsilon &= t_b - t_a \\ \epsilon &= t_{i+1} - t_i \\ t_0 &= t_a \qquad t_N = t_b \\ x_0 &= x_a \qquad x_N = x_b \end{aligned} \tag{2.19}$$

The resulting equation is

$$K(b,a) \sim \int \cdots \iint \phi[x(t)]\, dx_1\, dx_2 \cdots dx_{N-1} \tag{2.20}$$

We do not integrate over x_0 or x_N because these are the fixed end points x_a and x_b. This equation corresponds formally to Eq. (2.17). In the present case we can obtain a more representative sample of the complete set of all possible paths between a and b by making ϵ smaller. However, just as in the case of the Riemann integral, we cannot proceed to the limit of this process because the limit does not exist. Once again we must provide some normalizing factor which we expect will depend upon ϵ.

Unfortunately, to define such a normalizing factor seems to be a very difficult problem and we do not know how to do it in general. But we do know how to give the definition for all situations which so far seem to have practical value. For example, take the case where the lagrangian is given by Eq. (2.2). The normalizing factor turns out to be A^{-N}, where

$$A = \left(\frac{2\pi i\hbar\epsilon}{m}\right)^{1/2} \tag{2.21}$$

We shall see later (in Sec. 4-1) how this result is obtained. With this factor the limit exists° and we may write

$$K(b,a) = \lim_{\epsilon \to 0} \frac{1}{A} \int \cdots \iint e^{(i/\hbar)S[b,a]}\, \frac{dx_1}{A}\, \frac{dx_2}{A} \cdots \frac{dx_{N-1}}{A} \tag{2.22}$$

where

$$S[b,a] = \int_{t_a}^{t_b} L(\dot{x}, x, t)\, dt \tag{2.23}$$

is a line integral taken over the trajectory passing through the points x_i with straight sections between, as in Fig. 2-3.

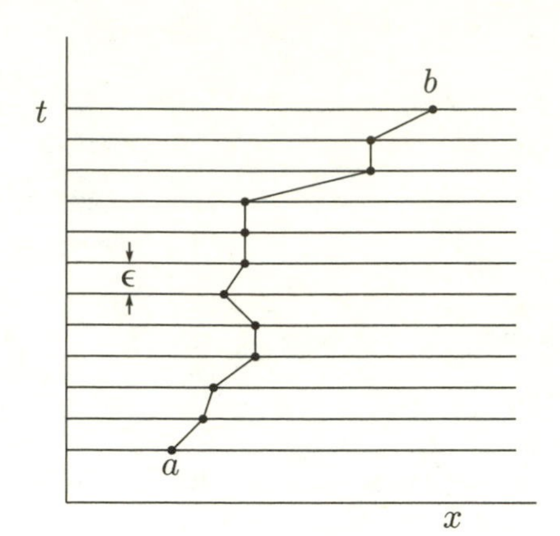

Fig. 2-3 The sum over paths is defined as a limit, in which at first the path is specified by giving only its coordinate x at a large number of specified times separated by very small time intervals ϵ. The path sum is then an integral over all these specific coordinates. Then to achieve the correct measure, the limit is taken as ϵ approaches 0.

It is possible to define the path in a somewhat more elegant manner. Instead of straight lines between the points i and $i+1$, we could use sections of the classical orbit. Then we could say that S is the minimum value of the integral of the lagrangian over all the paths which go through the specified points (x_i, t_i). With this definition no recourse is made to arbitrary straight lines.

The Path Integral. There are many ways to define a subset of all the paths between a and b. The particular definition we have used here may not be the best for some mathematical purposes. For example, suppose the lagrangian depends upon the acceleration of x. In the way we have constructed the path, the velocity is discontinuous at the various points (x_i, t_i); that is, the acceleration is infinite at these points. It is possible that this situation would lead to trouble. However, in the few such examples with which we have had experience the substitution

$$\ddot{x} = \frac{1}{\epsilon^2}(x_{i+1} - 2x_i + x_{i-1}) \tag{2.24}$$

has been adequate. There may be other cases where no such substitution is available or adequate, and the present definition of a sum over all paths is just too awkward to use.

A similar situation arises in ordinary integration, where sometimes the Riemann definition, Eq. (2.18), is not adequate and recourse must be had to some other definition, such as that of Lebesgue. The need to redefine the method of integration does not destroy the concept of integration. So we feel that the possible awkwardness of the special definition of the sum over all paths (as given in Eq. 2.22) may eventually require new definitions to be formulated. Nevertheless, the concept of the sum over all paths, like the concept of an ordinary integral, is independent of a special definition and valid in spite of the failure of such definitions. Thus we shall write the sum over all paths in a less restrictive notation as

$$K(b,a) = \int_a^b e^{(i/\hbar)S[b,a]}\,\mathcal{D}x(t) \tag{2.25}$$

which we shall call a *path integral.* The identifying notation in this expression is the script $\mathcal{D}$. Only rarely shall we return to the form given in Eq. (2.22).

Problem 2-6 The class of functionals for which path integrals can be defined is surprisingly varied. So far we have considered functionals such as that given in Eq. (2.15). Here we shall consider quite a different type. This latter type of functional arises in a one-dimensional relativistic problem. Suppose a particle moving in one dimension can go only forward or backward at the velocity of light. For convenience, we shall define the units such that the velocity of light, the mass of the particle, and Planck's constant are all unity. Then in the xt plane all trajectories shuttle back and forth with slopes of $\pm 45°$, as in Fig. 2-4. The amplitude for such a path can be defined as follows: Suppose time is divided into small equal steps of length ϵ. Suppose reversals of path direction can occur only at the boundaries of these steps, i.e., at $t = t_a + n\epsilon$, where n is an integer. For this relativistic problem the amplitude to go along such a path is different from the amplitude defined in Eq. (2.15). The correct definition for the present case is

$$\phi = (i\epsilon)^R \tag{2.26}$$

where R is the number of reversals, or corners, along the path.

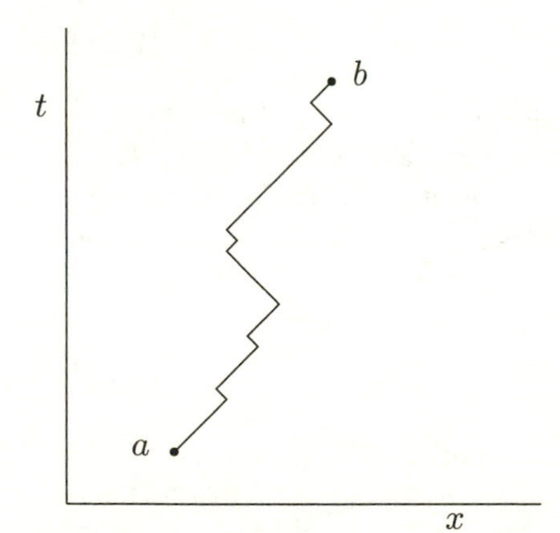

Fig. 2-4 The path of a relativistic particle traveling in one dimension is a zigzag of straight segments. The slope of the segments is constant in magnitude and differs only in sign from zig to zag. The amplitude for a particular path, as well as the kernel to go from a to b, depends on the number of corners R along a path, as shown by Eqs. (2.26) and (2.27).

As a problem, the reader may use this definition to calculate the kernel $K(b,a)$ by adding together the contribution for the paths of one corner, two corners, etc. Thus determine

$$K(b,a) = \sum_R N(R)(i\epsilon)^R \tag{2.27}$$

where $N(R)$ is the number of paths possible with R corners. It is best to calculate four separate K's, namely, the amplitude $K_{++}(b,a)$ of starting at the point a with a positive velocity and coming into the point b with a positive velocity, the amplitude $K_{+-}(b,a)$ of starting at the point a with a negative velocity and coming into the point b with a positive velocity, and the amplitudes $K_{-+}(b,a)$ and $K_{--}(b,a)$ defined in a similar fashion.

Next suppose the unit of time is defined as $\hbar/mc^2$. If the time interval is very long $[t_b - t_a \gg \hbar/mc^2]$ and the average velocity is small $[x_b - x_a \ll c(t_b - t_a)]$, show that the resulting kernel is approximately the same as that for a nonrelativistic free particle (given in Eq. 3.3), except for a factor $\exp\{-(i/\hbar)mc^2(t_b - t_a)\}$. The definition given here for the amplitude, and the resulting kernel, is correct for a relativistic theory of a free particle moving in one dimension. The result is equivalent to the Dirac equation for that case.

2-5 EVENTS OCCURRING IN SUCCESSION

The Rule for Two Events. In this section we shall derive an important law for the composition of amplitudes for events which occur successively in time. Suppose t_c is some time between t_a and t_b. Then the action along any path between a and b can be written as

$$S[b,a] = S[b,c] + S[c,a] \tag{2.28}$$

This follows from the definition of the action as an integral in time and also from the fact that L does not depend on derivatives higher than the velocity. (Otherwise, we would have to specify values of velocity and perhaps higher derivatives at point c.) Using Eq. (2.25) to define the kernel, we can write

$$K(b,a) = \int_a^b e^{(i/\hbar)S[b,c]+(i/\hbar)S[c,a]}\,\mathcal{D}x(t) \tag{2.29}$$

It is possible to split any path into two parts. The first part would have the end points x_a and $x_c = x(t_c)$, and the second part would have the end points x_c and x_b, as shown in Fig. 2.5. It is possible to integrate

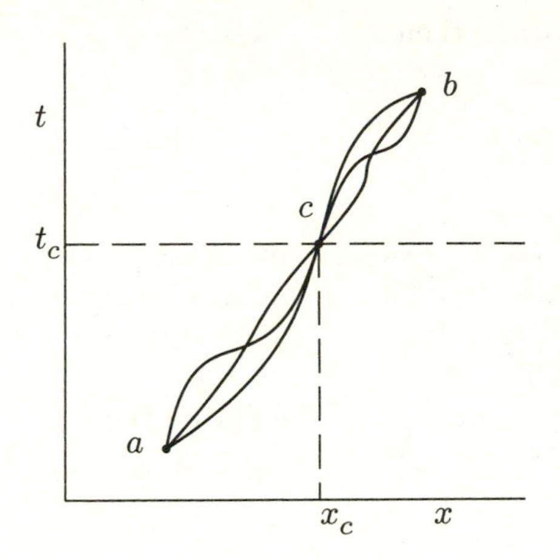

Fig. 2-5 One way the sum over all paths can be taken is by first summing over paths which go through the point at x_c and time t_c and later summing over the points x_c. The amplitude on each path that goes from a to b via c is a product of two factors: (1) an amplitude to go from a to c and (2) an amplitude to go from c to b. This is therefore valid also for the sum over all paths through c: the total amplitude to go from a to b via c is $K(b,c)K(c,a)$. Thus summing over the alternatives (values of x_c), we get for the total amplitude to go from a to b, Eq. (2.31).

over all paths from a to c, then over all paths from c to b, and finally integrate the result over all possible values of x_c. In performing the first step of this integration $S[b,c]$ is constant. Thus the result can be written as

$$K(b,a) = \int_{-\infty}^{\infty}\int_{c}^{b} e^{(i/\hbar)S[b,c]}K(c,a)\,\mathcal{D}x(t)\,dx_c \tag{2.30}$$

where integrations must now be carried out not only over paths between c and b but also over the variable end point x_c. In the next step we carry out the integration over all paths between some point with an arbitrary x_c and the point b. All that is left is an integral over all possible values of x_c. Thus

$$K(b,a) = \int_{-\infty}^{\infty} K(b,c)K(c,a)\,dx_c \tag{2.31}$$

Perhaps the argument is clearer starting from Eq. (2.22). Select one of the discrete times as t_c. Thus let $t_c = t_k$ and $x_c = x_k$. First carry out all the integrations over those x_i such that $i < k$. This will introduce the factor in the integral $K(c,a)$. Next carry out the integrals over all those x_i such that $i > k$. This introduces the factor $K(b,c)$. All that is left is an integral over x_c. The result can be written as Eq. (2.31).

This result can be summarized in the following way. All alternative paths from a to b can be labeled by specifying the position x_c through which they pass at time t_c. Then the kernel for a particle going from a to b can be computed from the rules:

1. The kernel to go from a to b is the sum, over all possible values of x_c, of amplitudes for the particle to go from a to c and then to b.

2. The amplitude to go from a to c and then to b is the kernel to go from a to c times the kernel to go from c to b.

Thus we have the rule: *Amplitudes for events occurring in succession in time multiply.*

Extension to Several Events. There are many applications for the important rule, and several will be developed in succeeding chapters. Here we shall show the application wherein we follow an alternative route in deriving the equation for the kernel, Eq. (2.22).

It is perfectly possible to make two divisions in all the paths: one at t_c and the other at, say, t_d. Then the kernel for a particle going from a to b can be written

$$K(b,a) = \int_{-\infty}^{\infty}\int_{-\infty}^{\infty} K(b,c)K(c,d)K(d,a)\,dx_d\,dx_c \tag{2.32}$$

This means that we look at a particle which goes from a to b as if it went first from a to d, then from d to c, and finally from c to b. The amplitude to follow such a path is the product of the kernels for each part of the path. The kernel taken over all such paths that go from a to b is obtained by integrating this product over all possible values of x_d and x_c.

We can continue this process until we have the time scale divided into N intervals. The result is

$$K(b,a) = \int_{x_{N-1}} \cdots \int_{x_2}\int_{x_1} K(b,N-1)K(N-1,N-2)\cdots \times K(i+1,i)\cdots K(1,a)\,dx_1\,dx_2\cdots dx_{N-1} \tag{2.33}$$

This means that we can define the kernel in a manner different from that given in Eq. (2.22). In this alternative definition the kernel for a particle to go between two points separated by an infinitesimal time interval ϵ is

$$K(i+1,i) = \frac{1}{A}\exp\left\{\frac{i}{\hbar}\epsilon L\left(\frac{x_{i+1}-x_i}{\epsilon}, \frac{x_{i+1}+x_i}{2}, \frac{t_{i+1}+t_i}{2}\right)\right\} \tag{2.34}$$

which is correct to first order in ϵ. Then by the rules for multiplying the amplitudes of events which occur successively in time, we have

$$\phi[x(t)] = \lim_{\epsilon\to 0}\prod_{i=0}^{N-1} K(i+1,i) \tag{2.35}$$

for the amplitude of a complete path. Then, using the rule that amplitudes for alternative paths add, we arrive at a definition for $K(b,a)$.

It can be seen that the resulting expression is actually the same as Eq. (2.22).

2-6 SOME REMARKS

In the relativistic theory of the electron we shall not find it possible to express the amplitude for a path as $e^{iS/\hbar}$, or in any other simple way. However, the laws for combining amplitudes still work (with some small modifications). The amplitude for a trajectory still exists. As a matter of fact, it is still given by Eq. (2.35). The only difference is that $K(i+1, i)$ is not so easily expressed in a relativistic theory as it is in Eq. (2.34). The complications arise from the necessity to consider spin and the possibility of the production of pairs of electrons and positrons.

In nonrelativistic systems with a larger number of variables, and even in the quantum theory of electromagnetic field, not only do the laws for combining amplitudes still hold but the amplitude itself follows the rules set down in this chapter. That is, each motion of a variable has an amplitude whose phase is $1/\hbar$ times the action associated with the motion.

We shall take up these more complicated examples in later chapters.

3

Developing the Concepts with Special Examples

IN this chapter we shall develop the kernels governing some special types of motion. We shall explore the physical meaning of the mathematical results in order to develop some physical intuition about motion under quantum-mechanical laws. The wave function will be introduced and its relation to the kernel will be described. This represents the first step in connecting the present approach to quantum mechanics with more traditional approaches.

We shall also introduce some special mathematical methods for computing the sum over all paths. The idea of a sum over all paths was described in Chap. 2 with the help of a particular computational method. Although that method may clarify the concept, it is an awkward tool with which to work. The simpler methods introduced in this chapter will be of great use in our future work.

Thus the present chapter has three purposes: deepening our understanding of quantum-mechanical principles, beginning the connection between our present approach and other approaches, and introducing some useful mathematical methods.

3-1 THE FREE PARTICLE

The Path Integral. The method used in Chap. 2 to describe a sum over all paths will be used here to compute the kernel for a free particle. The lagrangian for a free particle is

$$L = \frac{m}{2}\dot{x}^2 \tag{3.1}$$

Thus with the help of Eqs. (2.21) to (2.23) the kernel for a free particle (distinguished by the subscript 0) is

$$K_0(b,a) = \lim_{\epsilon\to 0}\left(\frac{m}{2\pi i\hbar\epsilon}\right)^{N/2} \times \int\cdots\int \exp\left\{\frac{im}{2\hbar\epsilon}\sum_{i=1}^{N}(x_i - x_{i-1})^2\right\} dx_1\cdots dx_{N-1} \tag{3.2}$$

This represents a set of gaussian integrals, i.e., integrals of the form $\int e^{-ax^2}\,dx$ or $\int e^{-ax^2+bx}\,dx$. Since the integral of a gaussian is again a gaussian, we may carry out the integrations on one variable after the other. After the integrations are completed, the limit may be taken. The result is

$$K_0(b,a) = \left(\frac{m}{2\pi i\hbar(t_b - t_a)}\right)^{1/2} \exp\left\{\frac{im(x_b - x_a)^2}{2\hbar(t_b - t_a)}\right\} \tag{3.3}$$

The calculation is carried out as follows. Notice first that

$$\left(\frac{m}{2\pi i\hbar\epsilon}\right)^{2/2}\int_{-\infty}^{\infty}\exp\left\{\frac{im}{2\hbar\epsilon}[(x_2-x_1)^2+(x_1-x_0)^2]\right\}dx_1 = \left(\frac{m}{2\pi i\hbar\cdot 2\epsilon}\right)^{1/2}\exp\left\{\frac{im}{2\hbar\cdot 2\epsilon}(x_2-x_0)^2\right\} \tag{3.4}$$

Next we multiply this result by

$$\left(\frac{m}{2\pi i\hbar\epsilon}\right)^{1/2}\exp\left\{\frac{im}{2\hbar\epsilon}(x_3-x_2)^2\right\} \tag{3.5}$$

and integrate again, this time over x_2. The result is similar to that of Eq. (3.4), except that $(x_2-x_0)^2$ becomes $(x_3-x_0)^2$ and the expression 2ϵ is replaced by 3ϵ in two places. Thus we get

$$\left(\frac{m}{2\pi i\hbar\cdot 3\epsilon}\right)^{1/2}\exp\left\{\frac{im}{2\hbar\cdot 3\epsilon}(x_3-x_0)^2\right\}$$

In this way a recursion process is established which, after $n-1$ steps, gives

$$\left(\frac{m}{2\pi i\hbar\cdot n\epsilon}\right)^{1/2}\exp\left\{\frac{im}{2\hbar\cdot n\epsilon}(x_n-x_0)^2\right\}$$

Since $n\epsilon = t_n - t_0$, it is easy to see that the result after $N-1$ steps is identical with Eq. (3.3).

There is an alternative procedure. Equation (3.4) can be used to integrate over all variables x_i for which i is odd (assuming N is even). The result is an expression formally like Eq. (3.2), but with half as many variables of integration. The remaining variables are defined at points in time spaced a distance 2ϵ apart. Hence, at least in the case that N is of the form 2^k, Eq. (3.3) results from k steps of this kind.

Problem 3-1 The probability that a particle arrives at the point b is by definition proportional to the absolute square of the kernel $K(b,a)$. For the free-particle kernel of Eq. (3.3) this is

$$P(b)\,dx = \frac{m}{2\pi\hbar(t_b-t_a)}\,dx \tag{3.6}$$

Clearly this is a relative probability, since the integral over the complete range of x diverges. What does the particular normalization mean? Show that this corresponds to a classical picture in which a particle starts from the point a with all momenta equally likely. Show that the corresponding relative probability that the momentum of the particle lies in the range dp is $dp/2\pi\hbar$.

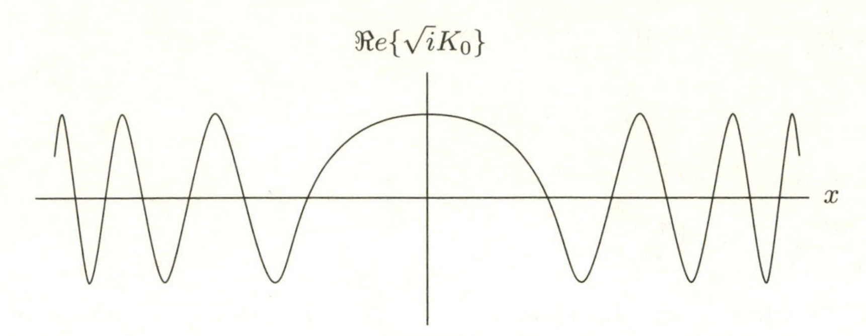

Fig. 3-1 The real part of $\sqrt{i}$ times amplitude to arrive at various distances x from the origin after a given time t. The imaginary part (not shown) is an analogous wave 90° out of phase, so that the absolute square of the amplitude is constant. The wavelength is short at large x, namely, where a classical particle could arrive only if it moved with high velocity. Generally, the wavelength and classical momentum are inversely related (see Eq. 3.10).

Momentum and Energy. We now study some of the implications of the free-particle kernel. For convenience let the point a represent the origin in both time and space. The amplitude to go to some other point $b = (x, t)$ is

$$K_0(x, t; 0, 0) = \left(\frac{m}{2\pi i\hbar t}\right)^{1/2} \exp\left\{\frac{imx^2}{2\hbar t}\right\} \tag{3.7}$$

If time is fixed, the amplitude varies with distance as shown in Fig. 3-1, in which the real part of $\sqrt{i}K_0(x, t,; 0, 0)$ is plotted.

We see that as we get farther from the origin the oscillations become more and more rapid. If x is so large that many oscillations have occurred, then the distance between successive nodes is nearly constant, at least for the next few oscillations. That is, the amplitude behaves much like a sine wave of slowly varying wavelength λ. Changing x by λ must increase the phase of the amplitude by 2π. That is,

$$2\pi = \frac{m(x+\lambda)^2}{2\hbar t} - \frac{mx^2}{2\hbar t} = \frac{mx\lambda}{\hbar t} + \frac{m\lambda^2}{2\hbar t} \tag{3.8}$$

Neglecting the quantity λ^2 relative to $x\lambda$ (that is, assuming $x \gg \lambda$), we find

$$\lambda = \frac{2\pi\hbar}{m(x/t)} \tag{3.9}$$

From a classical point of view a particle which moves from the origin to x in the time interval t has a velocity x/t and a momentum mx/t. From the quantum-mechanical point of view, when the motion can be adequately described by assigning a classical momentum to the particle of $p = mx/t$, then the amplitude varies in space with the wavelength

$$\lambda = \frac{h}{p} \tag{3.10}$$

We may show this relation still more generally. Suppose we have some large piece of apparatus, such as a magnetic analyzer, which is supposed to bring particles of a given momentum p to a given point. We shall show that, whenever the apparatus is large enough that classical physics offers a good approximation, then the amplitude for a particle to arrive at the prescribed point varies rapidly in space with a wavelength equal to h/p. For as we have seen, in such a situation, the kernel is approximated by

$$K \sim \exp\left\{\frac{i}{\hbar}S_{cl}(b,a)\right\} \tag{3.11}$$

Changes in the position of the final point x_b cause variations in the classical action. If this action is large compared to $\hbar$ (the semiclassical approximation), the kernel will oscillate very rapidly with changes in x_b. The change in phase per unit displacement of the end point is

$$k = \frac{1}{\hbar}\frac{\partial S_{cl}}{\partial x_b} \tag{3.12}$$

but $\partial S_{cl}/\partial x_b$ is the classical momentum of the particle when it arrives at the point x_b (see Prob. 2-4). Thus $p = \hbar k$. This quantity k, the phase change per unit distance of a wave, is called the wave number, and it is very convenient to use. Since the wavelength is the distance over which the phase changes by 2π, then $k = 2\pi/\lambda$. Equation (3.12) is de Broglie's formula relating the momentum to the wave number of a wave, $p = \hbar k$.

Next, let us study the time dependence of the free-particle kernel given by Eq. (3.7). Suppose we hold the distance fixed and vary the time. The variation of the real part of $\sqrt{i}$ times the kernel is shown in Fig. 3-2. Both frequency and amplitude change with t.

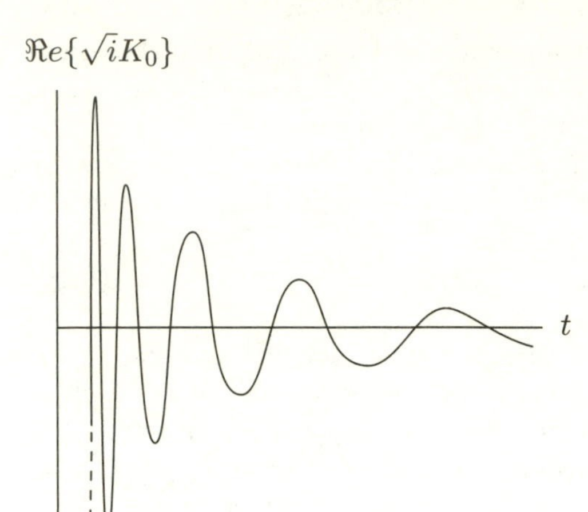

Fig. 3-2 The amplitude to find the particle at a given point varies with time. The real part of $\sqrt{i}K_0$ is plotted here. The frequency of the oscillations is proportional to the energy that a classical particle would have to have to arrive at the point in question within the time interval t.

Suppose t is very large and neglect the change in amplitude with variations of t. The period of oscillation T is defined as the time required to decrease the phase by 2π. Thus

$$2\pi = \frac{mx^2}{2\hbar t} - \frac{mx^2}{2\hbar(t+T)} = \frac{mx^2}{2\hbar t^2}\left(\frac{T}{1+T/t}\right) \tag{3.13}$$

By introducing the angular frequency $\omega = 2\pi/T$, and assuming $t \gg T$, we can write this equation as

$$\omega \approx \frac{m}{2\hbar}\left(\frac{x}{t}\right)^2 \tag{3.14}$$

Since $(m/2)(x/t)^2$ is the classical energy of a free particle, this equation says

$$\text{Energy} = \hbar\omega \tag{3.15}$$

This relation, like the one relating momentum and wavelength, holds for any apparatus which can be adequately described by classical physics; and, like the previous relation, it can be obtained from a more general argument.

Referring to Eq. (3.11), any variation of the time t_b of an end point will cause a rapid oscillation of the kernel. The resulting frequency is

$$\omega = -\frac{1}{\hbar}\frac{\partial S_{cl}}{\partial t_b} \tag{3.16}$$

The quantity $-\partial S_{cl}/\partial t_b$ is interpreted classically as the energy E (refer to Prob. 2-5). Thus

$$\omega = \frac{E}{\hbar} \tag{3.17}$$

In this way the concepts of momentum and energy are extended to quantum mechanics by the following rules:

1. If the amplitude varies in space as e^{ikx}, we say that the particle has momentum $\hbar k$.
2. If the amplitude varies in time as $e^{-i\omega t}$, we say that the particle has energy $\hbar\omega$.

We have just shown that these rules agree with the usual definitions of energy and momentum in the classical limit.

Problem 3-2 Show by substitution that the free-particle kernel $K_0(b,a)$ satisfies the differential equation

$$\frac{\partial K_0(b,a)}{\partial t_b} = -\frac{i}{\hbar}\left[-\frac{\hbar^2}{2m}\frac{\partial^2 K_0(b,a)}{\partial x_b^2}\right] \tag{3.18}$$

whenever t_b is greater than t_a.

3-2 DIFFRACTION THROUGH A SLIT°

The Conceptual Experiment. We can learn more about the physical interpretation of quantum mechanics and its relation to classical mechanics by considering another, somewhat more complicated, example. Suppose a free particle is liberated at $t = 0$ from the origin and then, at an interval of time T later, we observe that it is at a certain point X. Classically, we would say that the particle has had a velocity $V = X/T$. The implication would be that if a particle were left alone to continue for another interval of time t', it would move an additional distance $x' = Vt'$. To analyze this quantum-mechanically, we shall attempt to solve the following problem:

At $t = 0$ the particle starts from the origin $x = 0$. After an interval T we shall suppose that it is known that the particle is within the distance $\pm b$ of X. We ask: After an additional interval t', what is the probability of finding the particle at an additional displacement x' from the position X? The net amplitude to arrive at this position x' at the time $T + t'$ can be considered as the sum of contributions from every trajectory that goes from the origin to the final point, provided that that trajectory lies in the interval $\pm b$ from X at the time T.

We shall calculate this in a moment, but first it is worth remarking on what kind of an experiment we are contemplating here. How can we know that the particle passes the point X within the interval $\pm b$? One way would be to make an observation of the particle at the time

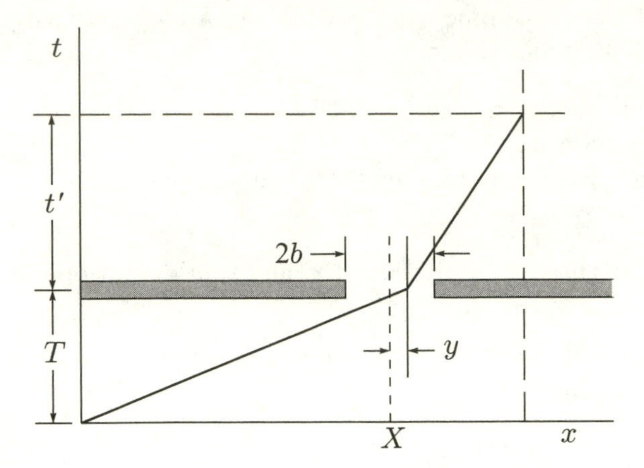

Fig. 3-3 A particle starting at $x = 0$ when $t = 0$ is determined to pass between $X - b$ and $X + b$ at $t = T$. We wish to calculate the probability of finding the particle at some point x at a time t' seconds later, i.e., when $t = T + t'$. According to classical laws, the particle would have to be between $(X - b)(1 + t'/T)$ and $(X + b)(1 + t'/T)$, that is, between the rectilinear extensions of the original slit. However, quantum-mechanical laws show that such particles have nonzero probability of appearing outside these classical limits.

We cannot approach this problem by a single application of the free-particle law of motion, since the particle is actually constrained by the slit. So we break the problem up into two successive free-particle motions. The first takes the particle from $x = 0$ at $t = 0$ to $x = X + y$ at $t = T$, where $|y| \leq b$. The second takes the particle from $x = X + y$ at $t = T$ to x at $t = T + t'$. The overall amplitude is an integral of the product of these two free-particle kernels, as shown in Eq. (3.19).

T to see if it is within the interval $\pm b$ This would be the most natural way to proceed, but it is somewhat more difficulty to analyze in detail (because of the complicated interaction between the particle and the observing mechanism) than another way of doing the experiment.

Suppose we look, say, with very strong light, everywhere all along the x axis *except* within $\pm b$ from the point X at the time T. If we find the particle, we discontinue the experiment. We consider only those cases in which a thorough investigation of the region, *except* for the region $\pm b$, shows no particle is present. That is, all trajectories which pass outside the limits $\pm b$ from X are rejected. The experimental situation is diagramed in Fig. 3-3. The amplitude then can be written as

$$\psi(x') = \int_{-b}^{b} K(X + x', T + t'; X + y, T) K(X + y, T; 0, 0)\, dy \tag{3.19}$$

This expression is written down in accordance with the rule for combining amplitudes of events occurring in succession in time (Sec. 2-5).

The first event is that the particle goes from the origin to the slit. The second event is that the particle proceeds from the slit to the point x' further on. The slit has a finite width, and passage through each elemental interval of the slit represents and alternative way of proceeding along the complete path. Thus we must integrate over the width of the slit. All particles which miss the slit are captured and removed from the experiment. Amplitudes for such particles are not included. All the particles which get through the slit move as free particles with kernels given by Eq. (3.3). Thus the amplitude is

$$\psi(x') = \int_{-b}^{b} \left(\frac{m}{2\pi i\hbar t'}\right)^{1/2} \exp\left\{\frac{im(x'-y)^2}{2\hbar t'}\right\} \times \left(\frac{m}{2\pi i\hbar T}\right)^{1/2} \exp\left\{\frac{im(X+y)^2}{2\hbar T}\right\} dy \tag{3.20}$$

This integral can be expressed in terms of Fresnel integrals. Such a representation contains the physical results we are after, but in an obscure way owing to the mathematical complexity of the Fresnel integral form. Rather than confuse the physical results by mathematical complexity, we shall set up a different, but analogous, expression which leads to a simpler mathematical form.

The Gaussian Slit. Suppose we introduce a function $G(y)$ as a factor in the integrand. If this function is defined as

$$G(y) = \begin{cases} 1 & \text{for } -b \leq y \leq b \\ 0 & \text{for } |y| > b \end{cases}$$

the limits of integration can be extended to infinity without any change in the result. Then

$$\psi(x') = \frac{m}{2\pi i\hbar\sqrt{t'T}} \int_{-\infty}^{\infty} G(y) \exp\left\{\frac{im}{2\hbar}\left[\frac{(x'-y)^2}{t'} + \frac{(X+y)^2}{T}\right]\right\} dy \tag{3.21}$$

Instead of this, suppose we define $G(y)$ to be a gaussian function, thus:

$$G(y) = e^{-y^2/2b^2} \tag{3.22}$$

This function has the shape shown in Fig. 3-4. The *effective width* of such a curve is related to the parameter b. For this particular function, approximately two-thirds of the area under the curve lies between $-b$ and $+b$.

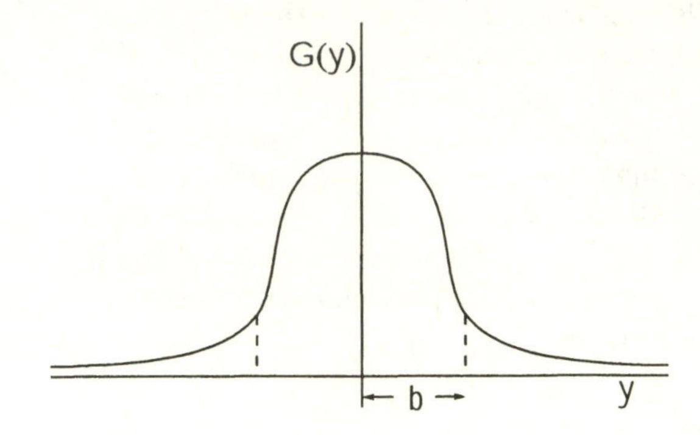

Fig. 3-4 The form of the gaussian function $G(y) = e^{-y^2/2b^2}$. The curve has the same shape as the normal distribution with a standard deviation b.

We do not know how to design metal parts for our imaginary experiment which will produce such a gaussian slit. However, there is no conceptual difficulty. We now have a situation in which the particles at time T are distributed along the x axis with a relative amplitude proportional to the function $G(y)$. (The relative probability is proportional to $[G(y)]^2$.) If the particles move classically, we would expect, after a succeeding interval of time t', to find them similarly distributed along the x axis with a new center a distance x'_1 beyond X and an increased width parameter b_1 given by

$$x'_1 = X\frac{t'}{T} \qquad b_1 = b\left(1 + \frac{t'}{T}\right) \tag{3.23}$$

as shown in Fig. 3-5.

With such a gaussian slit the equation for the amplitude is

$$\psi(x') = \frac{m}{2\pi i\hbar\sqrt{t'T}} \int_{-\infty}^{\infty} \exp\left\{\frac{im}{2\hbar}\left[\frac{x'^2}{t'} + \frac{X^2}{T}\right] + \frac{im}{\hbar}\left[-\frac{x'}{t'} + \frac{X}{T}\right]y + \left[\frac{im}{2\hbar}\left(\frac{1}{t'} + \frac{1}{T}\right) - \frac{1}{2b^2}\right]y^2\right\} dy \tag{3.24}$$

This integral is of the form

$$\int_{-\infty}^{\infty} \exp\{\alpha y^2 + \beta y\}\, dy = \sqrt{\frac{-\pi}{\alpha}} \exp\left\{-\frac{\beta^2}{4\alpha}\right\} \qquad \text{for } \Re e\{\alpha\} \leq 0 \tag{3.25}$$

which is integrated by completing the square in the exponent. Thus the amplitude becomes

$$\psi(x') = \frac{m}{2\pi i\hbar\sqrt{t'T}} \left(\frac{-\pi}{(im/2\hbar)(1/t' + 1/T) - 1/2b^2}\right)^{1/2} \times \exp\left\{\frac{im}{2\hbar}\left(\frac{x'^2}{t'} + \frac{X^2}{T}\right) + \frac{(m^2/4\hbar^2)(-x'/t' + X/T)^2}{(im/2\hbar)(1/t' + 1/T) - 1/2b^2}\right\} \tag{3.26}$$

The classical velocity to get from the origin to the center of the slit is $V = X/T$. When we use this as a substitution and rearrange some of the terms, the expression for the amplitude becomes

$$\psi(x') = \sqrt{\frac{m}{2\pi i\hbar}} \left(T + t' + it'T\frac{\hbar}{mb^2}\right)^{-1/2} \times \exp\left\{\frac{im}{2\hbar}\left(\frac{x'^2}{t'} + V^2T\right) + \frac{(m^2/2\hbar^2t'^2)(x' - Vt')^2}{(im/\hbar)(1/t' + 1/T) - 1/b^2}\right\} \tag{3.27}$$

We shall consider first the relative probability for the particle to arrive at various points along the x axis. This probability is proportional to the absolute square of the amplitude. The absolute value of an exponent with an imaginary argument is 1. So, by rationalizing the second factor and the denominator in the last exponent of Eq. (3.27), we obtain

$$P(x') = \frac{m}{2\pi\hbar}\frac{b}{T\Delta x}\exp\left\{-\frac{(x' - Vt')^2}{(\Delta x)^2}\right\} \tag{3.28}$$

where we have used the substitution

$$(\Delta x)^2 = b^2\left(1 + \frac{t'}{T}\right)^2 + \frac{\hbar^2 t'^2}{m^2b^2} = b_1^2 + \left(\frac{\hbar t'}{mb}\right)^2 \tag{3.29}$$

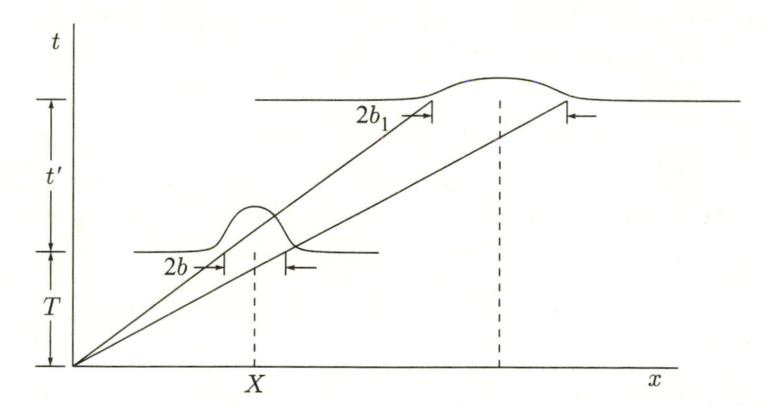

Fig. 3-5 The paths of particles moving through a gaussian slit. If the particles obeyed classical laws of motion, as shown here, then the distribution at time $T + t'$ would have the same form as the distribution at time T. The difference would be only a spreading out proportional to the time of flight. The characteristic width of the distribution would increase from $2b$ to $2b_1$, where $b_1 = b(1 + t'/T)$. For quantum-mechanical motion, the actual spreading is greater than this.

As we expected, the distribution is a gaussian centered about the point $x_1' = Vt'$ of Eq. (3.23). However, the spread of the distribution, Δx, is larger than the classical expected value b_1 of Eq. (3.23). This can be interpreted in the following manner. Suppose a_1 and a_2 are two independent random variables whose root-mean-square deviations about their average values are respectively α_1 and α_2. Then if $a_3 = a_1 + a_2$, the rms deviation of a_3 about its average value is $\alpha_3 = (\alpha_1^2 + \alpha_2^2)^{1/2}$. Now, the rms deviation in a particular distribution is a measure of the spread, or width, of the distribution. As a matter of fact, for the gaussian distribution $e^{-x^2/2b^2}$ the rms value is b.

Thus in the present case we find that the quantum-mechanical system acts as if it had an extra random variable x_2 whose rms deviation is

$$\Delta x_2 = \frac{\hbar t'}{mb} \tag{3.30}$$

It is this extra deviation Δx_2, or spreading, rather than the apparent extra variable x_2, which has physical significance. That this term is quantum-mechanical in nature is clear from the inclusion of the constant $\hbar$. Such a term is important for particles of small mass and for narrow slits.

Thus quantum mechanics tells us that for small particles, passage through a narrow slit makes the future position uncertain. This uncertainty Δx_2 is proportional to the time interval t' between passage through the slit and the next observation of position. If we introduce the classical notion of velocity, we can say that passage through a slit causes a velocity uncertainty whose size is

$$\delta v = \frac{\hbar}{mb} \tag{3.31}$$

We could take the width parameter $2b$ of the slit as a measure of the uncertainty of the position of the particle at the time it passed through the slit. If we call this uncertainty δx and write the product mv as the momentum p, then Eq. (3.31) becomes

$$\delta p \, \delta x = 2\hbar \tag{3.32}$$

Once more we have arrived at a statement of the uncertainty principle. It tells us that, although classically the velocity might be known, the future position has an additional uncertainty as though a random momentum had been generated by passing through a slit of width δx. If classical concepts are used to describe the results of quantum mechanics qualitatively, then we would say that knowledge of position creates uncertainty in momentum.

What about the factors that appear in front of the exponent in Eq. (3.28)? If we integrate this expression over the complete range of x from $-\infty$ to $+\infty$, the result is

$$P(\text{any } x) = \frac{m}{2\pi\hbar T} b\sqrt{\pi} \tag{3.33}$$

This must be the probability that the particle gets through the slit, since the integration includes those particles and only those particles which did get through. But we have another way of obtaining this result. Suppose we take the absolute square of the kernel $K(X+y,T;0,0)$, which comprises the second half of the integrand in Eq. (3.20). This is just the probability per unit distance that the particle arrives at the point $X+y$ in the slit. That is,

$$P(X+y)\,dy = \frac{m}{2\pi\hbar T}\,dy \tag{3.34}$$

This probability is independent of the position along the slit. Thus, if we were to multiply it by the width of the slit, we would obtain the total probability for the particle to arrive at the slit. This implies that the effective width of our gaussian slit is $\sqrt{\pi}\,b$. Had we used the original sharp-edged slit, we would have found the effective width to be $2b$.

Problem 3-3 By squaring the amplitude given in Eq. (3.20) and then integrating over x, show that the probability of passage through the original sharp-edged slit is

$$P(\text{going through}) = \frac{m}{2\pi\hbar T} 2b \tag{3.35}$$

In the course of this problem the integral

$$\int_{-\infty}^{\infty} e^{iax}\,dx \tag{3.36}$$

will appear. This is the integral representation of the Dirac delta function of a.†

Thus the quantum-mechanical results agree with the idea that the probability that a particle goes through a slit is equal to the probability that the particle arrives at the slit.

Momentum and Energy. Next we shall verify again that, when the momentum is definite, the amplitude varies as e^{ikx}. We return to a detailed study of the amplitude given in Eq. (3.27). This time we shall try to arrange conditions in our experiment so that the particle velocity after passing through the slit is known as accurately as possible.

†See Eq. (A.9) in the table of integrals in the Appendix and L.I. Schiff, "Quantum Mechanics," 2nd ed., pp. 50–52, McGraw-Hill Book Company, New York, 1955.

Quite apart from any quantum-mechanical considerations, there is a classical uncertainty of b/T in the velocity. For any given slit width we can make this uncertainty negligible by choosing T very large. We can also make X extremely large so that the average velocity $X/T = V$ does not go to 0. Considering V and t' as constants, the expression for the amplitude in the limit of large T is

$$\psi(x') \approx \text{const} \left(1 + it'\frac{\hbar}{mb^2}\right)^{-1/2} \times \exp\left\{\frac{imx'^2}{2\hbar t'} + \frac{(m^2/2\hbar^2t'^2)(x' - Vt')^2}{im/\hbar t' - 1/b^2}\right\} \tag{3.37}$$

Next we arrange that the quantum-mechanical uncertainty in momentum $\hbar/b$ is very small. That is, we take b very large, so we can neglect $1/b^2$. Then the amplitude can be written

$$\psi(x') \approx \text{const} \quad \exp\left\{\frac{imV}{\hbar}x' - \frac{imV^2}{2\hbar}t'\right\} \tag{3.38}$$

This is an important result. It says that, if we have arranged things so that the momentum of a particle is known to be p, then the amplitude for the particle to arrive at the point x at the time t is

$$\psi(x) \approx \text{const} \quad \exp\left\{\frac{i}{\hbar}px - \frac{i}{\hbar}\frac{p^2}{2m}t\right\} \tag{3.39}$$

We notice that this is a wave of definite wave number $k = p/\hbar$. Furthermore, it has a definite frequency $\omega = p^2/2m\hbar$. This means we can say that a free particle of momentum p has a definite quantum-mechanical energy (defined as $\hbar$ times frequency) which is $p^2/2m$ just as in classical mechanics.

The probability of arriving at any particular x, which is proportional to the square of the amplitude, is in this case independent of x. Thus exact knowledge of velocity means no knowledge of position. In arranging the experiment to give an accurately known velocity we have lost our chances for an accurate prediction of position. We have already seen that the reverse is true. The existence of the quantum-mechanical spread, inversely proportional to the slit width $2b$, implies that an exact knowledge of position precludes any knowledge of velocity. So, if you know where it is, you cannot say how fast it is going; and, if you know how fast it is going, you cannot say where it is. This again illustrates the uncertainty principle.

3-3 RESULTS FOR A SHARP-EDGED SLIT

Leaving the limiting case, suppose we return to a situation in which the slit width and quantum-mechanical spread are comparable in size and the times and distances of travel are not extremely large. We have seen that a gaussian slit leads to a gaussian distribution. If we use that more realistic sharp-edged version and work out the resulting Fresnel integrals, the probability distribution at the time t' after passing through the slit

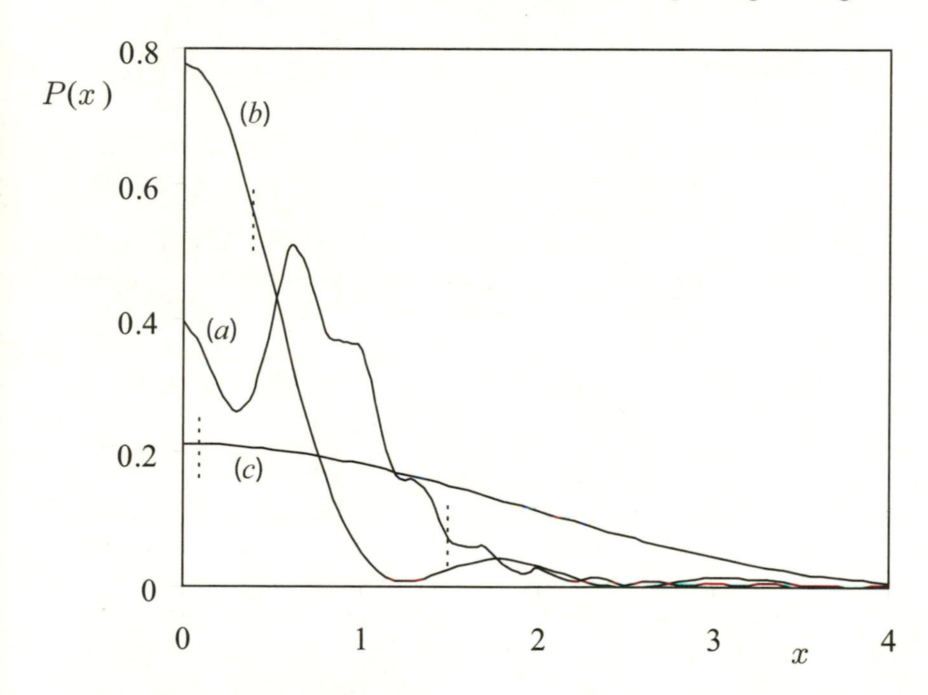

Fig. 3-6 The distribution of particles that have passed through sharp-edged slits of various widths. These distributions are symmetric about the mean position $V(T+t')$ so only the right-hand halves are shown. The classically-predicted width $b_1 = b(1 + t'/T)$ is indicated by a dashed vertical line. The three distributions differ in the ratio of the classical width b_1 to the quantum-mechanical spreading Δx_2: For curve (a), $b_1/\Delta x_2 = 15$; for curve (b), $b_1/\Delta x_2 = 1$; and for curve (c), $b_1/\Delta x_2 = 1/15$. In each case, the distribution spreads beyond the classical width. The rms width of the distribution is approximately equal to $[(b_1)^2 + (\Delta x_2)^2]^{1/2}$.

This distribution is expressed° by

$$P(x') = \frac{m}{2\pi\hbar(T+t')}\left(\frac{1}{2}[C(u_+) - C(u_-)]^2 + \frac{1}{2}[S(u_+) - S(u_-)]^2\right) \quad (3.40)$$

where

$$u_\pm = \frac{(x' - Vt') \pm b(1 + t'/T)}{\sqrt{(\pi\hbar t'/m)(1 + t'/T)}} \quad (3.41)$$

and $C(u)$ and $S(u)$ are the real and imaginary parts of the Fresnel integral. The first factor in this probability distribution is identical to the probability distribution of a free particle given in Eq. (3.6). The remaining factor contains a combination of real and imaginary Fresnel integrals.[1] It is this factor which is responsible for the variations shown in the curves of Fig. 3-6.

Thus for both slits the general result is the same. The most probable place to find the particle is within the classical projection of the slit. Beyond this there is the quantum-mechanical spreading.

We have treated this problem as if it were a combination of two separate motions. First the particle goes to the slit, and then it goes from the slit to the point of observation. The motion seems almost disjointed at the slit. It might be asked then, how does a particle with such a disjointed motion "remember" its velocity and head in the general direction predicted by classical physics? Or, to put it another way, how does making the slit narrower cause as "loss of memory" until, in the limit, all velocities are equally likely for the particle?

To understand this, let us investigate the amplitude to arrive at the slit. This is just the free-particle amplitude given by Eq. (3.3), with $x_a = 0$, $t_a = 0$, $x_b = X + y$, and $t_b = T$. As we move across the slit (vary y), both real and imaginary parts of the amplitude vary sinusoidally. As we have seen, the wavelength of this variation is connected to the momentum (refer to Eq. 3.10). The subsequent motion is a result of optical-like interference among these waves. The interference is constructive in the general direction predicted by classical physics and, in general, destructive in other directions.

If there are many wavelengths across the slit (i.e., the slit is very wide) the resulting interference pattern is quite sharp and the motion is approximately classical. But suppose the slit is made so narrow that not even one whole wavelength is included. There are no longer any oscillations to give an interference, and velocity information is lost. Thus

[1] Refer to p. 34 of E. Jahnke and F. Emde, "Tables of Functions," Dover Publications, Inc., New York, 1943.

in the limit as the slit width goes to zero all velocities are equally likely for the particle.

3-4 THE WAVE FUNCTION

We have developed the amplitude for a particle to reach a particular point in space and time by closely following its motion in getting there. However, it is often useful to consider the amplitude to arrive at a particular place without any special discussion of previous motion. Then we would say $\psi(x,t)$ is the total amplitude to arrive at (x,t) from the past in some (perhaps unspecified) situation. Such an amplitude has the same probability characteristics as those we have already studied; i.e., the probability of finding the particle at the point x and at the time t is $|\psi(x,t)|^2$. We shall call this kind of amplitude a wave function. The difference between this and the amplitudes we have studied before is just a matter of notation. One often hears the statement: The system is in the "state" ψ. This is just another way of saying: The system is described by the wave function $\psi(x,t)$.

Thus the kernel $K(x_b,t_b;x_a,t_a) = \psi(x_b,t_b)$ is actually a wave function. It is the amplitude to get to (x_b,t_b). The notation $K(x_b,t_b;x_a,t_a)$ gives us more information, in particular, that this is the amplitude for a special case in which the particle came from (x_a,t_a). Perhaps this information is of no interest to the problem, so that there is no point in keeping track of it. Then we just use the wave function notation $\psi(x_b,t_b)$.

Since the wave function is an amplitude, it satisfies the rules for combination of amplitudes for events occurring in succession in time. Thus since Eq. (2.31) is true for all points (x_a,t_a), we see that the wave function satisfies the integral equation

$$\psi(x_b,t_b) = \int_{-\infty}^{\infty} K(x_b,t_b;x_c,t_c)\psi(x_c,t_c)\,dx_c \tag{3.42}$$

This result can be stated in physical terms. The total amplitude to arrive at (x_b,t_b) [that is, $\psi(x_b,t_b)$] is the sum, or integral, over all possible values of x_c of the total amplitude to arrive at the point (x_c,t_c) [that is, $\psi(x_c,t_c)$] multiplied by the amplitude to go from c to b [that is, $K(x_b,t_b;x_c,t_c)$]. This means that the effects of all the past history of a particle can be expressed in terms of a single function. If we forget everything we knew about a particle except its wave function at a particular time, then we can calculate everything that can happen to that particle after that time. All of history's effect upon the future of the universe could be obtained from a single gigantic wave function.

Problem 3-4 Suppose a free particle has a definite momentum at the time $t = 0$ (that is, the wave function is $Ce^{i(p/\hbar)x}$). With the help of Eqs. (3.3) and (3.42), show that at some later time the particle has the same definite momentum (i.e., the wave function depends on x through the function $e^{i(p/\hbar)x}$) and varies in time as $e^{-(i/\hbar)(p^2/2m)t}$. This means that the particle has the definite energy $p^2/2m$.

Problem 3-5 Use the results of Prob. 3-2 and Eq. (3.42) to show that the wave function of a free particle satisfies the equation

$$\frac{\partial\psi}{\partial t} = -\frac{i}{\hbar}\left[-\frac{\hbar^2}{2m}\frac{\partial^2\psi}{\partial x^2}\right] \tag{3.43}$$

which is the Schrödinger equation for a free particle.

3-5 GAUSSIAN INTEGRALS

We are finished with the physical portion of this chapter, and we now proceed to mathematical considerations. We shall introduce some additional mathematical techniques which will help us to compute the sum over paths in certain situations.

The simplest path integrals are those in which all of the variables appear up to the second degree in an exponent. We shall call them gaussian integrals. In quantum mechanics this corresponds to a case in which the action S involves the path $x(t)$ up to and including the second power.

To illustrate how the method works in such a case, consider a particle whose lagrangian has the form

$$L = a(t)\dot{x}^2 + b(t)\dot{x}x + c(t)x^2 + d(t)\dot{x} + e(t)x + f(t) \tag{3.44}$$

The action is the integral of this function with respect to time between two fixed end points. (Actually, in this form the lagrangian is a little more general than necessary. The factor $\dot{x}$ could be removed from those terms in which it is linear through an integration by parts, but this fact is unimportant for our present purpose.) We wish to determine

$$K(b,a) = \int_a^b \exp\left\{\frac{i}{\hbar}\int_{t_a}^{t_b} L(\dot{x},x,t)\,dt\right\}\mathcal{D}x(t) \tag{3.45}$$

the integral over all paths which go from (x_a, t_a) to (x_b, t_b).

Of course, it is possible to carry out the integral over all paths in the way which was first described (in Sec. 2-4) by dividing the region into short time elements, and so on. That this will work follows from the fact that the integrand is an exponential of a quadratic form in the variables $\dot{x}$ and x. Such integrals can always be carried out. But we shall not go through this tedious calculation, since we can determine the most important characteristics of the kernel in the following manner.

Let $\bar{x}(t)$ be the classical path between the specified end points. This is the path which is an extremum for the action S. In the notation we have been using

$$S_{cl}[b,a] = S[\bar{x}(t)] \tag{3.46}$$

We can represent $x(t)$ in terms of $\bar{x}(t)$ and a new function $y(t)$:

$$x(t) = \bar{x}(t) + y(t) \tag{3.47}$$

That is to say, instead of defining a point on the path by its distance $x(t)$ from an arbitrary coordinate axis, we measure instead the deviation $y(t)$ from the classical path, as shown in Fig. 3-7.

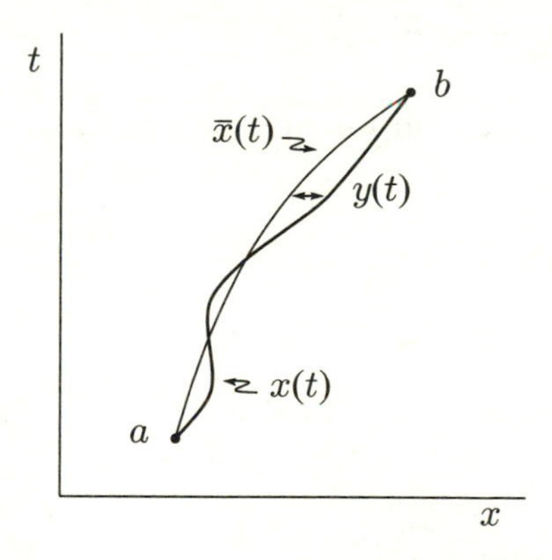

Fig. 3-7 The difference between the classical path $\bar{x}(t)$ and some possible alternative path $x(t)$ is the function $y(t)$. Since the paths must both reach the same end points, $y(t_a) = y(t_b) = 0$. In between these end points $y(t)$ can take any form. Since the classical path is completely fixed, any variation in the alternative path $x(t)$ is equivalent to the associated variation in the difference $y(t)$. Thus, in a path integral, the path differential $\mathcal{D}x(t)$ can be replaced by $\mathcal{D}y(t)$, and the path $x(t)$ by $\bar{x}(t) + y(t)$. In this form $\bar{x}(t)$ is a constant for the integration over paths. Furthermore, the new path variable $y(t)$ is restricted to take the value 0 at both end points. This substitution leads to a path integral independent of end-point positions.

At each t the variables x and y differ by the constant $\bar{x}$. (Of course, this is a different constant for each value of t.) Therefore, clearly, $dx_i = dy_i$ for each specific point t_i in the subdivision of time. In general, we may say $\mathcal{D}x(t) = \mathcal{D}y(t)$.

The integral for the action can be written

$$S[x(t)] = S[\bar{x}(t) + y(t)] = \int_{t_a}^{t_b} [a(t)(\dot{\bar{x}}^2 + 2\dot{\bar{x}}\dot{y} + \dot{y}^2) + \cdots] \, dt \tag{3.48}$$

If all the terms which do not involve y are collected, the resulting integral is just $S[\bar{x}(t)] = S_{cl}$. If all the terms which contain y as a linear factor are collected, the resulting integral vanishes. This could be proved by actually carrying out the integration (some integration by parts would be involved); however, such a calculation is unnecessary, since we already know the result is true. The function $\bar{x}(t)$ is determined by this very requirement. That is, $\bar{x}(t)$ is so chosen that there is no change in S, to first order, for variations of the path around $\bar{x}(t)$. All that remains are the second-order terms in y. These can be easily picked out, so that we can write

$$S[x(t)] = S_{cl}[b, a] + \int_{t_a}^{t_b} [a(t)\dot{y}^2 + b(t)\dot{y}y + c(t)y^2] \, dt \tag{3.49}$$

The integral over paths does not depend upon the classical path, so the kernel can be written

$$K(b, a) = e^{(i/\hbar)S_{cl}[b,a]} \int_0^0 \exp\left\{ \frac{i}{\hbar} \int_{t_a}^{t_b} [a(t)\dot{y}^2 + b(t)\dot{y}y + c(t)y^2] \, dt \right\} \mathcal{D}y(t) \tag{3.50}$$

Since all paths $y(t)$ start from and return to the point $y = 0$, the integral over paths can be a function only of times at the end points. This means that the kernel can be written as

$$K(b, a) = e^{(i/\hbar)S_{cl}[b,a]} F(t_b, t_a) \tag{3.51}$$

so K is determined except for a function of t_b and t_a. In particular, the kernel's dependence upon the spatial variables x_b and x_a is completely worked out. It should be noted that the dependence of the kernel upon the coefficients of the linear terms $d(t)$ and $e(t)$ and the remaining coefficient $f(t)$ is also completely worked out.

This seems to be characteristic of various methods of doing path integrals; a great deal can be worked out by some general methods, but often a multiplying factor is not fully determined. It must be determined by some other known property of the solution, as, for example, by Eq. (2.31).

It is interesting to note that the approximate expression $K \sim e^{iS_{cl}/\hbar}$ is exact for the case that S is a quadratic form.

Problem 3-6 Since the free-particle lagrangian is quadratic, show that (Prob. 2-1)

$$K(b,a) = F(t_b, t_a) \exp\left\{\frac{im(x_b - x_a)^2}{2\hbar(t_b - t_a)}\right\} \tag{3.52}$$

and give an argument to show that F can depend only on the difference $F(t_b - t_a)$.

Problem 3-7 Further information about this function F can be obtained from the property expressed by Eq. (2.31). First notice that the results of Prob. 3-6 imply that $F(t_b - t_a)$ can be written as $F(t)$, where t is the time interval $t_b - t_a$. By using this form for F in Eq. (3.52) and substituting into Eq. (2.31), express $F(t+s)$ in terms of $F(t)$ and $F(s)$, where $t = t_b - t_c$ and $s = t_c - t_a$. Show that if F is written as

$$F(t) = \left(\frac{m}{2\pi i\hbar t}\right)^{1/2} f(t) \tag{3.53}$$

the new function $f(t)$ must satisfy

$$f(t+s) = f(t)f(s) \tag{3.54}$$

This means that $f(t)$ must be of the form

$$f(t) = e^{at} \tag{3.55}$$

where a may be complex, that is, $a = \alpha + i\beta$. It is difficult to obtain more information about the function $f(t)$ from the principles we have so far laid down. However, the special choice of the normalizing factor A defined in Eq. (2.21) implies that $f(\epsilon) = 1$ to first order in ϵ. This corresponds to setting a in Eq. (3.55) equal to 0. The resulting value of $F(t)$ is in agreement with Eq. (3.3).

It is clear from this example how the important properties of path integrals may be easily obtained even though the integrand may be a complicated function. So long as the integrand is an exponential function which contains the path variables only up to the second order, a solution that will be complete except possibly for some simple multiplying factors can be obtained. This is true regardless of the number of variables. Thus, for example, a path integral of the form

$$\int_k^l \cdots \int_c^d \int_a^b \exp\{E[x(t), y(t), \ldots, z(t)]\}\, \mathcal{D}x(t)\, \mathcal{D}y(t) \cdots \mathcal{D}z(t) \tag{3.56}$$

contains as its important factor $e^{E_{cl}}$, where E_{cl} is the extremum of E subject to the boundary conditions. The only restriction is that in terms of the variables x, y, and so on, E is a function of the second degree. The remaining factor is a function of the times at the end points of the paths. For most of the path integrals which we shall study, the important information is contained in the exponential term rather then in the latter factor. In fact in most cases we shall not even find it necessary to evaluate this latter factor. This method of solving path integrals will be used frequently in the succeeding chapters.

3-6 MOTION IN A POTENTIAL FIELD

One simple application comes in the classical limiting case in which the action S is very large compared to Planck's constant $\hbar$. As we have already pointed out for this situation, the kernel K is approximately proportional to $e^{iS_{cl}/\hbar}$. We can now see more mathematically the basis of this approximation. Only those paths quite near to the classical path $\bar{x}(t)$ are important, so suppose we make the substitution $x(t) = \bar{x}(t) + y(t)$. Now if the particle is moving through the potential $V(x)$, we can write

$$V(x) = V(\bar{x} + y) = V(\bar{x}) + yV'(\bar{x}) + \frac{y^2}{2}V''(\bar{x}) + \frac{y^3}{6}V'''(\bar{x}) + \cdots \quad (3.57)$$

where the prime indicates differentiation with respect to x and all derivatives are evaluated along the classical path $\bar{x}$. Only small values of y are important, so suppose V is a sufficiently smooth function that we can neglect terms of order y^3 and higher. This means that we assume that y^3V''' and all higher-order terms are negligible compared to the terms kept.

Under this assumption the integrand can be expressed as a quadratic form in y. In fact, since $\bar{x}$ makes S extreme,

$$S = S_{cl} + \text{terms second order in } y.$$

The important term in the result is $e^{iS_{cl}/\hbar}$, where now, of course, S_{cl} contains the potential $V(\bar{x})$ along the classical path. The remaining integral over y goes from 0 to 0 and is of the form of the last factor in Eq. (3.50). It provides a smooth function as a factor to $e^{iS_{cl}/\hbar}$.

The result is true in situations other than the classical limiting case. For example, suppose $V(x)$ is a quadratic function of x. Then the solution is exact, since the expansion of $V(x)$ as in Eq. (3.57) contains no powers higher than the second. Some examples of this type are given in the problems. As another example, suppose $V(x)$ is a slowly varying function. In particular, if the third and higher derivatives are extremely small, the result given above is a very accurate approximation. This particular case is called the WKB approximation in quantum mechanics.

There are other situations in which the approximation is good. Suppose the total time interval for the motion is very short. If a particle moves along a path differing greatly from the classical path, it must have a very large extra velocity (to go out from the initial point and then return to the final point in the allotted time interval). The extra kinetic energy is proportional to the square of this large velocity, and the action contains a term roughly proportional to the kinetic energy multiplied by the time interval (thus, the square of the velocity multiplied by the time interval). The action for such paths will be very large, and the phase of the amplitude will vary greatly for closely neighboring paths. In this case again it is reasonable to drop the higher-order terms in the expansion of $V(x)$.

Problem 3-8 For a harmonic oscillator the lagrangian is

$$L = \frac{m}{2}\dot{x}^2 - \frac{m\omega^2}{2}x^2 \tag{3.58}$$

Show that the resulting kernel is (see Prob. 2-2)

$$K = F(T)\exp\left\{\frac{im\omega}{2\hbar\sin\omega T}[(x_b^2 + x_a^2)\cos\omega T - 2x_bx_a]\right\} \tag{3.59}$$

where $T = t_b - t_a$. Note that the multiplicative function $F(T)$ has not been explicitly worked out. It can be obtained by other means, and for the harmonic oscillator° it is (see Sec. 3-11)

$$F(T) = \left(\frac{m\omega}{2\pi i\hbar\sin\omega T}\right)^{1/2} \tag{3.60}$$

Problem 3-9 Find the kernel for a particle in a constant external field f where the lagrangian is

$$L = \frac{m}{2}\dot{x}^2 + fx \tag{3.61}$$

The result is

$$K = \left(\frac{m}{2\pi i\hbar T}\right)^{1/2} \exp\left\{\frac{i}{\hbar}\left[\frac{m(x_b - x_a)^2}{2T} + \frac{fT(x_b + x_a)}{2} - \frac{f^2T^3}{24\,m}\right]\right\} \tag{3.62}$$

where $T = t_b - t_a$.

Problem 3-10° The lagrangian for a particle of charge e and mass m in a constant external magnetic field B, in the z direction, is

$$L = \frac{m}{2}(\dot{x}^2 + \dot{y}^2 + \dot{z}^2) + \frac{\mathrm{e}B}{2c}(x\dot{y} - \dot{x}y) \tag{3.63}$$

Show that the resulting kernel is

$$K = \left(\frac{m}{2\pi i\hbar T}\right)^{3/2} \left(\frac{\omega T/2}{\sin(\omega T/2)}\right) \exp\left\{\frac{im}{2\hbar}\left[\frac{(z_b - z_a)^2}{T} + \left(\frac{\omega/2}{\tan(\omega T/2)}\right)[(x_b - x_a)^2 + (y_b - y_a)^2] + \omega(y_b x_a - x_b y_a)\right]\right\} \tag{3.64}$$

where $T = t_b - t_a$ and $\omega = \mathrm{e}B/mc$.

Problem 3-11 Suppose the harmonic oscillator of Prob. 3-8 is driven by an external force $f(t)$. The lagrangian is

$$L = \frac{m}{2}\dot{x}^2 - \frac{m\omega^2}{2}x^2 + f(t)x \tag{3.65}$$

Show that the resulting kernel is (with $T = t_b - t_a$)

$$K = \left(\frac{m\omega}{2\pi i\hbar \sin\omega T}\right)^{1/2} e^{iS_{cl}/\hbar}$$

where

$$S_{cl} = \frac{m\omega}{2\sin\omega T}\bigg[(x_b^2 + x_a^2)\cos\omega T - 2x_b x_a + \frac{2x_b}{m\omega}\int_{t_a}^{t_b} f(t)\sin\omega(t - t_a)\,dt + \frac{2x_a}{m\omega}\int_{t_a}^{t_b} f(t)\sin\omega(t_b - t)\,dt - \frac{2}{m^2\omega^2}\int_{t_a}^{t_b}\int_{t_a}^{t} f(t)f(s)\sin\omega(t_b - t)\sin\omega(s - t_a)\,ds\,dt\bigg] \tag{3.66}$$

This last result is of great importance in many advanced problems. It has particular applications in quantum electrodynamics because the electromagnetic field can be represented as a set of forced harmonic oscillators.

Problem 3-12 If the wave function for a harmonic oscillator is (at $t = 0$)

$$\psi(x, 0) = \exp\left\{-\frac{m\omega}{2\hbar}(x-a)^2\right\} \tag{3.67}$$

then, using Eq. (3.42) and the results of Prob. 3-8, show that

$$\psi(x, T) = \exp\left\{-\frac{i\omega T}{2} - \frac{m\omega}{2\hbar}[x^2 - 2axe^{-i\omega T} + a^2\cos(\omega T)e^{-i\omega T}]\right\} \tag{3.68}$$

and find the probability distribution $|\psi|^2$.

3-7 SYSTEMS WITH MANY VARIABLES[1]

Suppose a system has several degrees of freedom. A kernel for such a system can be represented by the form of Eq. (2.25), where the symbol $x(t)$ now represents several coordinates rather than just one.

We take as a first example a particle moving in three dimensions. The path is defined by giving three functions $x(t)$, $y(t)$, and $z(t)$. The action for a free particle, for example, is

$$\frac{m}{2}\int_{t_a}^{t_b}[\dot{x}(t)^2 + \dot{y}(t)^2 + \dot{z}(t)^2]\,dt$$

The kernel to go from some initial point (x_a, y_a, z_a) at time t_a to a final point (x_b, y_b, z_b) at time t_b is

$$K(x_b, y_b, z_b, t_b; x_a, y_a, z_a, t_a) = \int_a^b \exp\left\{\frac{i}{\hbar}\int_{t_a}^{t_b}\frac{m}{2}[\dot{x}^2(t) + \dot{y}^2(t) + \dot{z}^2(t)]\,dt\right\}\mathcal{D}x(t)\,\mathcal{D}y(t)\,\mathcal{D}z(t) \tag{3.69}$$

The differential is written as $\mathcal{D}x(t)\,\mathcal{D}y(t)\,\mathcal{D}z(t)$. If the time is divided into intervals ϵ, the position at the time t_i is given by three variables x_i, y_i, z_i and the integral over all variables is dx_i, dy_i, dz_i for each i in an expression like Eq. (2.22). (More generally, if we represent the

[1] R.P. Feynman, Space-Time Approach to Non-relativistic Quantum Mechanics, *Rev. Mod. Phys.*, vol. 20, p. 371, 1948.

position by a vector $\mathbf{x}$ in some s-dimensional space, the differential at each time interval is the volume element $d^s\mathbf{x}_i$ and the corresponding path differential is $\mathcal{D}^s\mathbf{x}$.)

If the definition of Eq. (2.22) is used, then the normalizing constant A (Eq. 2.21) must be included for each variable in each time interval. Thus if the total time interval is broken up into N steps of length ϵ, the factor A^{-3N} must be included in the integral.

Another situation involving several variables is that of two interacting systems. Suppose one system consists of a particle of mass m and coordinate x while the other system is a particle of mass M and coordinate X. Suppose these two systems interact through a potential $V(x, X)$. The resulting action is

$$S[x(t), X(t)] = \int_{t_a}^{t_b} \left[\frac{m}{2}\dot{x}^2 + \frac{M}{2}\dot{X}^2 - V(x, X)\right] dt \tag{3.70}$$

so that the kernel is

$$K(x_b, X_b, t_b; x_a, X_a, t_a) = \int_a^b \int_a^b \exp\left\{\frac{i}{\hbar} S[x(t), X(t)]\right\} \mathcal{D}x(t)\, \mathcal{D}X(t) \tag{3.71}$$

One might understand this generalization of Eq. (2.25) mathematically. Thus one might consider the motion of a point in some abstract two-dimensional space of coordinates x, X. However, it is much easier to think of it physically as representing the motion of two separate particles whose coordinates are respectively x and X. Then K is the amplitude that the particle of mass m goes from the point in space-time (x_a, t_a) to (x_b, t_b) and the particle of mass M goes from (X_a, t_a) to (X_b, t_b). The kernel is then the sum of an amplitude taken over all possible paths of both particles between their respective start and end points. The amplitude for any particular pair of paths (i.e., both $x(t)$ and $X(t)$ are specified) is $e^{iS/\hbar}$, where S is the action defined in Eq. (3.70). Mathematically, the amplitude is a function of two independent functions $x(t)$ and $X(t)$, and the integral is over both of the variable functions.

3-8 SEPARABLE SYSTEMS

Suppose we have a situation in which two particles are present, both moving in one or perhaps more dimensions. Let the vector $\mathbf{x}$ represent the coordinates of one particle and the vector $\mathbf{X}$ represent the coordinates of the other, as in the paragraph above, except that now we extend the picture to a three-dimensional space. It may happen that the resulting action can be separated into two parts, as

$$S[\mathbf{x}, \mathbf{X}] = S_x[\mathbf{x}] + S_X[\mathbf{X}] \tag{3.72}$$

where S_x involves only the paths $\mathbf{x}(t)$ and S_X involves only the paths $\mathbf{X}(t)$. This is the situation when the two particles do not interact.

In this case the kernel becomes the product of one factor depending on $\mathbf{x}$ alone and another depending on $\mathbf{X}$ alone. Thus

$$\begin{aligned} &K(\mathbf{x}_b, \mathbf{X}_b, t_b; \mathbf{x}_a, \mathbf{X}_a, t_a) \\ &= \int_a^b \int_a^b \exp\left\{\frac{i}{\hbar}\left[S_x[\mathbf{x}] + S_X[\mathbf{X}]\right]\right\} \mathcal{D}^3\mathbf{x}(t)\, \mathcal{D}^3\mathbf{X}(t) \\ &= \int_a^b \exp\left\{\frac{i}{\hbar}S_x[\mathbf{x}]\right\} \mathcal{D}^3\mathbf{x}(t) \int_a^b \exp\left\{\frac{i}{\hbar}S_X[\mathbf{X}]\right\} \mathcal{D}^3\mathbf{X}(t) \\ &= K_x(\mathbf{x}_b, t_b; \mathbf{x}_a, t_a) K_X(\mathbf{X}_b, t_b; \mathbf{X}_a, t_a) \end{aligned} \tag{3.73}$$

Here K_x is the amplitude computed as if only the particle of coordinates $\mathbf{x}$ were present, and K_X is defined similarly. Thus in a situation involving two independent noninteracting systems, the kernel for an event involving both systems is the product of two independent kernels. These are the kernels for each particle to carry out its individual portion of the overall event.

The wave function in a situation involving several particles is defined in a straightforward manner by analogy with the corresponding kernel as $\psi(\mathbf{x}, \mathbf{X}, \ldots, t)$. It is interpreted as the amplitude that, at time t, one particle is at the point $\mathbf{x}$, another particle is at the point $\mathbf{X}$, etc. The absolute square of the wave function is the probability that one particle is at point $\mathbf{x}$ per unit volume, another particle is at the point $\mathbf{X}$ per unit volume, etc. Equation (3.42), which holds for the one-dimensional case, can be immediately extended to read

$$\begin{aligned} &\psi(\mathbf{x}, \mathbf{X}, \ldots, t) = \\ &\iint \cdots K(\mathbf{x}, \mathbf{X}, \ldots, t; \mathbf{x}', \mathbf{X}', \ldots, t')\psi(\mathbf{x}', \mathbf{X}', \ldots, t')\, d^3\mathbf{x}'\, d^3\mathbf{X}' \cdots \end{aligned} \tag{3.74}$$

Where $d^3\mathbf{x}'$ is the product of as many differentials as there are coordinates in $\mathbf{x}'$ space.

In case two independent particles are represented by the sets of coordinates $\mathbf{x}$ and $\mathbf{X}$, then the kernel K is the product of one function of $\mathbf{x}$ and t and another of $\mathbf{X}$ and t, as mentioned above. However, this does not imply that, in general, ψ is such a product. In the special case that ψ is at some particular time a product of a function of $\mathbf{x}$ and another of $\mathbf{X}$ (thus $\psi = f(\mathbf{x})g(\mathbf{X})$), then it will remain so for all time. Each factor will change as it would for the partial system alone, since the kernel K represents the independent motion of two particles. But this is a special case. Just because the particles are independent now does not mean that they always were. There may have been some interaction in the past, which would imply that ψ is not a simple product.

Even though the action S does not appear as a simple product in the original coordinate system, there is often a transformation (such as that of center-of-gravity and internal coordinates) which will make it separable. Since the same form for the action is used in quantum mechanics as in classical physics, any transformation which will separate a classical system will also separate the corresponding quantum-mechanical system. Thus a part of the great body of work in classical physics can be applied directly to quantum mechanics. Such transformations are very important. It is hard to deal with a system consisting of several variables. Separation of variables permits us to reduce a complex problem to a number of simpler problems.

3-9 THE PATH INTEGRAL AS A FUNCTIONAL

When a problem contains more than one variable and a separation is not possible, the analysis is generally very difficult. Later on we shall discuss some approximations which can be applied to this case. Here we shall describe one very powerful tool which can sometimes be applied. Consider the kernel given by Eq. (3.71). This can be written out in detail as

$$K(b,a) = \int_a^b \int_a^b \exp\left\{\frac{i}{\hbar}\int_{t_a}^{t_b} \frac{m}{2}\dot{x}^2\,dt + \frac{i}{\hbar}\int_{t_a}^{t_b} \frac{M}{2}\dot{X}^2\,dt - \frac{i}{\hbar}\int_{t_a}^{t_b} V(x,X,t)\,dt\right\} \mathcal{D}x(t)\,\mathcal{D}X(t) \tag{3.75}$$

First, suppose we carry out the integral over the paths $X(t)$. The result can be written formally as

$$K(b,a) = \int_a^b \exp\left\{\frac{i}{\hbar}\int_{t_a}^{t_b} \frac{m}{2}\dot{x}^2\,dt\right\} T[x(t)]\,\mathcal{D}x(t) \tag{3.76}$$

where

$$T[x(t)] = \int_a^b \exp\left\{\frac{i}{\hbar}\int_{t_a}^{t_b} \left[\frac{M}{2}\dot{X}^2 - V(x,X,t)\right] dt\right\} \mathcal{D}X(t) \tag{3.77}$$

These results are interpreted in the following manner. Integrating over all paths available to the X particle produces a *functional* T. A functional is a number whose value depends on specifying a complete function. For example, the area under a curve is a functional of the curve: $A = \int f(y)\,dy$. To find it, a function (the curve) must be specified. We write a functional as $A[f(y)]$ to indicate that A depends on the function $f(y)$. We do not write $A(f(y))$, for that might be interpreted as a function of a function, i.e., that A just depends what value f takes at some specified point y. This is not the case. $A[f(y)]$ depends on the entire shape of the function $f(y)$. It does not depend on y in any way.

The functional defined in Eq. (3.77) is the amplitude that the X particle alone goes between its end points X_a and X_b under the influence of a potential V. This potential, which depends upon both x and X, is computed assuming x is held to be a fixed path as X changes. Thus it is the potential for the X particle when the x particle is moving along a specific trajectory. Clearly, this amplitude T depends upon the trajectory chosen for $x(t)$, so we write it as a functional of $x(t)$. Then the total amplitude is obtained by summing over all paths a functional consisting of the product of T and the free-particle kernel for $x(t)$.

Thus the amplitude K, like all others, is a sum over the amplitudes of all possible alternatives. Each of these amplitudes is a product of two lesser amplitudes. The first of these is the amplitude T that the X particle goes between its given end points when x has a specified trajectory. The second is the amplitude that x has that specified trajectory. The final sum over alternatives becomes the sum over all possible trajectories of x. It is important to understand this concept clearly, for it includes one of the fundamental principles of quantum electrodynamics, a subject which will be taken up in a later chapter.

Of course it is not practical to use this method unless the integral T can actually be worked out, either exactly or approximately, for the possible values of the trajectory $x(t)$. As we have seen (in Prob. 3-11) one exact case it that in which X is a harmonic oscillator. This is a very important practical case. For example, when a particle interacts with a quantized field, the field is an oscillator.

3-10 INTERACTION OF A PARTICLE AND A HARMONIC OSCILLATOR

We shall consider in more detail the interaction of a particle and a harmonic oscillator. Let the coordinate of the particle be x and that of the oscillator be X. The action can be written as

$$S[x,X] = S_0[x] + \int_{t_a}^{t_b} g(x(t),t)X(t)\,dt + \int_{t_a}^{t_b} \frac{M}{2}(\dot{X}^2 - \omega^2 X^2)\,dt \quad (3.78)$$

where S_0 is the action of the particle in the absence of the oscillator. In the discussion above (Sec. 3-9) we assumed that this action corresponded to that for a free particle. This assumption is not necessary. The motion of x could be complicated by the existence of a potential depending upon x and t only. Thus, for example, the action S_0 might be

$$S_0[x] = \int_{t_a}^{t_b} \left[\frac{m}{2}\dot{x}^2 - V(x,t)\right] dt \quad (3.79)$$

The second term in Eq. (3.78) represents the interaction between the particle and the oscillator. Note that this term is linear in X. Omission of a dependence upon $\dot{X}$ does not imply any loss in generality, since if such a term were to occur, it could be removed by an integration by parts. We can call the coefficient g the coupling coefficient. Its dependence upon $x(t)$ is indicated, but it could also depend upon other variables, such as $\dot{x}(t)$. Since the analysis we are presenting is general, it is not important to write down the exact form of g. The last term in Eq. (3.78) is, of course, the action of the oscillator alone. By combining this with the second term, the functional T of Eq. (3.77) can be written as

$$T[x(t)] = \int_a^b \exp\left\{\frac{i}{\hbar}\int_{t_a}^{t_b}\left[g(x(t),t)X(t) + \frac{M}{2}(\dot{X}^2 - \omega^2 X^2)\right] dt\right\} \mathcal{D}X(t) \quad (3.80)$$

Now as far as X is concerned, the situation is just that of a forced harmonic oscillator. The forcing function $g(x(t),t)$ is some special function of t, say, $f(t)$. Thus the path integral is the same as that considered in Prob. 3-11, with $f(t)$ replaced by $g(x(t),t)$ and the final and initial coordinate values (x_b, x_a) replaced by (X_b, X_a).

For illustrative purposes, to simplify the expressions somewhat, we take the special case in which the oscillator initially and finally is at the origin, so $X_b = X_a = 0$. (The general case is just as easily handled.) Then according to Prob. 3-11 in this case we have

$$T = \left(\frac{M\omega}{2\pi i\hbar \sin\omega T}\right)^{1/2} \exp\left\{\frac{-i}{\hbar M\omega \sin\omega T}\right. \quad (3.81)$$
$$\left.\times \int_{t_a}^{t_b}\int_{t_a}^{t} g(x(t),t)g(x(s),s)\sin\omega(t_b - t)\sin\omega(s - t_a)\,ds\,dt\right\}$$

Therefore, the kernel for the present situation can be written

$$K(b,a) = \left(\frac{M\omega}{2\pi i\hbar\sin\omega T}\right)^{1/2} \tag{3.82}$$
$$\times \int_a^b \exp\left\{\frac{i}{\hbar}\left[S_0[x] - \frac{1}{M\omega\sin\omega T}\right.\right.$$
$$\left.\left.\times \int_{t_a}^{t_b}\int_{t_a}^{t} g(x(t),t)g(x(s),s)\sin\omega(t_b-t)\sin\omega(s-t_a)\,ds\,dt\right]\right\}\mathcal{D}x(t)$$

with a similar (but more complicated) expression for arbitrary X_a, X_b.

This is a more complicated path integral than any we have had to solve so far. It is not possible to proceed further with the solution until various methods of approximation have been developed in succeeding chapters. Note that the integrand of this path integral can still be thought of as being of the form $e^{iS/\hbar}$, but now S is no longer a function of only $\dot{x}$, x, and t. Instead, S contains a product of variables defined at two different times, s and t. The separation of past and future can no longer be made. This happens because the variable x at some previous time affects the oscillator which, at some later time, reacts back to affect x. No wave function $\psi(x,t)$ can be defined to give the amplitude that the particle is at some particular place x at a particular time t. Such an amplitude would be insufficient for continuing calculations into the future, since at any time one must also know what the oscillator is doing.

3-11 EVALUATION OF PATH INTEGRALS BY FOURIER SERIES

Consider the path integral for the harmonic oscillator problem (Prob. 3-8). This is

$$K(b,a) = \int_a^b \exp\left\{\frac{i}{\hbar}\int_{t_a}^{t_b}\frac{m}{2}(\dot{x}^2 - \omega^2 x^2)\,dt\right\}\mathcal{D}x(t) \tag{3.83}$$

Using the methods of Sec. 3-5 this path integral can be reduced to a product of two functions, as in Prob. 3-8. The more important of these two functions depends upon the classical orbit for a harmonic oscillator and is given in Eq. (3.59). The remaining function depends upon the time interval only and is written down in Eq. (3.60). This latter function can be written as

$$F(T) = \int_0^0 \exp\left\{\frac{i}{\hbar}\int_0^T \frac{m}{2}(\dot{y}^2 - \omega^2 y^2)\,dt\right\}\mathcal{D}y(t) \tag{3.84}$$

We shall solve this, at least to within a factor independent of ω, by a method which illustrates still another way of handling path integrals. Since all paths $y(t)$ go from 0 at $t = 0$ to 0 at $t = T$, such paths can be written as a Fourier sine series with a fundamental period of T. Thus

$$y(t) = \sum_{n=1}^{\infty} a_n \sin \frac{n\pi t}{T} \tag{3.85}$$

It is possible then to specify a path through the coefficients a_n instead of through the function values y at any particular value of t. This is a linear transformation whose jacobian J is a dimensionless constant, obviously independent of ω, m, and $\hbar$.

Of course, it is possible to evaluate this jacobian directly. However, here we shall avoid the evaluation of J by collecting all factors which are independent of ω (including J) into a single constant factor. We can always recover the correct factor at the end, since we know the value for $\omega = 0$, namely $F(T) = (m/2\pi i\hbar T)^{1/2}$ (a free particle).

The integral for the action can be written in terms of the Fourier series of Eq. (3.85). Thus the kinetic-energy term becomes

$$\begin{aligned} \frac{m}{2}\int_0^T \dot{y}^2\,dt &= \frac{m}{2}\sum_{n=1}^{\infty}\sum_{m=1}^{\infty}\frac{n\pi}{T}\frac{m\pi}{T}a_n a_m \int_0^T \cos\frac{n\pi t}{T}\cos\frac{m\pi t}{T}\,dt \\ &= \frac{m}{2}\frac{T}{2}\sum_{n=1}^{\infty}\left(\frac{n\pi}{T}\right)^2 a_n^2 \end{aligned} \tag{3.86}$$

and similarly the potential-energy term is

$$\frac{m\omega^2}{2}\int_0^T y^2\,dt = \frac{m\omega^2}{2}\frac{T}{2}\sum_{n=1}^{\infty} a_n^2 \tag{3.87}$$

On the assumption that the time T is divided into discrete steps of length ϵ (asin Eq. 2.19) so that there are only a finite number N of coefficients a_n, the path integral becomes

$$\begin{aligned} F(T) = J\frac{1}{A}\int_{-\infty}^{\infty}\cdots\int_{-\infty}^{\infty}\int_{-\infty}^{\infty}\exp\left\{\frac{im}{2\hbar}\frac{T}{2}\sum_{n=1}^{N}\left[\left(\frac{n\pi}{T}\right)^2-\omega^2\right]a_n^2\right\} \\ \times\frac{da_1}{A}\frac{da_2}{A}\cdots\frac{da_N}{A} \end{aligned} \tag{3.88}$$

Since the exponent can be separated into factors, the integral over each coefficient a_n can be done separately. The result of one such integration is

$$\int_{-\infty}^{\infty}\exp\left\{\frac{im}{2\hbar}\frac{T}{2}\left(\frac{n^2\pi^2}{T^2}-\omega^2\right)a_n^2\right\}\frac{da_n}{A} = \left(\frac{2}{\epsilon T}\right)^{1/2}\left(\frac{n^2\pi^2}{T^2}-\omega^2\right)^{-1/2} \tag{3.89}$$

Thus the path integral is proportional to

$$\prod_{n=1}^{N}\left(\frac{n^2\pi^2}{T^2}-\omega^2\right)^{-1/2}=\prod_{n=1}^{N}\left(\frac{n^2\pi^2}{T^2}\right)^{-1/2}\prod_{n=1}^{N}\left(1-\frac{\omega^2T^2}{n^2\pi^2}\right)^{-1/2} \tag{3.90}$$

The first product does not depend on ω and combines with the jacobian and other factors we have collected into a single constant. The second factor has the limit $[(\sin\omega T)/\omega T]^{-1/2}$ as $N\to\infty$, that is, as $\epsilon\to 0$. Thus

$$F(T)=C\left(\frac{\sin\omega T}{\omega T}\right)^{-1/2} \tag{3.91}$$

where C is independent of ω. But for $\omega=0$ our integral is that for a free particle, for which we have already found that

$$F(T)=\left(\frac{m}{2\pi i\hbar T}\right)^{1/2} \tag{3.92}$$

Hence for the harmonic oscillator we have

$$F(T)=\left(\frac{m\omega}{2\pi i\hbar\sin\omega T}\right)^{1/2} \tag{3.93}$$

which is to be substituted in Eq. (3.59) to obtain the complete solution.

Problem 3-13 By keeping track of all the constants, show that the jacobian satisfies

$$J\left(\frac{\pi}{\sqrt{2}}\right)\left(\frac{2T}{\pi^2\epsilon}\right)^{(N+1)/2}\left(\prod_{n=1}^{N}\frac{1}{n}\right)\to 1 \qquad \text{as } N\to\infty \tag{3.94}$$

4

The Schrödinger Description of Quantum Mechanics

THE path integrals which we have discussed so far (with the exception of Eq. 3.82) have integrands which are exponentials of actions with the property

$$S[b,a] = S[b,c] + S[c,a] \tag{4.1}$$

Such path integrals can be analyzed in terms of the properties of integral equations which can be deduced from them. We have already seen this in Chap. 2 (e.g., Eq. 2.31) and Chap. 3 (e.g., Eq. 3.42).

A still more convenient method is to reduce the path integrals to differential equations if possible. This possibility exists in quantum mechanics and is, in fact, the most convenient way to present that theory. It is in almost every case easier to solve the differential equation than it is to evaluate the path integral directly. The conventional presentation of quantum mechanics is based on this differential equation, called the Schrödinger equation. Here we shall derive this equation from our formulation. We shall not solve this equation for a large number of examples, because such solutions are presented in a detailed and satisfactory fashion in other books on quantum mechanics.[1]

In this chapter our purpose is twofold: (1) For the reader primarily interested in quantum mechanics our aim is to connect the path integral formulation with other formulations which are found in the standard literature and textbooks so that he can continue his study in those books and can learn to translate back and forth between the two different languages. (2) For the reader primarily interested in path integrals this chapter will show a technique which is available for a certain class of path integrals to reduce these path integrals to differential equations. This technique is best shown by the particular example of quantum mechanics which we shall develop here.

4-1 THE SCHRÖDINGER EQUATION

The Differential Equation Form. The reason that we can develop a differential equation is that the relationship of Eq. (4.1) is correct for any values of the points a, b, and c. For example, the time t_b can be only an infinitesimal time ϵ greater than the time t_c. This will permit us to relate the value of a path integral at one time to its value a short time later. In this manner we can obtain a differential equation for the path integral.

[1]For example, see L.I. Schiff, "Quantum Mechanics," 2nd ed., McGraw-Hill Book Company, New York, 1955.

We have already found that as a consequence of Eq. (4.1) we can define a wave function. Furthermore, we know that the equation

$$\psi(x_b, t_b) = \int_{-\infty}^{\infty} K(x_b, t_b; x_a, t_a)\psi(x_a, t_a)\, dx_a \tag{4.2}$$

gives the wave function at a time t_b in terms of the wave function at a time t_a. In order to obtain the differential equation that we seek, we apply this relationship in the special case that the time t_b differs only by an infinitesimal interval ϵ from t_a. The kernel $K(b,a)$ is proportional to the exponential of $i/\hbar$ times the action for the interval t_a to t_b. For a short interval ϵ the action is approximately ϵ times the lagrangian for this interval. That is, using the same approximation at that of Eq. (2.34), we have

$$\psi(x, t+\epsilon) = \frac{1}{A}\int_{-\infty}^{\infty} \exp\left\{\frac{i}{\hbar}\epsilon L\left(\frac{x-y}{\epsilon}, \frac{x+y}{2}\right)\right\}\psi(y,t)\, dy \tag{4.3}$$

We shall now apply this to the special case of a particle moving in one dimension subject to a potential energy $V(x,t)$, i.e., that for which $L = (m/2)\dot{x}^2 - V(x,t)$. In this case Eq. (4.3) becomes

$$\psi(x, t+\epsilon) = \frac{1}{A}\int_{-\infty}^{\infty} \exp\left\{\frac{i}{\hbar}\frac{m(x-y)^2}{2\epsilon}\right\} \times \exp\left\{-\frac{i}{\hbar}\epsilon V\left(\frac{x+y}{2}, t\right)\right\}\psi(y,t)\, dy \tag{4.4}$$

The quantity $(x-y)^2/\epsilon$ appears in the exponent of the first factor. It is clear that if y is appreciably different from x, this quantity is very large and the exponential consequently oscillates very rapidly as y varies. When this factor oscillates rapidly, the integral over y gives a very small value (because of the smooth behavior of the other factors). Only if y is near x (where the exponential changes more slowly) do we get important contributions. For this reason we make the substitution $y = x + \eta$ with the expectation that appreciable contributions to the integral will occur only for small η. We obtain

$$\psi(x, t+\epsilon) = \frac{1}{A}\int_{-\infty}^{\infty} \exp\left\{\frac{im\eta^2}{2\hbar\epsilon}\right\} \times \exp\left\{-\frac{i}{\hbar}\epsilon V\left(x+\frac{\eta}{2}, t\right)\right\}\psi(x+\eta,t)\, d\eta \tag{4.5}$$

The phase of the first exponential changes from 0 to of 1 radian when η changes from 0 to $\sqrt{2\hbar\epsilon/m}$, so most of the integral is contributed by values of η in this order.

We may expand ψ in a power series. We need only keep terms of order ϵ. This implies keeping second-order terms in η. The term $\epsilon V(x+\eta/2, t)$ may be replaced by $\epsilon V(x,t)$ because the error is of higher order than ϵ. Expanding the left-hand side to first order in ϵ and second order in η, we obtain

$$\psi(x,t) + \epsilon\frac{\partial\psi}{\partial t} = \frac{1}{A}\int_{-\infty}^{\infty} \exp\left\{\frac{im\eta^2}{2\hbar\epsilon}\right\} \times \left[1 - \frac{i}{\hbar}\epsilon V(x,t)\right]\left[\psi(x,t) + \eta\frac{\partial\psi}{\partial x} + \frac{\eta^2}{2}\frac{\partial^2\psi}{\partial x^2}\right] d\eta \tag{4.6}$$

If we take the leading term on the right-hand side, we have the quantity $\psi(x,t)$ multiplied by the integral

$$\frac{1}{A}\int_{-\infty}^{\infty} \exp\left\{\frac{im\eta^2}{2\hbar\epsilon}\right\} d\eta = \frac{1}{A}\left(\frac{2\pi i\hbar\epsilon}{m}\right)^{1/2} \tag{4.7}$$

On the left-hand side we have just $\psi(x,t)$. In order that both sides agree in the limit ϵ approaches 0, it is necessary that A be so chosen that the expression of Eq. (4.7) equals 1. That is,

$$A = \left(\frac{2\pi i\hbar\epsilon}{m}\right)^{1/2} \tag{4.8}$$

as we have stated previously (see Eq. 2.21). This is a way of obtaining the quantity A in more complicated problems also. The A must be so chosen that the equation is correct to zero order in ϵ. Otherwise, no limit will exist as ϵ approaches 0 in the original path integral.

In order to evaluate the right-hand side of Eq. (4.6), we shall have to use two integrals

$$\frac{1}{A}\int_{-\infty}^{\infty} \eta\exp\left\{\frac{im\eta^2}{2\hbar\epsilon}\right\} d\eta = 0 \tag{4.9}$$

and

$$\frac{1}{A}\int_{-\infty}^{\infty} \eta^2\exp\left\{\frac{im\eta^2}{2\hbar\epsilon}\right\} d\eta = \frac{i\hbar\epsilon}{m} \tag{4.10}$$

Writing out the right-hand side of Eq. (4.6) gives

$$\psi + \epsilon\frac{\partial\psi}{\partial t} = \psi - \frac{i}{\hbar}\epsilon V\psi + \frac{i\hbar\epsilon}{2m}\frac{\partial^2\psi}{\partial x^2} \tag{4.11}$$

This will be true to order ϵ if $\psi(x,t)$ satisfies the differential equation

$$\frac{\partial\psi}{\partial t} = -\frac{i}{\hbar}\left[-\frac{\hbar^2}{2m}\frac{\partial^2\psi}{\partial x^2} + V(x,t)\psi\right] \tag{4.12}$$

This is the Schrödinger equation for our problem of a particle moving in one dimension. Corresponding equations in more complicated situations can be worked out in the same way, as demonstrated by the following problems.

Problem 4-1 Show that for a single particle moving in three dimensions in a potential energy $V(\mathbf{x}, t)$ the Schrödinger equation is

$$\frac{\partial \psi(\mathbf{x}, t)}{\partial t} = -\frac{i}{\hbar}\left[-\frac{\hbar^2}{2m}\nabla^2\psi(\mathbf{x}, t) + V(\mathbf{x}, t)\psi(\mathbf{x}, t)\right] \tag{4.13}$$

This equation was discovered by Schrödinger in 1925 and formed the central feature of the development of quantum mechanics thereafter.

The Operator Form. The equations which result from various problems corresponding to different forms for the lagrangian can all be written for convenience in the form

$$\frac{\partial \psi}{\partial t} = -\frac{i}{\hbar} H\psi \tag{4.14}$$

Here H does not represent a number but indicates an operation on ψ. It is called the hamiltonian operator. For example, in Eq. (4.12) this operation is

$$H = -\frac{\hbar^2}{2m}\frac{\partial^2}{\partial x^2} + V(x, t) \tag{4.15}$$

Such an equation with operators on both sides means this: If any function $f(x)$ is written after each operator on each side, the equation will be true. That is, Eq. (4.15) symbolizes the statement: The relation

$$Hf(x) = -\frac{\hbar^2}{2m}\frac{\partial^2 f(x)}{\partial x^2} + V(x, t)f(x) \tag{4.16}$$

holds for any function $f(x)$.

Problem 4-2 For a particle of charge e in a magnetic field the lagrangian is

$$L = \frac{m}{2}\dot{\mathbf{x}}^2 + \frac{\mathrm{e}}{c}\dot{\mathbf{x}}\cdot\mathbf{A}(\mathbf{x}, t) - \mathrm{e}\phi(\mathbf{x}, t) \tag{4.17}$$

where $\dot{\mathbf{x}}$ is the velocity vector, c is the velocity of light, and $\mathbf{A}$ and ϕ are the vector and scalar potentials. Show that the corresponding Schrödinger equation is

$$\frac{\partial \psi}{\partial t} = -\frac{i}{\hbar}\left[\frac{1}{2m}\left(\frac{\hbar}{i}\boldsymbol{\nabla} - \frac{\mathrm{e}}{c}\mathbf{A}\right)\cdot\left(\frac{\hbar}{i}\boldsymbol{\nabla} - \frac{\mathrm{e}}{c}\mathbf{A}\right)\psi + \mathrm{e}\phi\,\psi\right] \tag{4.18}$$

Thus the hamiltonian is

$$H = \frac{1}{2m}\left(\frac{\hbar}{i}\boldsymbol{\nabla} - \frac{\mathrm{e}}{c}\mathbf{A}\right)\cdot\left(\frac{\hbar}{i}\boldsymbol{\nabla} - \frac{\mathrm{e}}{c}\mathbf{A}\right) + \mathrm{e}\phi \tag{4.19}$$

Problem 4-3 Show that the complex conjugate function ψ^*, defined as the function ψ with every i changed to $-i$, satisfies

$$\frac{\partial\psi^*}{\partial t} = +\frac{i}{\hbar}(H\psi)^* \tag{4.20}$$

The notation for operators can be described by giving a number of examples. For example, the operator x means multiplication by x, the operator x^2 means multiplication by x^2, the operator $V(x)$ (some function of x) means multiplication by $V(x)$, the operator $\partial/\partial x$ means partial differentiation with respect to x, $\partial\psi/\partial x$, etc.

If A and B are operators, then the operator AB means that we first apply B and then A, that is, $AB\psi$ means $A(B\psi)$. Thus, for example, the operator $x(\partial/\partial x)$ means x times $\partial\psi/\partial x$. On the other hand, the operator $(\partial/\partial x)x$ means the partial derivative with respect to x of $x\psi$, or

$$\frac{\partial}{\partial x}(x\psi) = x\frac{\partial\psi}{\partial x} + \psi$$

We see that in general the operator AB and the operator BA are *not* identical.

We further define the operator $A+B$ by the rule that $A+B$ operating on ψ is $A\psi + B\psi$. For example, the previous equation can be written as an equation among operators as follows:

$$\frac{\partial}{\partial x}x = x\frac{\partial}{\partial x} + 1 \tag{4.21}$$

the meaning being that $(\partial/\partial x)xf = x(\partial/\partial x)f + f$ for any function f.

Problem 4-4 Show

$$\frac{\partial^2}{\partial x^2}x = x\frac{\partial^2}{\partial x^2} + 2\frac{\partial}{\partial x} \tag{4.22}$$

and therefore that, for the H of Eq. (4.15),

$$Hx - xH = -\frac{\hbar^2}{m}\frac{\partial}{\partial x} \tag{4.23}$$

This operator notation is used a great deal in the conventional formulations of quantum mechanics.

The Schrödinger Equation for the Kernel. Since $K(b, a)$, thought of as a function of the variables b, is a special wave function (namely, that for a particle which starts at a), we see that K must also satisfy a Schrödinger equation. Thus for the case specified by Eq. (4.15)

$$\frac{\partial}{\partial t_b}K(b,a) = -\frac{i}{\hbar}\left[-\frac{\hbar^2}{2m}\frac{\partial^2}{\partial x_b^2}K(b,a) + V(b)K(b,a)\right] \tag{4.24}$$

for $t_b > t_a$. In general we have

$$\frac{\partial}{\partial t_b}K(b,a) = -\frac{i}{\hbar}H_b K(b,a) \qquad \text{for } t_b > t_a \tag{4.25}$$

wherein the operator H_b operates on the b variables only.

Problem 4-5 Using the relation

$$K(b,a) = \int_{-\infty}^{\infty} K(b,c)K(c,a)\,dx_c \tag{4.26}$$

with $t_c - t_a = \epsilon$, an infinitesimal, show that if t_b is greater than t_a, the kernel K satisfies

$$\frac{\partial}{\partial t_a}K(b,a) = +\frac{i}{\hbar}H_a^* K(b,a) \tag{4.27}$$

where H_a now operates on the a variables only.

The function $K(b, a)$ defined by a path integral in Eq. (2.25) is defined only for $t_b > t_a$. The function is not defined if $t_b < t_a$. It will prove to be very convenient in later work (e.g., Chap. 6) to define $K(b, a)$ to be zero for $t_b < t_a$. (With this convention Eq. (4.2), for example, is valid only if $t_b > t_a$.) With the condition

$$K(b,a) = 0 \qquad \text{for } t_b < t_a \tag{4.28}$$

it is evident that Eq. (4.25) is satisfied also for $t_b < t_a$ (in a trivial fashion, of course, since $K = 0$). But this equation is not satisfied at the point $t_b = t_a$, because $K(b, a)$ is discontinuous at $t_b = t_a$.

Problem 4-6 Show that $K(b, a) \to \delta(x_b - x_a)$ as $t_b \to t_a + 0$.

From the result of Prob. 4-6 we see that the derivative of K with respect to t_b gives a delta function in the time multiplied by the height of the jump, $\delta(x_b - x_a)$. Hence $K(b, a)$ satisfies

$$\frac{\partial}{\partial t_b}K(b,a) = -\frac{i}{\hbar}H_b K(b,a) + \delta(x_b - x_a)\delta(t_b - t_a) \tag{4.29}$$

This equation plus the boundary condition of Eq. (4.28) could serve to define $K(b,a)$ if one were to have started out from the Schrödinger equation as the fundamental definition in quantum mechanics. It is clear that the quantity $K(b,a)$ is a kind of Green's function for the Schrödinger equation.

The Conservation of Probability. The hamiltonian operator given by Eq. (4.15) has the interesting property that, if f, g are any functions which fall off to zero at infinity,

$$\int_{-\infty}^{\infty} (Hg)^* f\,dx = \int_{-\infty}^{\infty} g^*(Hf)\,dx \tag{4.30}$$

The meaning of the symbols is this. On the left we are to take g, operate on it with H (forming Hg), and then take the complex conjugate. The result is then multiplied by f and integrated over all space. The result is the same as taking Hf, multiplying by the complex conjugate of g, and integrating. It is easily verified that this is true by integrating the term $\int (Hg)^* f\,dx$ (by parts, where necessary).

For our example in Eq. (4.15) we have for the left side of Eq. (4.30)

$$-\frac{\hbar^2}{2m}\int_{-\infty}^{\infty} \frac{d^2g^*}{dx^2} f\,dx + \int_{-\infty}^{\infty} Vg^*f\,dx = \tag{4.31}$$
$$-\frac{\hbar^2}{2m}\left[\frac{dg^*}{dx}f - g^*\frac{df}{dx}\right]_{-\infty}^{+\infty} - \frac{\hbar^2}{2m}\int_{-\infty}^{\infty} g^*\frac{d^2f}{dx^2}\,dx + \int_{-\infty}^{\infty} Vg^*f\,dx$$

(integrating by parts twice). If f, g, fall off at infinity, the integrated parts vanish and Eq. (4.30) is established. An operator which has the property given by Eq. (4.30) is called *hermitian.* In all cases of quantum mechanics the hamiltonian is hermitian. For more general cases than that considered above the integration over our one-dimensional variable x becomes an integration (or sum) over all the variables of the system.

If we put f and g equal to $\psi(x,t)$, we get

$$\int_{-\infty}^{\infty} (H\psi)^*\psi\,dx = \int_{-\infty}^{\infty} \psi^*(H\psi)\,dx \tag{4.32}$$

and if ψ satisfies the wave equation (4.14), this becomes

$$\int_{-\infty}^{\infty} \frac{\partial\psi^*}{\partial t}\psi\,dx + \int_{-\infty}^{\infty} \psi^*\frac{\partial\psi}{\partial t}\,dx = \frac{d}{dt}\left(\int_{-\infty}^{\infty} \psi^*\psi\,dx\right) = 0 \tag{4.33}$$

That is, $\int \psi^*\psi\,dx$ is a constant independent of time. This is easily interpreted. For if ψ is suitably normalized, $\psi^*\psi$ is the probability of being found at x; so the integral is the probability of being found somewhere, which is certainty (or 1) and is constant. Of course, as far as the wave equation is concerned ψ can be multiplied by any constant and still be a solution. Then $\psi^*\psi$ is multiplied by the square of the constant, and the integral is this constant squared.

It is fundamental to our definition of ψ as probability amplitude that the integral of $\psi^*\psi$ is constant. In terms of the kernel this means that if f is the wave function at time t_a, then at time t_b it has the same square integral. That is, if

$$\psi(b) = \int_{-\infty}^{\infty} K(b,a) f(a)\, dx_a \tag{4.34}$$

then

$$\int_{-\infty}^{\infty} \psi^*(b)\psi(b)\, dx_b = \int_{-\infty}^{\infty} f^*(a) f(a)\, dx_a \tag{4.35}$$

or

$$\int_{-\infty}^{\infty}\int_{-\infty}^{\infty}\int_{-\infty}^{\infty} K^*(b;x_a',t_a)K(b;x_a,t_a)f^*(x_a')f(x_a)\, dx_a\, dx_a'\, dx_b = \int_{-\infty}^{\infty} f^*(x_a)f(x_a)\, dx_a \tag{4.36}$$

For this to be true for arbitrary f we must have

$$\int_{-\infty}^{\infty} K^*(b;x_a',t_a)K(b;x_a,t_a)\, dx_b = \delta(x_a' - x_a) \tag{4.37}$$

That is, in order to interpret ψ as a probability amplitude, the kernel must satisfy Eq. (4.37). We have derived this by means of the Schrödinger equation. It would be nicer to demonstrate this and other properties, such as Eq. (4.38) and Prob. 4-7, directly in terms of the path integral definition of K instead of coming through the differential equation. It is possible, of course, but it is not so simple or neat as a derivation of such a fundamental relation should be. One can verify Eq. (4.37) as follows: For a small interval with $t_b = t_a + \epsilon$, Eq. (4.37) follows directly from the expression $e^{i\epsilon L/\hbar}$ for this interval. By induction, the complete Eq. (4.37) results. One disadvantage of the approach to quantum mechanics through path integrals is the fact that relations involving ψ^* or K^* are not self-evident.

By changing the variable name in Eq. (4.37) from a to c, multiplying by $K(c,a)$, and integrating over x_c, we find

$$\int_{-\infty}^{\infty} K^*(b,c)K(b,a)\, dx_b = K(c,a) \tag{4.38}$$

where, as usual, $t_b > t_c > t_a$. Compare this to $\int K(b,c)K(c,a)\,dx_c = K(b,a)$. We may describe the second relation this way: Starting at t_a, $K(c,a)$ gives us the amplitude at the later time t_c. If we wish to go to a still later time t_b, we can do so by using the kernel $K(b,c)$. On the other hand, if having the amplitude at t_b we want to *work back* to find it at an earlier time $t_c < t_b$, we can do this by using the function $K^*(b,c)$, according to Eq. (4.38). That is, $K^*(b,c)$ undoes the work of $K(c,b)$.

Problem 4-7 Show that $\int K^*(b,a)K(b,c)\,dx_b = K^*(c,a)$, where our usual convention of $t_b > t_c > t_a$ holds.

4-2 THE TIME-INDEPENDENT HAMILTONIAN

Steady States of Definite Energy. The special case that the hamiltonian H is independent of time is of great practical importance. This corresponds to the case that the action S does not depend on the time explicitly; e.g., the potentials $\mathbf{A}$ and ϕ, and the potential energy V, do not contain t. In this case the kernel cannot depend upon the absolute time but instead is a function only of the interval $t_b - t_a$. As a consequence, there exist wave functions that depend periodically on the time.

It is easiest to see what happens by studying the differential equation. Starting from the Schrödinger equation (4.14), we try a special solution of the form $\psi(x,t) = \phi(x)f(t)$, a function of position only multiplied by a function of time only. Substitution gives us the relation

$$\phi(x)f'(t) = -\frac{i}{\hbar}[H\phi(x)]f(t) \tag{4.39}$$

or

$$\frac{f'(t)}{f(t)} = -\frac{i}{\hbar}\frac{H\phi(x)}{\phi(x)} \tag{4.40}$$

The left-hand side of this equation does not depend upon x, whereas the right-hand side does not depend upon t. Because they are always equal neither side can depend upon either variable t or x. That is, each side is a constant. Let us call the constant $-(i/\hbar)E$. Then $f'(t) = -(i/\hbar)Ef(t)$, or $f(t) = f_0 e^{-(i/\hbar)Et}$ where f_0 is an arbitrary constant factor. Thus the special solution is of the form

$$\psi(x,t) = \phi(x)e^{-(i/\hbar)Et} \tag{4.41}$$

where $\phi(x)$ satisfies

$$H\phi(x) = E\phi(x) \tag{4.42}$$

That is, for this special solution the wave function oscillates with the same definite frequency at every point in space. We saw (Eq. 3.17) that the frequency with which a wave function oscillates corresponds, in classical physics, to the energy. Therefore, we say that when the wave function is of this special form, the state has a definite energy E. For each value of E a different particular function $\phi(x)$, a solution of Eq. (4.42), must be sought.

The probability that a particle is at x is the absolute square of the wave function $\psi(x)$, or $|\psi(x)|^2$. In view of Eq. (4.41) this is equal to $|\phi(x)|^2$ and does not depend upon the time. That is, the probability of finding the particle at any location is independent of the time. We say under these circumstances that the system is in a stationary state — stationary in the sense that there is no variation in the probabilities as a function of time.

This situation is somewhat related to the uncertainty principle; for in a situation in which we know that the energy is exactly E we must be completely uncertain of the time. This is consonant with the idea that the properties of an atom in a specific state are absolutely independent of the time, so that at any time we would obtain the same result.

Suppose that E_1 is a possible energy for which Eq. (4.42) has a solution $\phi_1(x)$ and that E_2 is another value for energy for which this equation has some other solution $\phi_2(x)$. Then we know two special solutions of the Schrödinger equation, namely,

$$\psi_1(x,t) = \phi_1(x)e^{-(i/\hbar)E_1 t} \quad \text{and} \quad \psi_2(x,t) = \phi_2(x)e^{-(i/\hbar)E_2 t} \tag{4.43}$$

Since the Schrödinger equation is linear, it is clear that if ψ is a solution, then so is $c\psi$. Furthermore, if ψ_1 is a solution and ψ_2 is a solution, then the sum $\psi_1 + \psi_2$ is also a solution. Evidently, then, the function

$$\psi(x,t) = c_1\phi_1(x)e^{-(i/\hbar)E_1 t} + c_2\phi_2(x)e^{-(i/\hbar)E_2 t} \tag{4.44}$$

is also a solution of the Schrödinger equation.

As a matter of fact, it can be shown that if all of the possible values of E and the corresponding functions $\phi(x)$ are worked out, any solution $\psi(x,t)$ of Eq. (4.14) can be written as a linear combination of these special solutions of definite energy.

The total probability to be anywhere is constant, as we showed in Sec. 4-1. This must be true no matter what the values of c_1 and c_2, so that, using Eq. (4.44) for $\psi(x,t)$, we have

$$\begin{aligned}\int \psi^*\psi\,dx = c_1^*c_1\int\phi_1^*\phi_1\,dx + c_1^*c_2e^{+(i/\hbar)(E_1-E_2)t}\int\phi_1^*\phi_2\,dx \\ + c_1c_2^*e^{-(i/\hbar)(E_1-E_2)t}\int\phi_1\phi_2^*\,dx + c_2^*c_2\int\phi_2^*\phi_2\,dx\end{aligned} \tag{4.45}$$

Since this must give a constant result, the time-varying terms (i.e., terms including $e^{\pm(i/\hbar)(E_1-E_2)t}$) must vanish for all possible choices of c_1 and c_2. This means that

$$\int_{-\infty}^{\infty} \phi_1^*(x)\phi_2(x)\,dx = \int_{-\infty}^{\infty} \phi_1(x)\phi_2^*(x)\,dx = 0 \tag{4.46}$$

When two functions $f(x)$ and $g(x)$ satisfy $\int f^*(x)g(x)\,dx = 0$, we say they are orthogonal. Thus Eq. (4.46) says that two stationary states of different energies are orthogonal.

In Sec. 5-2 we shall learn an interpretation for expressions such as $\int f^*(x)g(x)\,dx$, and we shall find that Eq. (4.46) records the fact that if a particle is known to have energy E_1 (and hence a wave function $\psi_1 = e^{-(i/\hbar)E_1t}\phi_1$), then the amplitude that it is found to have a different energy E_2 (i.e., wave function $\psi_2 = e^{-(i/\hbar)E_2t}\phi_2$), must be zero.

Problem 4-8 Show from the fact that H is hermitian that E is real. (*Hint:* Choose $f = g = \phi$ in Eq. (4.30).)

Problem 4-9 Show from the fact that H is hermitian that Eq. (4.46) holds. (*Hint:* Choose $f = \phi_2$, $g = \phi_1$ in Eq. (4.30).)

Linear Combinations of Steady-state Functions. Suppose that our functions corresponding to the set of energy levels E_n are not only orthogonal but also normalized, i.e., that the integral of the absolute square over all x is 1. Then we shall have

$$\int_{-\infty}^{\infty} \phi_n^*(x)\phi_m(x)\,dx = \delta_{n,m} \tag{4.47}$$

where $\delta_{n,m}$, the Kronecker delta, is defined by $\delta_{n,m} = 0$ if $n \neq m$ and $\delta_{n,n} = 1$. Many functions can be expressed as a linear combination of such $\phi_n(x)$'s. In particular, any function which is likely to arise as a wave function can be so expressed. That is,

$$f(x) = \sum_{n=1}^{\infty} a_n\phi_n(x) \tag{4.48}$$

The coefficients a_n are easily obtained: multiply Eq. (4.48) by $\phi_m^*(x)$ and integrate over all x to obtain

$$\int_{-\infty}^{\infty} \phi_m^*(x)f(x)\,dx = \sum_{n=1}^{\infty} a_n \int_{-\infty}^{\infty} \phi_m^*(x)\phi_n(x)\,dx = a_m \tag{4.49}$$

That is,

$$a_n = \int_{-\infty}^{\infty} \phi_n^*(x)f(x)\,dx \tag{4.50}$$

Thus we have the identity

$$f(x) = \sum_{n=1}^{\infty} \phi_n(x) \int_{-\infty}^{\infty} \phi_n^*(y) f(y)\, dy = \int_{-\infty}^{\infty} \left[\sum_{n=1}^{\infty} \phi_n(x) \phi_n^*(y) \right] f(y)\, dy \tag{4.51}$$

This shows that, in terms of the Dirac delta function,

$$\delta(x-y) = \sum_{n=1}^{\infty} \phi_n(x) \phi_n^*(y) \tag{4.52}$$

It is possible to express the kernel $K(b,a)$ in terms of these functions $\phi_n(x)$ and energy values E_n. We do so by the following consideration. Let us ask this: If $f(x)$ is the known wave function at the time t_a, what is the wave function at time t_b? It can be written at any time t as

$$\psi(x,t) = \sum_{n=1}^{\infty} c_n e^{-(i/\hbar)E_n t} \phi_n(x) \tag{4.53}$$

for it is a solution of the Schrödinger equation, and any solution can be written in this form. But at the time t_a we have

$$f(x) = \psi(x,t_a) = \sum_{n=1}^{\infty} c_n e^{-(i/\hbar)E_n t_a} \phi_n(x) = \sum_{n=1}^{\infty} a_n \phi_n(x) \tag{4.54}$$

since we can always express $f(x)$ in the form of Eq. (4.48). So we conclude

$$c_n = a_n e^{+(i/\hbar)E_n t_a} \tag{4.55}$$

Putting this into Eq. (4.53), we have

$$\psi(x,t_b) = \sum_{n=1}^{\infty} c_n e^{-(i/\hbar)E_n t_b} \phi_n(x) = \sum_{n=1}^{\infty} a_n e^{-(i/\hbar)E_n (t_b - t_a)} \phi_n(x) \tag{4.56}$$

Now using Eq. (4.50) for the coefficient a_n, we obtain

$$\begin{aligned} \psi(x,t_b) &= \sum_{n=1}^{\infty} \phi_n(x) e^{-(i/\hbar)E_n(t_b-t_a)} \int_{-\infty}^{\infty} \phi_n^*(y) f(y)\, dy \\ &= \int_{-\infty}^{\infty} \sum_{n=1}^{\infty} \phi_n(x) \phi_n^*(y) e^{-(i/\hbar)E_n(t_b-t_a)} f(y)\, dy \end{aligned} \tag{4.57}$$

This final expression determines the wave function at time t_b completely in terms of $f(x)$, the wave function at time t_a. Previously we represented this relation by the equation

$$\psi(x,t_b) = \int_{-\infty}^{\infty} K(x,t_b;y,t_a) f(y)\, dy \tag{4.58}$$

Comparing Eqs. (4.57) and (4.58), we finally obtain the desired expression for the kernel $K(b,a)$,

$$K(x_b,t_b;x_a,t_a) = \begin{cases} \displaystyle\sum_{n=1}^{\infty} \phi_n(x_b)\phi_n^*(x_a)e^{-(i/\hbar)E_n(t_b-t_a)} & \text{for } t_b > t_a \\ 0 & \text{for } t_b < t_a \end{cases} \tag{4.59}$$

This expression for $K(b,a)$ is very useful for translating expressions to more conventional representations. It expresses the kernel, which was originally a path integral, entirely in terms of solutions of the differential equation (4.42).

Problem 4-10 Verify that $K(b,a)$ as expressed in Eq. (4.59) satisfies the Schrödinger equation (4.29).

Problem 4-11 Show that for free particles in three dimensions the solutions

$$\phi_{\mathbf{p}}(\mathbf{x}) = e^{i(\mathbf{p}/\hbar)\cdot\mathbf{x}} \tag{4.60}$$

go with the energy $E_{\mathbf{p}} = p^2/2m$. Consider the vector $\mathbf{p}$ as an index n and note the orthogonality. That is, as long as $\mathbf{p} \neq \mathbf{p}'$,

$$\int \phi_{\mathbf{p}}^*(\mathbf{x})\phi_{\mathbf{p}'}(\mathbf{x})\,d^3\mathbf{x} = 0 \qquad \text{even if } E_{\mathbf{p}} = E_{\mathbf{p}'} \tag{4.61}$$

Therefore the free-particle kernel must be

$$K_0(\mathbf{x}_b,t_b;\mathbf{x}_a,t_a) = \sum_{\mathbf{p}} e^{i(\mathbf{p}/\hbar)\cdot(\mathbf{x}_b-\mathbf{x}_a)}e^{-(i/\hbar)(p^2/2m)(t_b-t_a)} \tag{4.62}$$

Since the $\mathbf{p}$'s are distributed over a continuum, the sum over the "indices" $\mathbf{p}$ is really equivalent to an integral over the values of $\mathbf{p}$, namely

$$\sum_{\mathbf{p}}(\quad) = \int(\quad)\frac{d^3\mathbf{p}}{(2\pi\hbar)^3} \tag{4.63}$$

Therefore, we find that the free-particle kernel is given by

$$K_0(\mathbf{x}_b,t_b;\mathbf{x}_a,t_a) = \int e^{i(\mathbf{p}/\hbar)\cdot(\mathbf{x}_b-\mathbf{x}_a)}e^{-(i/\hbar)(p^2/2m)(t_b-t_a)}\frac{d^3\mathbf{p}}{(2\pi\hbar)^3} \tag{4.64}$$

Problem 4-12 Carry out the integral in Eq. (4.64) by completing the square. Show that the correct free-particle kernel (i.e., the three-dimensional version of Eq. 3.3) results.

4-3 NORMALIZING THE FREE-PARTICLE WAVE FUNCTIONS

The derivation of the kernel for a free particle, as given in Prob. 4-11, is unsatisfactory for two related reasons. First, the idea of a sum over distinct states n used in Eq. (4.59) is not satisfactory if the states lie in a continuum, as they do for a free particle where any $\mathbf{p}$ is allowed. Second, the plane-wave functions for free particles, although orthogonal, cannot be normalized; that is, $\int_{-\infty}^{\infty} \phi^* \phi \, dx = \int_{-\infty}^{\infty} 1 \, dx = \infty$, so the condition of Eq. (4.47), used in deriving Eq. (4.59), is not satisfied. Both of these points can be remedied together in a perfectly straightforward mathematical way. Starting all the way back when we expressed an arbitrary function as a sum of eigenfunctions,

$$f(x) = \sum_n a_n \phi_n(x) \tag{4.65}$$

we allow part, or all, of the states to lie in a continuum, so that the sum over n must be replaced partly by an integral. With mathematical care one can find the correct expression for K analagous to Eq. (4.59) but applying also when the states are in a continuum.

Normalizing in a Box. Many physicists prefer another, less rigorous approach. They modify the original problem in a way that (from physical reasoning) will not essentially modify the result yet will leave all the states separate in energy and all the simple sums as simple sums. In our example this may be accomplished as follows. We are studying the amplitude that in a finite time a particle goes from $\mathbf{x}_a$ at time t_a to $\mathbf{x}_b$ at time t_b. Now if these two points are some finite distance apart and the time is not extremely long, surely it can make no appreciable difference to the amplitude whether the electron is really free or is instead confined to some enormous box of volume "Vol" with walls very, very far from $\mathbf{x}_a$ and $\mathbf{x}_b$. The amplitude could be affected only if the particle could run out to the walls and back in the time $t_b - t_a$; but if the walls are far enough away, there is no appreciable amplitude for this.

In is always possible that this assumption fails for some special-shaped walls such that, for example, $\mathbf{x}_b$ is at a focus of waves from $\mathbf{x}_a$ reflected at the walls. From time to time someone lets an error creep in by replacing a system in empty space with one at the center of a large spherical box. The fact that this system remains at the exact center of a perfect sphere may have an effect (like the spot of light at the center of the shadow of a perfectly circular object) which does not vanish as the sphere radius goes to infinity. For another shape, or a system off-center to the sphere, the surface effect would vanish.

Take first the case of one dimension. In empty space the space-dependent wave functions are $e^{i(p/\hbar)x}$ (any p, positive or negative). If, instead, the range of x is limited to $-L/2$ to $+L/2$, say, what are the functions $\phi(x)$? The answer depends on the boundary conditions defining $\phi(x)$ at $x = -L/2$ and $x = +L/2$. The easiest conditions to understand physically are those for walls which offer very high repulsive potentials to the particle, thus confining it (i.e., perfect reflectors). They correspond to $\phi(x) = 0$ at $x = -L/2$ and $x = +L/2$. The solutions of the wave equation

$$-\frac{\hbar^2}{2m}\frac{\partial^2 \phi}{\partial x^2} = E\phi(x) \tag{4.66}$$

in the range $|x| < L/2$ are, for $E = p^2/2m = \hbar^2 k^2/2m$,

$$e^{ikx} \qquad \text{and} \qquad e^{-ikx}$$

or any linear combination. Neither e^{ikx} nor e^{-ikx} can satisfy the boundary conditions, but with $k = n\pi/L$ (n an integer) satisfactory solutions are given by half the sum (which is $\cos(kx)$) for n odd and $i/2$ times the difference (which is $\sin(kx)$) for n even, as diagrammed in Fig. 4-1. Thus the states are sines and cosines and the energy levels are separated (i.e. not in a continuum).

If the solutions are written as

$$\sqrt{\frac{2}{L}}\cos(kx) \qquad \text{and} \qquad \sqrt{\frac{2}{L}}\sin(kx)$$

then they are normalized, since

$$\int_{-L/2}^{+L/2} \left(\sqrt{\frac{2}{L}}\cos(kx)\right)^2 dx = 1 \tag{4.67}$$

A sum over states is a sum over n. If we consider, say, the sine wave functions (thus, even values of n) for very large L but not large x (walls far from the point of interest), the successive functions differ by only a

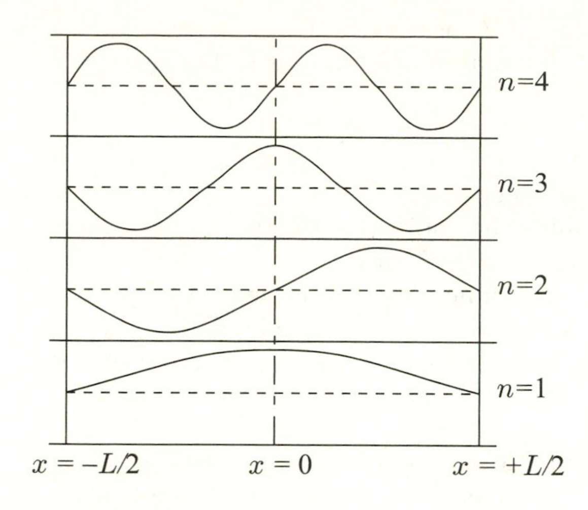

Fig. 4-1 The form of the one-dimensional wave functions which have been normalized in a box. The first four are shown. The corresponding energy levels are $E_1 = \hbar^2\pi^2/2mL^2$, $E_2 = 4E_1$, $E_3 = 9E_1$, and $E_4 = 16E_1$. The magnitude of the energy in absolute terms, which depends on the size of our fictitious box, is not important for more realistic problems. Rather, it is the relation between the energy levels of the various states which has significance.

small amount. This difference

$$\begin{aligned}
&\sqrt{\frac{2}{L}}\left[\sin\left(2\pi(n+1)\frac{x}{L}\right) - \sin\left(2\pi n\frac{x}{L}\right)\right] \\
&= \sqrt{\frac{2}{L}}\left[2\cos\left(2\pi\frac{2n+1}{2}\frac{x}{L}\right)\sin\left(2\pi\frac{1}{2}\frac{x}{L}\right)\right] \\
&\approx \sqrt{\frac{2}{L}}\frac{2\pi x}{L}\cos\left(2\pi(n+\tfrac{1}{2})\frac{x}{L}\right)
\end{aligned} \tag{4.68}$$

is approximately proportional to the small quantity x/L. So a sum on n can be replaced by an integral over $k = 2\pi n/L$. Since the successive allowed values of k (for sine functions) are spaced by $2\pi/L$, there are $(L/2\pi)\Delta k$ states in range Δk. All of this applies also to states with the cosine wave function, so that we may replace sums by integrals in our formulas with

$$\sum_{n=1}^{\infty}(\quad) \to \int_0^{\infty}(\quad)\frac{L}{2\pi}\,dk \tag{4.69}$$

and remember to add the result for the two kinds of wave functions, namely, $\sqrt{2/L}\,\cos(kx)$ and $\sqrt{2/L}\,\sin(kx)$.

It is often inconvenient to use $\sin(kx)$ and $\cos(kx)$ for the wave functions as we would like to use the linear combinations

$$e^{ikx} = \cos(kx) + i\sin(kx) \quad \text{and} \quad e^{-ikx} = \cos(kx) - i\sin(kx)$$

We were forced by our box to use sines and cosines and not the linear combination, because for a given k one, but not both, of the functions is a solution. But if we can disregard small errors arising from these small differences in k, we might still expect to be able to get the correct results from these new linear combinations. Normalized, they are

$$\sqrt{\frac{1}{L}}\,e^{ikx} \quad \text{and} \quad \sqrt{\frac{1}{L}}\,e^{-ikx}$$

Since the wave e^{-ikx} can be thought of as e^{ikx} but for negative values of k, our new procedure, including the addition of the two kinds of wave functions, becomes the following practical rule:

To deal with free-particle wave functions e^{ikx}, normalize them to a range of x of length L (i.e., use $\phi(x) = \sqrt{1/L}\,e^{ikx}$), and replace sums on states by integrals over k with the rule that the number of states with k in the range k to $k + dk$ is $(L/2\pi)\,dk$ and the range of k is $-\infty$ to $+\infty$.

Periodic Boundary Conditions. Sometimes this excursion into cosines and sines and back to exponentials is avoided by the following argument. The wall is artificial anyway, so its particular position and the particular boundary condition should not make any physical difference as long as it is far away. So instead of the physically simple conditions $\phi(x) = 0$ at $x = +L/2$ and at $x = -L/2$, let us use two others for which the solutions are indeed e^{ikx} directly. These are

$$\phi(x) \quad \left(\text{at } x = +\frac{L}{2}\right) = \phi(x) \quad \left(\text{at } x = -\frac{L}{2}\right) \tag{4.70}$$

and

$$\phi'(x) \quad \left(\text{at } x = +\frac{L}{2}\right) = \phi'(x) \quad \left(\text{at } x = -\frac{L}{2}\right) \tag{4.71}$$

These are called *periodic boundary conditions*, because the same ones would result by the requirement that $\phi(x)$ is periodic in x in all space with period $x = L$. It is readily verified that the functions $\sqrt{1/L}\,e^{ikx}$ are solutions, normalized to range L, provided $k = 2\pi n/L$ with n an integer: positive, negative, or zero. From this our rule follows directly.

In three dimensions we can see what happens by using a rectangular box of sides L_x, L_y, L_z in the three directions. Let us use periodic boundary conditions. That is, the magnitude and first derivatives of a wave function at a point on one face are respectively equal to the magnitude and first derivative at the corresponding point on the opposite face. The normalized wave function for a free particle is

$$\sqrt{\frac{1}{L_x}}\,e^{ik_x x}\sqrt{\frac{1}{L_y}}\,e^{ik_y y}\sqrt{\frac{1}{L_z}}\,e^{ik_z z} = \frac{1}{\sqrt{\text{Vol}}}e^{i\mathbf{k}\cdot\mathbf{x}} \tag{4.72}$$

where Vol $= L_x L_y L_z$ is the volume of the box and the allowed values of k_x are $2\pi n_x/L_x$ for n_x an integer, those of k_y are $2\pi n_y/L_y$ for n_y an integer, and those of k_z are $2\pi n_z/L_z$ for n_z an integer. Furthermore, the number of solutions with k_x in range dk_x, k_y in dk_y, and k_z in dk_z is

$$\frac{L_x}{2\pi}dk_x\,\frac{L_y}{2\pi}dk_y\,\frac{L_z}{2\pi}dk_z = \frac{\text{Vol}}{(2\pi)^3}d^3\mathbf{k} \tag{4.73}$$

That is, use plane waves normalized to volume "Vol": $\exp\{i\mathbf{k}\cdot\mathbf{x}\}/\sqrt{\text{Vol}}$. The number of states in range $d^3\mathbf{k}$ (differential volume of $\mathbf{k}$ space) is $\text{Vol}\,d^3\mathbf{k}/(2\pi)^3$.

Let us apply this to Prob. 4-11 and recall the connection between momentum and wave number $p = \hbar k$ brought out in Sec. 3-1. In Eq. (4.64) we must make two changes. First, since the wave functions used were $\exp\{i\mathbf{p}\cdot\mathbf{x}/\hbar\}$, whereas we should have used

$$\frac{1}{\sqrt{\text{Vol}}}\exp\left\{\frac{i\mathbf{p}\cdot\mathbf{x}}{\hbar}\right\}$$

there should be an additional factor $1/\text{Vol}$; for the product of two wave functions was involved. Second, the symbol

$$\sum_{\mathbf{p}}(\quad) \qquad \text{must be replaced by} \qquad \text{Vol}\int(\quad)\frac{d^3\mathbf{p}}{(2\pi\hbar)^3}$$

This justifies what was done in Prob. 4-11.

It is noted that the "Vol" factors cancel out, as indeed they must; for as $\text{Vol} \to \infty$ the kernel $K(b,a)$ must be independent of the size of the box.

Some Remarks on Mathematical Rigor. The reader may have one of two reactions on seeing how the volumes "Vol" cancel at the end of this calculation. One might be: How nicely it cancels out as it should, for the walls have no effect. The other might be: Why do it in the complicated and "dirty" nonrigorous manner, putting in walls which make no difference, etc., when all this can be done much more elegantly and rigorously mathematically without the need of walls, etc? It depends on whether you are physically minded or mathematically minded. There are many misunderstandings between mathematicians and physicists on the place of mathematical rigor in physics, so perhaps a word as to the value of each method (the box or mathematical rigor) may be in order.

There is, of course, the more trivial point: Which is most familiar — which takes the least new knowledge? Most physicists have seen this argument about how to count the states in a box before.

Another point is that the mathematically rigorous solution may not be physically rigorous. That is, the box may in fact exist. It may not be a rectangular box, but it is not often that experiments are done under the stars. Rather they are done in a room. Although it is physically reasonable that the walls have no effect, it is true that the original problem is set up as an idealization. It is no more satisfactory idealization to move the walls to infinity than to replace them by perfect mirrors far away. The mathematical rigor is wasted in the first idealization, since the walls are not at infinity.

The box approach is just as rigorous, or rather just as nonrigorous. It has several advantages. For example, in finding that the volume cancels out we do learn that at least one aspect of the idealized walls, namely how far away they are, is unimportant. This discovery makes us more intuitively convinced that the actual disposition of the real environment may be unimportant. Finally, the formula derived is very useful when in fact we do have a finite sample. For example, in Chap. 8 we shall use it to count sound-wave modes in a large, rectangular block of material.

On the other hand, the advantage of the mathematically clean argument is the avoidance of much unnecessary detail that cancels out. Although, using the box approach, one may learn something about how the walls have no effect, one may be firmly convinced that this is true anyway and not wish to descend into details to see it again.

The normalization problem is a special example, but it illustrates the point. The physicist cannot understand the mathematician's care in solving an idealized physical problem. The physicist knows the real problem is much more complicated. It has already been simplified by intuition, which discards the unimportant and often approximates the remainder.

5

Measurements and Operators

So far we have described quantum-mechanical systems as if we intended to measure only the coordinates of position and time. Indeed, all measurements of quantum-mechanical systems could be made to reduce eventually to position and time measurements (e.g., the position of a needle on a meter or the time of flight of a particle). Because of this possibility a theory formulated in terms of position measurements is complete enough in principle to describe all phenomena. Nevertheless, it is convenient to try to answer directly a question involving, say, a measurement of momentum without insisting that the ultimate recording of the equipment must be a position measurement and without having to analyze in detail that part of the apparatus which converts momentum to a recorded position. Thus, in this chapter, instead of concentrating on the amplitude that a particle has a definite position, we shall develop the idea of an amplitude to find a definite momentum, energy, or other physical quantity.

In the first section of this chapter we shall show how a system may be described in terms of momentum and energy. The concepts learned here will be extended in the second section to describe in general various ways of representing the quantum-mechanical system. The transformation functions which enable us to go from one method of representation to another have many interesting properties. Among them is the concept of an operator, which was introduced in the preceding chapter and will be discussed further in the third section of this chapter.

5-1 THE MOMENTUM REPRESENTATION

The Momentum Amplitude. So far we have used the concept of probability in terms of the position of a particle, but suppose we wish to measure the momentum. Is there an amplitude $\phi(p)$ whose absolute square will give us the probability $P(p)$ that a measurement of momentum will show that the particle has momentum p? There is in fact such an amplitude, and we can easily find it.

Some ways of measuring momentum (or other physical quantities) correspond to measurements of position, and thus they can be analyzed if we know how to analyze coordinate measurements. For example, working in one dimension, suppose we have a particle whose position at $t = 0$ is localized within $\pm b$ of the origin of the x axis. The uncertainty b can be as large as desired so long as it is finite. We can measure the momentum of such a particle by a time-of-flight technique. That is, we can observe how far the particle has traveled (assuming no forces) by the time $t = T$. If the position is y, then the velocity is y/T and the momentum is $p = my/T$. The error in such a momentum measurement, $\pm mb/T$, can be made as small as desired by making T sufficiently large.

Suppose we analyze the momentum probability $P(p)$ as defined by such an experiment. The probability $P(p)\,dp$ that the momentum lies between p and $p + dp$ is the probability $P(y)\,dy$ that, if all the potentials affecting the particle are suddenly turned off, then after the time T the particle will be found between the points y and $y + dy$. Of course, this requires that we connect p with y by $p = my/T$. Assume the wave function of the particle is given by $f(x)$ at $t = 0$, and our problem is to find $P(p)$ directly in terms of $f(x)$.

The amplitude for the particle to arrive at y at the time $t = T$ is

$$\psi(y,T) = \int_{-\infty}^{\infty} K_0(y,T;x,0) f(x)\,dx \tag{5.1}$$

Upon substitution for the free-particle kernel K_0 (Eq. 3.3), this expression becomes

$$\begin{aligned}\psi(y,T) = {} & \left(\frac{m}{2\pi i\hbar T}\right)^{1/2} \\ & \times \exp\left\{\frac{imy^2}{2\hbar T}\right\}\int_{-\infty}^{\infty}\exp\left\{\frac{im(-2yx + x^2)}{2\hbar T}\right\} f(x)\,dx\end{aligned} \tag{5.2}$$

The absolute square of this amplitude gives the probability that the particle lies between y and $y + dy$. According to our definition, this is identical (in the limit $T \to \infty$) with the probability that the momentum of the particle lies between p and $p + dp$.

$$\begin{aligned}P(y)\,dy &= \frac{m\,dy}{2\pi\hbar T}\left|\int_{-\infty}^{\infty}\exp\left\{\frac{im(-2yx + x^2)}{2\hbar T}\right\} f(x)\,dx\right|^2 \\ &= P(p)\,dp \qquad \text{as } T \to \infty\end{aligned} \tag{5.3}$$

Then substituting $p = my/T$, and supposing that we pass to the limit of large T, there results

$$P(p)\,dp = \frac{dp}{2\pi\hbar}\left|\int_{-\infty}^{\infty}\exp\left\{\frac{-ipx}{\hbar} + \frac{imx^2}{2\hbar T}\right\} f(x)\,dx\right|^2 \tag{5.4}$$

We assumed earlier that, initially, the particle would be restricted to a region within $\pm b$ of the origin. This means that the initial wave function $f(x)$ drops to 0 for values of x larger in absolute magnitude than b. Now as T becomes large the quantity $imb^2/2\hbar T$ becomes negligibly small. Since there is no contribution to the integral of Eq. (5.4) for

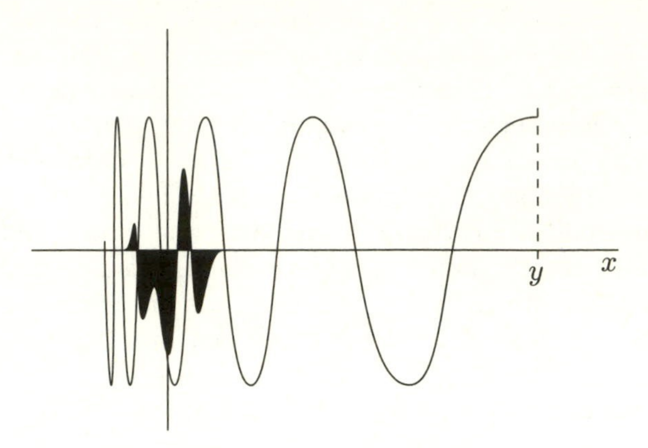

Fig. 5-1 The amplitude for a particle traveling freely to arrive at position y after time interval T is determined by the convolution of two functions. The first is the amplitude $f(x)$ for the particle to start at position x, as indicated by the shaded curve in the figure. The second, the amplitude to go from x to y, is the free particle kernel $K_0(y, T; x, 0)$, indicated by the sine wave of slowly changing wavelength. (Shown here is $\Re e\{\sqrt{i}K_0(y, T; x, 0)\}$ as a function of x for fixed y and T.) If the point y is far from the origin, compared to the distance $-b$ to $+b$ over which $f(x)$ is nonzero, the wave has an approximately constant wavelength near the origin. Its form there is approximately proportional to $\exp\{(-i/\hbar)(my/T)x\}$. The two functions are multiplied together and then integrated over x to find the amplitude for arrival at y. Since the particle has traveled approximately the distance y (again assuming $y \gg b$) in time T, this amplitude is equivalent to the amplitude that the particle has momentum $p = my/T$.

values of x greater in absolute magnitude than b, the probability $P(p)\,dp$ approaches $dp/2\pi\hbar$ times the absolute square° of the amplitude[1]

$$\phi(p) = \int_{-\infty}^{\infty} \exp\left\{\frac{-ipx}{\hbar}\right\} f(x)\,dy \tag{5.5}$$

An alternative explanation of this result is given in Fig. 5-1 and extended in Fig. 5-2.

The expression for the momentum amplitude given by Eq. (5.5) applies to a one-dimensional situation. It is easy to extend the definition

[1]Many writers prefer to account for the factor $1/2\pi\hbar$ in the definition of $\phi(p)$, where it appears as $1/\sqrt{2\pi\hbar}$. However, following the development of Sec. 4-3, we prefer to write it in the form we have used and remember that the differential element of momentum always includes the factor $1/2\pi\hbar$ in each dimension. For example, the differential element of momentum in three-dimensional momentum space is $d^3\mathbf{p}/(2\pi\hbar)^3$.

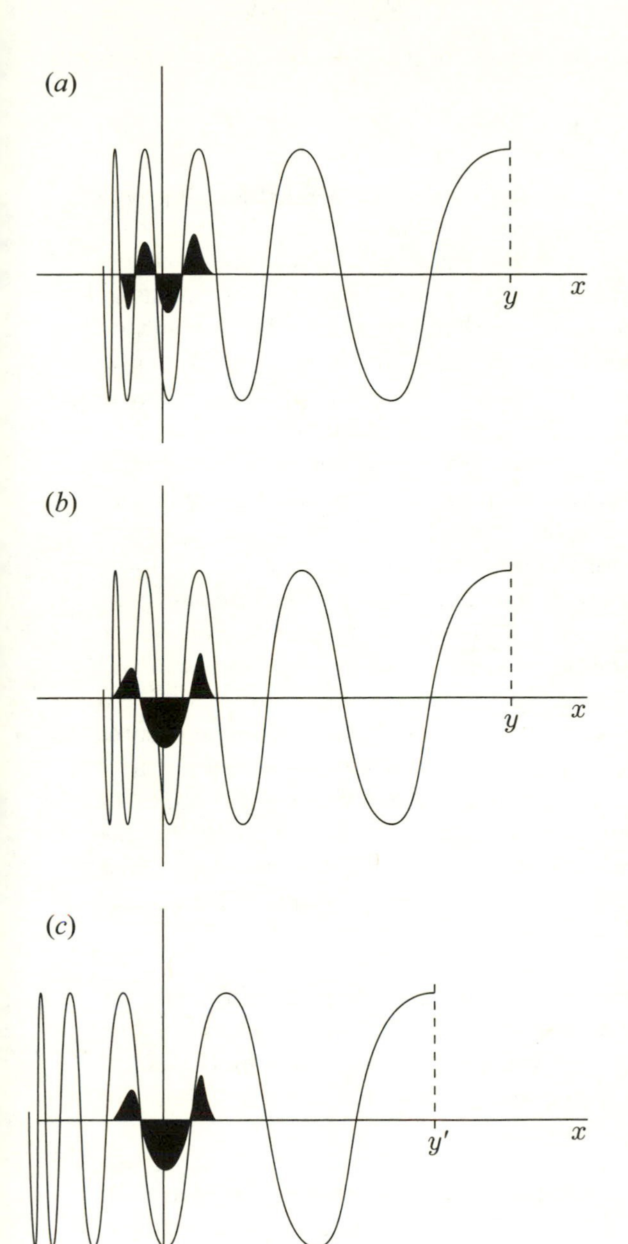

Fig. 5-2 If the amplitude $f(x)$ is roughly periodic with the same wavelength as the overlying kernel, as shown in (a), then the integral of the product of the two functions is large. That is, the probability that the momentum is $p = my/T$ is large.

On the other had, suppose the wavelengths differ for some new function $f'(x)$, as shown in (b). Then, when the product is taken, the contributions to the integral from different values of x tend to cancel each other. Now the probability that the momentum is my/T is small.

If a new position y' is chosen as a final point, as shown in (c), then a new region of the kernel curve overlies the space $-b$ to $+b$. For a correct choice of y', the wavelength of the kernel in this new region is the same as the wavelength of $f'(x)$ and a large probability results. That is, there is a large probability that such a particle has the new momentum value $p' = my'/T$.

to the three-dimensional case where the amplitude for the momentum is

$$\phi(\mathbf{p}) = \int \exp\left\{\frac{-i\mathbf{p}\cdot\mathbf{x}}{\hbar}\right\} f(\mathbf{x})\, d^3\mathbf{x} \tag{5.6}$$

where the wave function $f(\mathbf{x})$ is now assumed to be defined for all points in the three-dimensional coordinate space. This is the amplitude that the particle has the momentum $\mathbf{p}$ at the time $t = 0$. (Note that it is *not* defined for the time $t = T$. The time interval T is part of the measuring equipment, and it can be varied without changing the momentum amplitude.) The square of this amplitude, multiplied by the differential momentum element, gives the probability of finding the momentum in the interval (three-dimensional) $d^3\mathbf{p}/(2\pi\hbar)^3$ of momentum space.

We have analyzed a momentum measurement which is based on a time-of-flight technique. However, such an analysis can be applied to other techniques. The analysis of any technique for measuring momentum will give the same result for the momentum amplitude. For suppose we have two methods or techniques which purport to measure the same quantity, momentum. If one gives a different result than the other, we have to explain why one or the other apparatus is faulty. So if you will grant that the time-of-flight technique is an adequate way to define a momentum measurement, any other piece of equipment which measures momentum must give the same results $P(p)\,dp$ for the distribution of momenta if the system is in the state $f(x)$. Analysis of any equipment which measures momentum must give the same expression $\phi(p)$ for the amplitude for momentum p, within possibly an irrelevant constant phase difference (i.e., a factor $e^{i\delta}$ with δ constant). For example, consider the following problem.

Problem 5-1 Consider any piece of experimental equipment designed to measure momentum by means of a classical approximation, such as a magnetic field analyzer. Analyze the equipment by the methods outlined in the preceding paragraphs. Show that the same result for the momentum amplitude is obtained.

Transformation to Momentum Representation. We have called $\psi(\mathbf{x}, t)$ the amplitude for a particle to be at the point $\mathbf{x}$ at the time t. We have found that the momentum amplitude is given by

$$\phi(\mathbf{p}, t) = \int \exp\left\{\frac{-i\mathbf{p}\cdot\mathbf{x}}{\hbar}\right\} \psi(\mathbf{x}, t)\, d^3\mathbf{x} \tag{5.7}$$

We shall call this the amplitude that the particle has momentum $\mathbf{p}$ at the time t.

It is often useful to analyze problems in this momentum representation rather than in the coordinate representation, or, as it is often stated, in momentum space rather than in coordinate space. Actually, the transformation from one representation to the other is just a Fourier transform. Thus if we have the momentum representation and wish to find the coordinate representation, we use the inverse transform given by

$$\psi(\mathbf{x},t) = \int \exp\left\{\frac{+i\mathbf{p}\cdot\mathbf{x}}{\hbar}\right\} \phi(\mathbf{p},t)\, \frac{d^3\mathbf{p}}{(2\pi\hbar)^3} \tag{5.8}$$

We can describe this last formula in the same physical terms we have used to describe the structure of other amplitudes. The amplitude that the particle is at the position $\mathbf{x}$ is given by the sum over alternatives. In this case each alternative corresponds to the product of two terms. One of these is the amplitude that the momentum of the particle is $\mathbf{p}$, given by $\phi(\mathbf{p})$. The other term, $\exp\{i\mathbf{p}\cdot\mathbf{x}/\hbar\}$, *is the amplitude that if the momentum is* $\mathbf{p}$, *then the particle is at the position* $\mathbf{x}$. This second factor is not new to us, for we have discussed such an expression in Prob. 3-4.

Note that in the transform of Eq. (5.7) the exponent has a minus sign. Such a term can be described in a manner parallel to that used in the preceding paragraph. Thus we say that $\exp\{-i\mathbf{p}\cdot\mathbf{x}/\hbar\}$ *is the amplitude that if a particle is at position* $\mathbf{x}$, *it has the momentum* $\mathbf{p}$.

The Kernel in Momentum Representation. We have shown (Sec. 3-4) how a wave function at a particular time t_b can be obtained from the wave function at an earlier time t_a with the help of the kernel describing the motion of the particle in the intervening time. Thus

$$\psi(\mathbf{x}_b,t_b) = \int K(\mathbf{x}_b,t_b;\mathbf{x}_a,t_a)\psi(\mathbf{x}_a,t_a)\, d^3\mathbf{x}_a \tag{5.9}$$

It is possible to define a kernel in momentum space which would be used in a parallel expression. Thus the momentum amplitude at the time t_b can be derived from the momentum amplitude at an earlier time t_a by

$$\phi(\mathbf{p}_b,t_b) = \int \mathcal{K}(\mathbf{p}_b,t_b;\mathbf{p}_a,t_a)\phi(\mathbf{p}_a,t_a)\, \frac{d^3\mathbf{p}_a}{(2\pi\hbar)^3} \tag{5.10}$$

Substituting in Eq. (5.9) for $\psi(\mathbf{x}_a,t_a)$ the expression of Eq. (5.8) and taking the Fourier transform of $\psi(\mathbf{x}_b,t_b)$ to get $\phi(\mathbf{p}_b,t_b)$, as in Eq. (5.7), we see that the kernel in momentum representation is given in terms of the kernel in coordinate representation by the expression

$$\mathcal{K}(\mathbf{p}_b,t_b;\mathbf{p}_a,t_a) = \iint e^{-(i/\hbar)\mathbf{p}_b\cdot\mathbf{x}_b} K(\mathbf{x}_b,t_b;\mathbf{x}_a,t_a) e^{+(i/\hbar)\mathbf{p}_a\cdot\mathbf{x}_a}\, d^3\mathbf{x}_b\, d^3\mathbf{x}_a \tag{5.11}$$

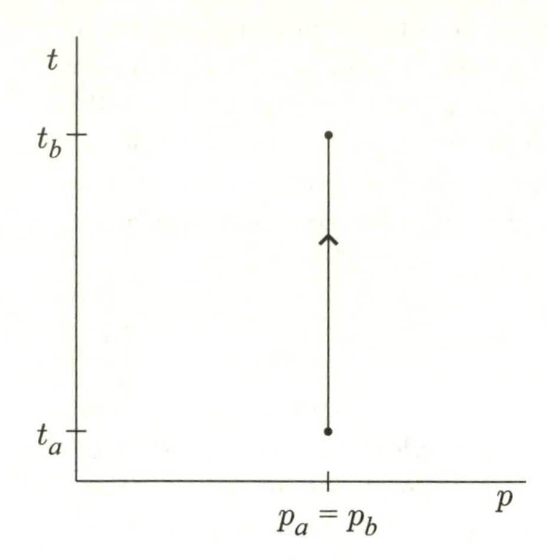

Fig. 5-3 The kernel for a free particle in momentum space is unlike the kernel in coordinate space. In momentum space, there is only one path which can carry the particle to the momentum value p_b at the time t_b. That single path must start at the momentum $p_a = p_b$. No other paths contribute to the kernel.

For example, the kernel describing the motion of a free particle in momentum space is found by using K_0 from Eq. (3.3) in Eq. (5.11). The result of the integration is

$$\mathcal{K}_0(\mathbf{p}_b, t_b; \mathbf{p}_a, t_a) = \begin{cases} (2\pi\hbar)^3\delta^3(\mathbf{p}_b - \mathbf{p}_a)\exp\left\{-\frac{i}{\hbar}\frac{|\mathbf{p}_a|^2}{2m}(t_b - t_a)\right\} & \text{for } t_b > t_a \\ 0 & \text{for } t_b < t_a \end{cases} \tag{5.12}$$

(The last line follows the convention of Eq. (4.28).) The occurrence of the Dirac delta function in this expression shows that the momentum of a free particle does not change, as diagrammed in Fig. 5-3. However, the phase of the momentum wave function changes continuously in accordance with the factor $e^{-(i/\hbar)Et}$, where $E = p^2/2m$. This result given by Eq. (5.12) can also be seen directly from Eq. (4.64).

This momentum-space kernel offers a much simpler representation of the free particle than does the coordinate-space kernel. Generally, when the particle is not free, but rather moves under the influence of a potential, the kernel in momentum representation loses its simplicity. But if the effect of the potential can be represented in a perturbation expansion, this simplicity is regained (Chap. 6).

The Energy-Time Transformation. For many applications, particularly in relativistic quantum mechanics, it is best to treat the variables of space and time in a symmetric manner. Then in transforming from coordinate representation to momentum representation we include a transformation from time to energy. Thus the complete transformation for a kernel is

$$k(\mathbf{p}_b, E_b; \mathbf{p}_a, E_a) = \iiint_{-\infty}^{\infty} \int_{t_a}^{\infty} e^{-(i/\hbar)\mathbf{p}_b\cdot\mathbf{x}_b} e^{+(i/\hbar)E_b t_b} K(\mathbf{x}_b, t_b; \mathbf{x}_a, t_a) \times e^{+(i/\hbar)\mathbf{p}_a\cdot\mathbf{x}_a} e^{-(i/\hbar)E_a t_a}\, dt_b\, dt_a\, d^3\mathbf{x}_b\, d^3\mathbf{x}_a \tag{5.13}$$

The energy E is not equal to $p^2/2m$, but is instead an extra independent variable (the coefficient of time) needed to define the kernel. Only if the system exists in the same energy state for an infinite time can an exact measurement of E be made to establish the relation between energy and momentum.

As an example, we shall work out the kernel for a free particle. For this case the integrals over $\mathbf{x}_b$ and $\mathbf{x}_a$ have already been worked out, with the results given in Eq. (5.12). Thus we are left with the integrals over t_b and t_a. Make the substitution $t_b = t_a + \tau$. Then the double integral can be written as

$$\int_{-\infty}^{\infty} e^{(i/\hbar)(E_b - E_a)t_a}\, dt_a \int_0^{\infty} e^{(i/\hbar)(E_b - p_a^2/2m)\tau}\, d\tau \tag{5.14}$$

The first of these two integrals is a representation of the Dirac delta function. In particular it is $2\pi\hbar\delta(E_b - E_a)$. The second integral is of the form

$$\int_0^{\infty} e^{i\omega\tau}\, d\tau \tag{5.15}$$

This latter integral arises often in quantum-mechanical problems. If ω is a real number, the integral does not converge. In order to carry out the present calculation, we shall replace ω with a complex number $\omega + i\epsilon$. When both ω and ϵ are real numbers, with $\epsilon > 0$, the integral has the value $i/(\omega + i\epsilon)$.

Now it would be possible to take the limit of this fraction as ϵ approaches 0 and interpret the result simply as i/ω. However, such an interpretation would lead to incorrect (or rather, incomplete) results in further work. The function we are evaluating is a kernel, and in future work it will often be integrated (multiplied by some other function) over values of ω or its equivalent. If ϵ were dropped from the expression, then such integrals would have a pole at $\omega = 0$, and we would be at a loss what to do.

It would not be correct to take just the principal part of the integral at such a pole. This would give the wrong result. In particular, such a result would imply that the inverse transform of the kernel would not give back the original coordinate representation kernel with which we started. Such a transform would differ from the correct kernel in that it would not be zero for values of time less than zero. One way to obtain the correct result from such integrals is to place the pole an infinitesimal distance above the real axis. This is accomplished by leaving ϵ in the expression.

If we rationalize the expression as

$$\frac{i}{\omega + i\epsilon} = \frac{i(\omega - i\epsilon)}{\omega^2 + \epsilon^2} = \frac{i\omega}{\omega^2 + \epsilon^2} + \frac{\epsilon}{\omega^2 + \epsilon^2} \tag{5.16}$$

we can interpret the first term on the right-hand side as i/ω and in further integrations use the principal part of an integral involving this term. The second term becomes $\pi\delta(\omega)$ as ϵ approaches 0, and it is to be interpreted as such in further integrations. That is, if a more precise mathematical definition is wanted, $i/(\omega + i\epsilon)$ should be replaced by P.P.$(i/\omega) + \pi\delta(\omega)$. This means that

$$\begin{aligned}\int_0^\infty e^{i\omega\tau}\, d\tau &= \lim_{\epsilon \to 0+} \frac{i}{\omega + i\epsilon} \\ &= \text{P.P.}\left(\frac{i}{\omega}\right) + \pi\delta(\omega)\end{aligned} \tag{5.17}$$

(This result is recorded in the Appendix as Eq. A.7.) In all expressions containing ϵ, a limit as $\epsilon \to 0+$ is implied.

Returning to the evaluation of the kernel, we replace ω with $(E_b - p_a^2/2m)/\hbar$ to find

$$k_0(\mathbf{p}_b, E_b; \mathbf{p}_a, E_a) = \frac{(2\pi\hbar)^4 \delta^3(\mathbf{p}_b - \mathbf{p}_a)\delta(E_b - E_a)\, i\hbar}{E_a - p_a^2/2m + i\epsilon} \tag{5.18}$$

The existence of the delta functions in this expression means that neither the energy E nor the momentum $\mathbf{p}$ changes during the motion of a free particle. These two quantities affect the motion of the particle as shown by the remaining pieces of this equation. That is, the amplitude for the motion from one point to another of a free particle with energy E and momentum $\mathbf{p}$ is proportional to $i/(E - p^2/2m + i\epsilon)$.

Earlier in this section it was mentioned that the energy E is not in general identical to $p^2/2m$, but is instead a separate variable. To understand the distinction, let us look at the kernel for a free particle, which is a wave-like function in time and space and wherein E is the coefficient of time and thus has the properties of a frequency. This kernel, given in Eq. (5.12), has the form shown in Fig. 5-4 when plotted against the time difference $T = t_b - t_a$.

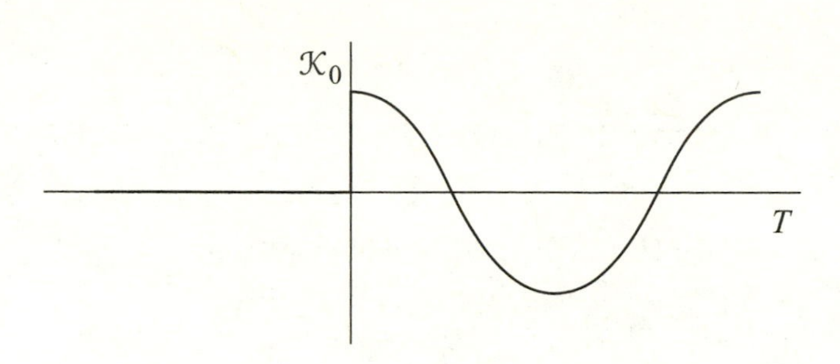

Fig. 5-4 The real part of the free particle kernel $\mathcal{K}_0$ as a function of time. The function is zero for negative times, then starts with a sharp jump at $T = 0$ and continues as a cosine wave of constant amplitude and frequency.

$\mathcal{K}_0$ is zero for T less than zero, and it suddenly begins to oscillate at $T = 0$. The transformation from time to energy representation is equivalent to a Fourier transformation. Since the wave has a sharp beginning (at $T = 0$), the Fourier transform contains components at all frequencies and thus at all energies. If the function extends over a long time interval (many periods), then one frequency begins to dominate in the Fourier transform. For the free particle this dominating frequency corresponds to the energy $E_0 = p^2/2m$.

It is for this reason that the free-particle kernel contains the factor

$$\frac{i}{E_a - p_a^2/2m + i\epsilon} = \text{P.P.}\left(\frac{i}{E_a - p_a^2/2m}\right) + \pi\delta\left(E_a - p_a^2/2m\right) \tag{5.19}$$

Here the first term on the right accounts for the transient effects that result from the sudden start at $T = 0$. The second term gives the steady-state behavior and shows that, if we wait long enough, the only energy found is the usual $p^2/2m$; but near $T = 0$ the energy is not given by this classical formula.

Problem 5-2 If we transform only the time and not the spatial variables, defining

$$\text{k}(x_b, E_b; x_a, E_a) = \iint e^{+(i/\hbar)E_b t_b} K(x_b, t_b; x_a, t_a) e^{-(i/\hbar)E_a t_a}\, dt_b\, dt_a \tag{5.20}$$

show that for a system with a time-independent hamiltonian H

$$\text{k}(x_b, E_b; x_a, E_a) = 2\pi\hbar^2 i\delta(E_b - E_a) \sum_n \frac{\phi_n(x_b)\phi_n^*(x_a)}{E_a - E_n + i\epsilon} \tag{5.21}$$

where E_n and $\phi_n(x)$ are the eigenvalues and eigenfunctions of H.

5-2 MEASUREMENT OF QUANTUM-MECHANICAL VARIABLES

The Characteristic Function. In the preceding section we have shown how an experiment designed to measure momentum leads to a definition of a probability distribution for the momentum. That is, from the results of a correctly designed experiment we can answer the question: What is the probability that the momentum of a particle is $\mathbf{p}$? From the existence of a probability function for momentum we were led to the discovery of a wave function or amplitude written in terms of momentum variables. In fact, we found that a system could be completely described and problems completely analyzed in a momentum-energy representation as well as the space-time representation which we have used heretofore.

These same results apply to physical variables other then momentum. If any physical quantity can be measured experimentally, a probability function can be associated with it. That is, if an experiment is capable of measuring some characteristic A associated with a system (e.g., the x component of momentum), then after repeating the experiment several times it will be possible to construct the probability function $P(a)$ which gives the probability that in any particular experiment the numerical values of A will be found to be equal to a.

In general, it is possible to associate a probability amplitude with such a probability function. This amplitude would be defined in terms of the measured variable, together with other variables necessary to complete the specification. Let us see what is involved by generalizing our example of a momentum measurement. First we shall take just one dimension, but the extension to several dimensions will be obvious. We ask: Does the system have the property G? For example, G might stand for the statement: The value of the quantity A is equal to a. We must have some way to answer this experimentally. So let us imagine some equipment can be designed so that, if it has the property G, the particle will pass through the equipment and arrive at a certain location y on some screen or meter.

The probability of this may be written

$$P(G) = \left| \int_{-\infty}^{\infty} K_{\text{exp}}(y,x) f(x)\, dx \right|^2 \tag{5.22}$$

where $f(x)$ is the wave function of the system to be measured, $K_{\text{exp}}(y,x)$ is the kernel for going through the particular experimental apparatus, and y is the position of arrival for particles with the property G. This probability has the alternative mathematical form

$$P(G) = \left| \int_{-\infty}^{\infty} g^*(x) f(x)\, dx \right|^2 \tag{5.23}$$

where we have defined

$$g^*(x) = K_{\text{exp}}(y, x) \tag{5.24}$$

(Defining this as the complex conjugate of a function is just for convenience, as we shall see later.) So° we can say

$$\psi(G) = \int_{-\infty}^{\infty} g^*(x) f(x)\, dx \tag{5.25}$$

is the amplitude that the system has the property G. This concept is further described in Fig. 5-5.

The property is defined by the function $g^*(x)$ for the following reason. Suppose that some other experiment with different equipment, and hence a different kernel $K_{\text{exp}'}(y', x)$, should be built to measure the same property. In this second experiment the particle arrives at y'. Then the probability of finding that the system has the property G is

$$\left| \int_{-\infty}^{\infty} K_{\text{exp}'}(y', x) f(x)\, dx \right|^2 \quad \text{or} \quad \left| \int_{-\infty}^{\infty} g'^*(x) f(x)\, dx \right|^2 \tag{5.26}$$

Since the property measured is the same, we must obtain the same result in every case for $P(G)$ as we did with the previous experiment.

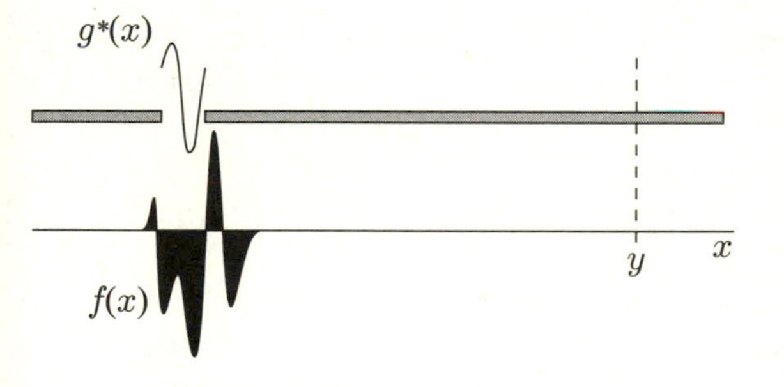

Fig. 5-5 A device designed to measure the property G is placed between the incoming particle (with wave function $f(x)$) and the final point y. The equipment modifies the kernel for the motion (compare Figs. 5-1 and 5-2), making it equal to $g^*(x)$. The product $g^*(x)f(x)$, integrated over x, is the amplitude to arrive at y after passing through the equipment.

That is to say, we must have

$$\left|\int_{-\infty}^{\infty} g'^*(x)f(x)\,dx\right|^2 = \left|\int_{-\infty}^{\infty} g^*(x)f(x)\,dx\right|^2 \tag{5.27}$$

for any arbitrary function $f(x)$. This means $g^*(x) = g'^*(x)$ within at least an unimportant constant phase factor $e^{i\delta}$. That is, all methods to determine the same property correspond (within a phase) to the same $g^*(x)$. For this reason we call $g^*(x)$ the *characteristic function* of the property G.

We may ask another question. What must the state $f(x)$ be so that it is sure to have the property G? (For example, what is the wave function for a particle whose momentum is definite?) That is, we wish to find an $f(x)$, say $F(x)$, so that the particle going through the apparatus will certainly arrive at y and at no other point $\breve{y}$. The amplitude to arrive at $\breve{y}$ should be proportional to $\delta(y - \breve{y})$ (that is, zero unless $\breve{y} = y$). Hence

$$\int_{-\infty}^{\infty} K_{\text{exp}}(\breve{y}, x)F(x)\,dx = \delta(y - \breve{y}) \tag{5.28}$$

This we can solve by the relation of the complex conjugate of a kernel to its inverse, discussed in Sec. 4-1. We have from Eq. (4.37)

$$\int_{-\infty}^{\infty} K_{\text{exp}}(\breve{y}, x)K^*_{\text{exp}}(y, x) = \delta(y - \breve{y}) \tag{5.29}$$

so that

$$F(x) = K^*_{\text{exp}}(y, x) = g(x) \tag{5.30}$$

That is, $g(x)$ is the wave function of a particle having the property G with certainly. We can say either (1) the particle has the property G or (2) the particle is in the state $g(x)$. So we find: If a particle is in a state $f(x)$, the amplitude that it will be found in a state $g(x)$ is

$$\psi(G) = \int_{-\infty}^{\infty} g^*(x)f(x)\,dx \tag{5.31}$$

For more dimensions, x becomes a space of several variables.

We might say loosely: The probability that the particle is in the state $g(x)$ is $|\int g^*(x)f(x)\,dx|^2$. This is all right if we know what we mean. The system is in state $f(x)$, so it is not in $g(x)$; but if a measurement is made to ask if it is *also* in $g(x)$, the answer will be affirmative with probability

$$P(G) = \left|\int_{-\infty}^{\infty} g^*(x)f(x)\,dx\right|^2 \tag{5.32}$$

A measurement which asks "Is the state $g(x)$?" will always have the answer yes if the wave function actually is $g(x)$. For all other wave functions, repetition of the experiment will result in yes some fraction P (between 0 and 1) of the tries. This is a central result for the probabilistic interpretation of the theory of quantum mechanics.

From all of this we deduce an interesting inverse relationship between a wave function and its complex conjugate. In accordance with the interpretation of Eq. (5.25), $g^*(x)$ is the amplitude that if a system is at position x, then it has the property G. (Such a statement is put mathematically by substituting a Dirac delta function for $f(x)$ in Eq. (5.31).) On the other hand, $g(x)$ is the amplitude that if the system has the property G, it is at position x. (This is just a way of giving the definition of a wave function.) One function gives the amplitude for: If A, then B. The other function gives the amplitude for: If B, then A. The inversion is accomplished simply by taking the complex conjugate.

Equation (5.31) can be interpreted as follows: The amplitude that a system has property G is (1) the amplitude $f(x)$ that it is at x times (2) the amplitude $g^*(x)$ that if it is at x, it has property G, with this product summed over the alternatives x.

Problem 5-3 Assume $\int_{-\infty}^{\infty} f^*(x)f(x)\,dx$, which is the probability that a particle of wave function $f(x)$ is somewhere, has been normalized to the value 1. Under this constraint, show that the state $f(x)$ which has the highest probability of having the property G is $f(x) = g(x)$.

Problem 5-4 Suppose the wave function for a system is $\psi(x)$ at time t_a. Suppose further that the behavior of the system is described by the kernel $K(x_b, t_b; x_a, t_a)$ for motions in the interval $t_b \geq t \geq t_a$. Show that the probability that the system is found to be in the state $\chi(x)$ at time t_b is given by the square of the integral

$$\int_{-\infty}^{\infty}\int_{-\infty}^{\infty} \chi^*(x_b)K(x_b, t_b; x_a, t_a)\psi(x_a)\,dx_a\,dx_b$$

We call this integral the *transition amplitude* to go from state $\psi(x)$ to state $\chi(x)$.

Measurements of Several Variables. In the considerations of the preceding section we assumed an ideal experiment, which means that no quantity besides A could be measured at the same time. That is, we do not allow that more than one $g(x)$ would give the same result, but assert that the maximum possible amount of information has been obtained from the system by a measurement of A.

Now in reality it is common for several variables to determine the state of a system. For example, if only the x component of momentum is measured in a three-dimensional system, no definite $g(x)$ can be defined. Both the wave functions $\exp\{ip_x x/\hbar\}$ and $\exp\{ip_x x/\hbar - ip_y y/\hbar\}$ give the same value p_x for the x component of momentum. So if only p_x is measured in a three-dimensional system, the particle could be moving with any component of momentum in the y direction and not change the outcome of the measurement. Nor need the particle come to some unique point in the measuring apparatus. All the particles which arrive at some line or set of points could have the same value for p_x.

Thus in general, we see that the wave function $g(x)$ defines the property G as follows: A state described by the wave function $g(x)$ is certain to have the property G. However, the converse is not necessarily true. That is, it is not certain that all sates having the property G are described by the wave function $g(x)$. Only if G includes a specification of all the quantities that may be simultaneously measured is the wave function completely defined by G. Even then there remains an undefined (and unimportant) constant phase factor $e^{i\delta}$.

It is easy to make the necessary extension of the characteristic function $g^*(x)$ when the ideal experiment requires the measurement of more than one variable. Thus suppose we have a set of quantities which we shall call A, B, C, ..., and which can all be simultaneously measured in an experiment: For example, the x component of momentum, the y component of momentum, etc. Suppose we can completely describe the state of a system by specifying the numerical values a, b, c, ... assigned to these quantities. That is, we completely describe the state by saying whether or not it has a certain property. In this case the property in question is that the value of A is a, the value of B is b, etc. Furthermore, suppose that no additional information (information not derivable from a knowledge of the numerical values of A, B, etc.) could be obtained simultaneously by any means.

Imagine we have an experimental setup capable of measuring all these quantities, i.e., capable of telling us whether or not the state has the property that the value of A is a, etc. We shall call the characteristic function of such a property

$$g^*(x) = \chi^*_{a,b,c,\ldots}(x) \tag{5.33}$$

This function is, of course, a function of the numerical values a, b, c, ... which the experiment is set up to measure, as well as the coordinate variable x.

Suppose the system is in state $f(x)$. Then the probability that the experiment would show that the value of A is a, the value of B is b, etc. (i.e., the probability that the state has the property in question), is

$$P(a,b,c,\ldots) = \left| \int \chi^*_{a,b,c,\ldots}(x) f(x)\, dx \right|^2 \tag{5.34}$$

Transformation Functions. Suppose the system is actually in the state $\chi_{a',b',c',\ldots}(x)$, that is, the value of A is a', etc. Then with our experiment the probability of finding the system in a state described by a, b, c, ... is zero unless $a = a'$, $b = b'$, $c = c'$, ... This means that, with suitable normalizing factors, we have

$$\int_{-\infty}^{\infty} \chi^*_{a,b,c,\ldots}(x)\chi_{a',b',c',\ldots}(x)\, dx = \delta(a-a')\delta(b-b')\delta(c-c')\cdots \tag{5.35}$$

The function $\chi_{a,b,c,\ldots}(x)$ is the amplitude that if the system is in the state described by a, b, c, ..., then it will be found at x. The function $\chi^*_{a,b,c,\ldots}(x)$, which we have called the characteristic function, is the amplitude that, if the system is at x, it will be found in the state specified by a, b, c, ...

If the system is in the state $f(x)$, then

$$F_{a,b,c,\ldots} = \int_{-\infty}^{\infty} \chi^*_{a,b,c,\ldots}(x) f(x)\, dx \tag{5.36}$$

is the amplitude to find the system in the state specified by A having the value a, B having the value b, etc.

The quantities $F_{a,b,c,\ldots}$ are just as good a representation of the state as the function $f(x,y,z,\ldots)$. In fact, if we know the function $F_{a,b,c,\ldots}$ we can reproduce the function $f(x,y,z,\ldots)$ by means of an inverse transformation.

The function $F_{a,b,c,\ldots}$ is called the A, B, C, ... representation of the state. (In the preceding section we had an example of this in the momentum representation.) The function $f(x,y,z,\ldots)$ is the customary coordinate representation, or x, y, z, ... representation, of the state. Transformations between the two are carried out with the help of the functions χ and χ^*. In particular, the function $\chi^*_{a,b,c,\ldots}(x,y,z,\ldots)$ is the transformation function going from the x, y, z, ... representation to the A, B, C, ... representation, while the function $\chi_{a,b,c,\ldots}(x,y,z,\ldots)$ is the transformation function going in the opposite direction. Thus the inverse of the transformation given by Eq. (5.36) is

$$f(x,y,z,\ldots) = \sum_a \sum_b \sum_c \cdots F_{a,b,c,\ldots}\chi_{a,b,c,\ldots}(x,y,z,\ldots) \tag{5.37}$$

This says that the amplitude to be found at x is the amplitude $F_{a,b,c,\ldots}$ to be found with $A = a$, $B = b$, ... times the amplitude $\chi_{a,b,c,\ldots}(x)$ to be at x if $A = a$, $B = b$, etc., summed over alternatives a, b, c, ...

Problem 5-5 Assume that the function $f(x, y, z, \ldots)$ can be represented by

$$f(x, y, z, \ldots) = \sum_a \sum_b \sum_c \cdots F'_{a,b,c,\ldots} \chi_{a,b,c,\ldots}(x, y, z, \ldots) \tag{5.38}$$

By substituting this relation into Eq. (5.36), and using the orthogonal properties of χ as defined by Eq. (5.35), show that $F'_{a,b,c,\ldots} = F_{a,b,c,\ldots}$.

Problem 5-6 Suppose A, B, and C are the three cartesian components of momentum p_x, p_y, p_z. What is the form of the function $\chi_{a,b,c}(x, y, z)$? Using the results of Sec. 5-2, verify the relations obtained in Sec. 5-1.

Problem 5-7 Suppose that the A, B, C, ... representation does not correspond to either coordinate representation or momentum representation, but instead is some third way of representing the state of the system. Suppose we know the function $\chi_{a,b,c,\ldots}(x, y, z, \ldots)$ which permits us to transform back and forth between coordinate representation and A, B, C, ... representation. Suppose further that we know the transformation function necessary to transform back and forth between coordinate representation and momentum representation. What then is the function necessary for the transformation between momentum representation and A, B, C, ... representation?

5-3 OPERATORS

Expected Values. We can develop a few further properties of these transformation functions. Let us try to answer this question: A system is in a state specified by the wave function $f(x)$, and the quantity A is measured. If the measurement is repeated many times, what is the average value which will be obtained for A? We shall denote this average value (sometimes called the *expected value*) by the symbol $\langle A \rangle$.

Suppose it is possible, in principle, to measure simultaneously several physical quantities A, B, C, ..., where a measurement of A could produce any one of a continuous or discrete set of values $\{a\}$, a measurement of B could produce any one of a continuous or discrete set of values $\{b\}$, etc. The probability of obtaining one particular set of values

a, b, c, ... is $|F_{a,b,c,\ldots}|^2$. So the probability of obtaining a particular value a in a measurement of A, irrespective of the values taken on by B, C, ... (for example, if B, C, ... were not measured at all) is

$$P(a) = \sum_b \sum_c \cdots |F_{a,b,c,\ldots}|^2 \tag{5.39}$$

In this equation summations are carried out over all possible values in the continuous or discrete sets of $\{b\}$, $\{c\}$, ...

The average, or expected, value resulting from a measurement of A is obtained by multiplying the probability of Eq. (5.39) by a and summing the result over all possible values of a. Thus

$$\langle A \rangle = \sum_a \sum_b \sum_c \cdots a|F_{a,b,c,\ldots}|^2 \tag{5.40}$$

The need for computing such expected values arises frequently in quantum-mechanical problems. It is useful to have available formulas which simplify such computations. This subject, the subject of operators, was discussed briefly in Sec. 4-1. Now we shall develop a few additional results. However, nowhere in this book shall we attempt a really thorough study of operator calculus, since several excellent works along this line are already available.[1]

The Operator. Let us try to express the expected value of A directly in terms of the original wave function $f(x)$. Note first that the absolute square of $F_{a,b,c,\ldots}$ can be written as

$$|F_{a,b,c,\ldots}|^2 = F^*_{a,b,c,\ldots} F_{a,b,c,\ldots} \tag{5.41}$$

Then, using Eq. (5.36), we can write

$$\begin{aligned}\langle A \rangle &= \sum_a \sum_b \sum_c \cdots a \int_{-\infty}^{\infty} \chi_{a,b,c,\ldots}(x) f^*(x)\, dx \int_{-\infty}^{\infty} \chi^*_{a,b,c,\ldots}(x') f(x')\, dx' \\ &= \int_{-\infty}^{\infty} f^*(x) R(x)\, dx \end{aligned}\tag{5.42}$$

In the second line of this equation we have made use of the substitution

$$R(x) = \int_{-\infty}^{\infty} G_A(x, x') f(x')\, dx' \tag{5.43}$$

where we have written

$$G_A(x, x') = \sum_a \sum_b \sum_c \cdots a\chi_{a,b,c,\ldots}(x) \chi^*_{a,b,c,\ldots}(x') \tag{5.44}$$

[1] For example, see P.A.M. Dirac, "The Principles of Quantum Mechanics," Clarendon Press, Oxford, 1947.

Equation (5.43) says that the function $R(x)$ results from the function $f(x)$ as the result of an integration performed with the help of a linear integral operator $G_A(x, x')$ associated with the quantity A. Often an equation like Eq. (5.43) is symbolized by the notation

$$R = \mathcal{A} f \tag{5.45}$$

where $\mathcal{A}$ stands for linear operator which operates on the function f. In the present case $\mathcal{A}$ stands for the operation displayed on the right-hand side of Eq. (5.43), that is, multiplication by the function G_A and integration. The operator $\mathcal{A}$ is associated with the physical quantity A. Using this notation, we can write

$$\langle A \rangle = \int_{-\infty}^{\infty} f^*(x) \mathcal{A} f(x)\, dx = \int_{-\infty}^{\infty} \int_{-\infty}^{\infty} f^*(x) G_A(x, x') f(x')\, dx'\, dx \tag{5.46}$$

Problem 5-8 Note that Eq. (5.44) implies $G_A^*(x, x') = G_A(x', x)$. With this in mind show that for any two wave functions $g(x)$ and $f(x)$, both of which approach 0 as x goes to $\pm\infty$,

$$\int_{-\infty}^{\infty} g^*(x) \mathcal{A} f(x)\, dx = \int_{-\infty}^{\infty} [\mathcal{A} g(x)]^* f(x)\, dx \tag{5.47}$$

Any operator, such as $\mathcal{A}$, for which Eq. (5.47) holds is called *hermitian* (see Eq. 4.30).

Problem 5-9 The transformation function between space representation and momentum representation is

$$\chi_{a,b,c}(\mathbf{x}) = e^{(i/\hbar)\mathbf{p}\cdot\mathbf{x}} \tag{5.48}$$

(see Prob. 5-6). Choose the physical quantity A as the momentum p_x in the x direction. Show that the function G_A is

$$G_{p_x}(x, x') = \frac{\hbar}{i} \delta'(x - x') \delta(y - y') \delta(z - z') \tag{5.49}$$

where $\delta'(x) = \dfrac{d}{dx}\delta(x)$. With this result determine the operator corresponding to the x component of momentum and show that the expected value of this component of momentum can be written as

$$\langle p_x \rangle = \int_{-\infty}^{\infty} f^*(x) \frac{\hbar}{i} \frac{\partial f}{\partial x}\, dx \tag{5.50}$$

Problem 5-10 Suppose the quantity A corresponds to the x coordinate of position. Show that the correct formula for the expected value of x results when the function $G_A(x, x')$ is taken to be

$$G_x(x, x') = x\delta(x - x')\delta(y - y')\delta(z - z') \tag{5.51}$$

and the operator corresponding to x is simply multiplication by x, that is,

$$\mathcal{X}f(x) = xf(x) \tag{5.52}$$

Eigenfunctions and Eigenvalues. The wave function $\chi_{a,b,c,\ldots}(x)$, as discussed in Sec. 5-2, shows a particularly simple behavior when subjected to the operation $\mathcal{A}$. Thus

$$\mathcal{A}\chi_{a,b,c,\ldots}(x) = a\chi_{a,b,c,\ldots}(x) \tag{5.53}$$

Problem 5-11 Show that this last result is true.

When a function χ satisfies an equation such as (5.53), we say that χ is an eigenfunction of the operator $\mathcal{A}$ associated with the eigenvalue a.

If two physical quantities can be simultaneously measured, then the operators associated with these quantities, $\mathcal{A}$ and $\mathcal{B}$, for example, satisfy an interesting relationship, namely, $\mathcal{A}(\mathcal{B}f) = \mathcal{B}(\mathcal{A}f)$. This relation says that the result of performing one operation after the other is the same regardless of the order in which the operations are performed. In this case the two operators are said to *commute:*

$$\mathcal{A} \cdot \mathcal{B} = \mathcal{B} \cdot \mathcal{A}$$

In general, we cannot expect the commutation relation to hold between operators, but in this special case it does. The reason for this is that if A and B are physical quantities which can be measured simultaneously, they can form part of a set A, B, C, ... of simultaneously measurable quantities with a single characteristic function $\chi_{a,b,c,\ldots}$. If the operator $\mathcal{B}$ is substituted for $\mathcal{A}$ and the value b is substituted for a in Eq. (5.53), the result is still valid, so

$$\mathcal{A}(\mathcal{B}\chi) = \mathcal{A}(b\chi) = b(\mathcal{A}\chi) = ba\chi = ab\chi \tag{5.54}$$

which is true, since a and b are just numbers. Now also

$$\mathcal{B}(\mathcal{A}\chi) = \mathcal{B}(a\chi) = a(\mathcal{B}\chi) = ab\chi \tag{5.55}$$

A comparison of these two equations proves the commutation of the operators $\mathcal{A}$ and $\mathcal{B}$ when acting upon any of the functions $\chi_{a,b,c,\ldots}$. Since both these operations are linear (i.e., they do not involve computations with higher powers of the function $\mathcal{X}$), the commutation relation must also apply to any linear combination of the $\mathcal{X}$ functions.

If the $\mathcal{X}$ functions constitute a "complete set" (which is typical) we can construct any function at all from such a linear combination. So the operation $\mathcal{AB}$ and the operation $\mathcal{BA}$ give the same result on any function; that is, they commute.

Problem 5-12 Show that the x coordinate of position and the x coordinate of momentum are not simultaneously measurable quantities.

There are situations in which a set of commuting mathematical operators $\mathcal{A}$, $\mathcal{B}$, $\mathcal{C}$, ... are already known and it is required to find the functions (the eigenfunctions) which are associated with them. This requires solving a set of equations such as

$$\mathcal{A}\mathcal{X} = a\mathcal{X} \qquad \mathcal{B}\mathcal{X} = b\mathcal{X} \qquad \mathcal{C}\mathcal{X} = c\mathcal{X} \qquad \cdots \tag{5.56}$$

For example, suppose the operators for momentum in the x, y, z directions p_x, p_y, p_z are given as $\frac{\hbar}{i}\frac{\partial}{\partial x}$, $\frac{\hbar}{i}\frac{\partial}{\partial y}$, $\frac{\hbar}{i}\frac{\partial}{\partial z}$. What are the eigenfunctions of this set of operators corresponding to a state in which p_x has the value a, p_y has the value b, and p_z has the value c? (These are, of course, the eigenvalues.) We must solve the equations

$$\frac{\hbar}{i}\frac{\partial \mathcal{X}}{\partial x} = a\mathcal{X} \qquad \frac{\hbar}{i}\frac{\partial \mathcal{X}}{\partial y} = b\mathcal{X} \qquad \frac{\hbar}{i}\frac{\partial \mathcal{X}}{\partial z} = c\mathcal{X} \tag{5.57}$$

and the solution is some arbitrary constant times $e^{(i/\hbar)(ax+by+cz)}$. This agrees with our previous knowledge that a particle with a definite momentum $\mathbf{p}$ has the wave function $e^{(i/\hbar)\mathbf{p}\cdot\mathbf{x}}$.

Interpretation of Energy Expansion. Various expressions involving $\phi_n(x)$ can be interpreted more completely now. For example, consider the expansion in Eq. (4.59) of the kernel in terms of the solutions $\phi_n(x)$ of a constant hamiltonian

$$K(x_b, t_b; x_a, t_a) = \sum_n \phi_n(x_b)\phi_n^*(x_a)e^{-(i/\hbar)E_n(t_b-t_a)} \tag{5.58}$$

We notice first that $\phi_n(x)$ is the amplitude that if we are in energy state n, we are at position x. Therefore, from our previous discussion (Sec. 5-2), $\phi_n^*(x)$ is the amplitude that if we are at x, we are in n. Now let us interpret Eq. (5.58) this way. The amplitude to get from position x_a at time t_a to position x_b at time t_b is the sum over alternatives. This time the alternatives will be divided into the various energy states in which the transition can be made. Thus we must sum over all of the energy states n the product of the following terms:

1. $\phi_n^*(x_a)$, which is the amplitude that if we are at x_a, then we are in the energy state n.
2. $e^{-(i/\hbar)E_n(t_b-t_a)}$, which is the amplitude to be in energy state n at the time t_b if we were in the energy state n at the time t_a.†
3. $\phi_n(x_b)$, which is the amplitude to be found at x_b when we are in the energy state n.

Problem 5-13 Discuss the possibility of interpreting $\phi_n(x)$ as a $\chi_{a,b,c,\ldots}(x)$ function discussed in Sec. 5-2. That is, say $\phi_n(x)$ is the transformation function to go from the x representation to a representation specified by n (energy representation).

†There is no amplitude to change the state. That is the importance of these particular states $\phi_n(x)$.

6

The Perturbation Method in Quantum Mechanics

IF a quantum-mechanical system is subjected to a potential energy which introduces only quadratic terms into the action, then we have seen in Sec. 3-5 how the resulting motion can be determined with the path integral method. However, many of the interesting potentials which arise in quantum-mechanical problems are not of this special type and cannot be handled so easily. In this chapter we shall develop a method of treating more complicated potentials. The method which we discuss, called the *perturbation expansion*, is most useful when the potential is comparatively weak (compared, for instance, to the kinetic energy of the system).

Although the perturbation expansion can be developed along strictly mathematical lines, it is capable of an interesting physical interpretation. This interpretation, which we shall also present, leads to a deeper understanding of quantum-mechanical behavior.

In the second section of this chapter we shall undertake a special application of the perturbation method. We shall consider the motion of an electron when it is scattered by an atom. In describing the scattering interaction we shall find useful the classical notion of a cross-sectional area which the atom presents to the impinging electron. Although this area is related to the actual size of the atom, we shall find that its complete description depends upon the quantum-mechanical aspects of the interacting system.

6-1 THE PERTURBATION EXPANSION

The Terms of the Expansion. Suppose a particle is moving in a potential $V(x,t)$. For the present, the motion will be restricted to one dimension. Then the kernel for motion between the points a and b is

$$K_V(b,a) = \int_a^b \exp\left\{\frac{i}{\hbar}\int_{t_a}^{t_b}\left(\frac{m}{2}\dot{x}^2 - V(x,t)\right)dt\right\}\mathcal{D}x(t) \tag{6.1}$$

The subscript notation K_V is used to remind us that the particle is in the potential V. The notation K_0 denotes the kernel for the motion of a free particle.

In some cases the kernel K_V can be determined by the methods already studied. For instance, in Sec. 3-6 we determined the kernel for the harmonic oscillator subject to an outside force $f(t)$. Here the potential was (see Eq. 3.65)

$$V(x,t) = \frac{m\omega^2}{2}x^2 - f(t)x \tag{6.2}$$

In general, we have found that if the potential is quadratic in x, the kernel can be determined exactly, whereas if it is sufficiently slowly varying, the semiclassical approximation is adequate. There are some other types of potentials which can be successfully treated with the help of Schrödinger's equation. Now we are studying a technique which is often useful if the effect of the potential is small.

Suppose the potential is small, or more precisely, suppose the time integral of the potential along a path is small compared to $\hbar$. Then the part of the exponential of Eq. (6.1) which depends upon $V(x,t)$ can be expanded as

$$\exp\left\{-\frac{i}{\hbar}\int_{t_a}^{t_b} V(x,t)\,dt\right\} = 1-\frac{i}{\hbar}\int_{t_a}^{t_b} V(x,t)\,dt+\frac{1}{2!}\left(-\frac{i}{\hbar}\int_{t_a}^{t_b} V(x,t)\,dt\right)^2+\cdots \tag{6.3}$$

which is defined along any particular path $x(t)$.

Using this expansion in Eq. (6.1) results in

$$K_V(b,a)=K_0(b,a)+K^{(1)}(b,a)+K^{(2)}(b,a)+\cdots \tag{6.4}$$

where

$$K_0(b,a)=\int_a^b \exp\left\{\frac{i}{\hbar}\int_{t_a}^{t_b}\frac{m}{2}\dot{x}^2\,dt\right\}\mathcal{D}x(t) \tag{6.5}$$

$$K^{(1)}(b,a)=-\frac{i}{\hbar}\int_a^b \exp\left\{\frac{i}{\hbar}\int_{t_a}^{t_b}\frac{m}{2}\dot{x}^2\,dt\right\}\int_{t_a}^{t_b} V(x(s),s)\,ds\,\mathcal{D}x(t) \tag{6.6}$$

$$K^{(2)}(b,a)=-\frac{1}{2\hbar^2}\int_a^b \exp\left\{\frac{i}{\hbar}\int_{t_a}^{t_b}\frac{m}{2}\dot{x}^2\,dt\right\} \times\int_{t_a}^{t_b} V(x(s),s)\,ds\int_{t_a}^{t_b} V(x(s'),s')\,ds'\,\mathcal{D}x(t) \tag{6.7}$$

and so forth. To avoid confusion in the integrals over V, we call the time variables s, s', etc.

Evaluation of the Terms. First consider the kernel $K^{(1)}$. We wish to interchange the order of integration over the variable x and the path $x(t)$. We write

$$K^{(1)}(b,a)=-\frac{i}{\hbar}\int_{t_a}^{t_b} F(s)\,ds \tag{6.8}$$

where

$$F(s)=\int_a^b \exp\left\{\frac{i}{\hbar}\int_{t_a}^{t_b}\frac{m}{2}\dot{x}^2\,dt\right\}V(x(s),s)\,\mathcal{D}x(t) \tag{6.9}$$

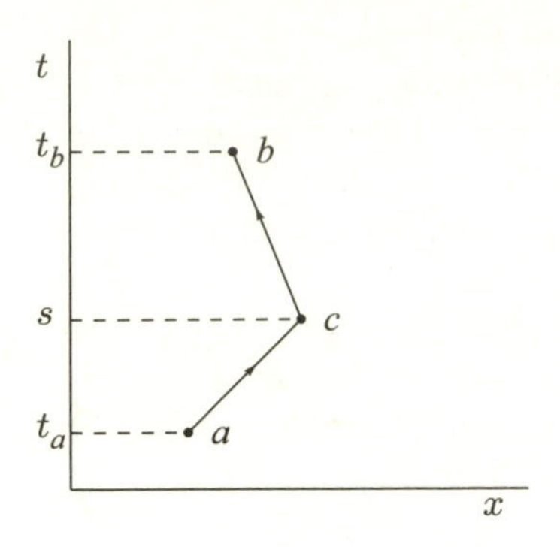

Fig. 6-1 A particle starts from a and moves as a free particle to c. Here it is acted upon, or scattered, by the potential $V(x(s), s) = V_c$. Thereafter, it moves as a free particle to b. The amplitude for such a motion is given in Eq. (6.10). If this amplitude is integrated over all possible positions of the point c, the result is the first-order term in the perturbation expansion.

The path integral $F(s)$ can be described as follows. It is the sum over all paths of the free-particle amplitude. However, each path is weighed by the potential $V(x(s), s)$, evaluated at time s. The only characteristic of the path $x(t)$ which is involved in this particular V is the position of the path at the particular time $t = s$. This means that before and after the time s the paths involved in $F(s)$ are the paths of an ordinary free particle. The situation is sketched in Fig. 6-1.

Using the same arguments which led to Eq. (2.31), we divide each path into two parts, one before the time $t = s$ and one after this time. To be specific, we shall assume that each path goes through the point x_c at this division time. Later on we shall integrate over all values of x_c. If we denote the point $x_c(s)$ by c (that is, $s = t_c$), then the sum over all such paths can be written as $K_0(b, c)K_0(c, a)$. This means that $F(s) = F(t_c)$ can be written as

$$F(t_c) = \int_{-\infty}^{\infty} K_0(b, c)V(x_c, t_c)K_0(c, a)\, dx_c \tag{6.10}$$

Substituting this into Eq. (6.8) gives [with $V(c) = V(x_c, t_c)$]

$$K^{(1)}(b, a) = -\frac{i}{\hbar}\int_{t_a}^{t_b}\int_{-\infty}^{\infty} K_0(b, c)V(c)K_0(c, a)\, dx_c\, dt_c \tag{6.11}$$

The path integral (6.6) has been evaluated as an ordinary integral (6.11).

Here the limits on the integral over x have been written as $\pm\infty$. In a practical problem the limits will be established by the potential (which in most cases drops to 0 when x becomes very large) or by the equipment, which restricts the range of x.

Interpretation of the Terms. Equation (6.11) is very important and very useful, so we shall develop a special interpretation to help think about it physically. We call the interaction between the potential and the particle a *scattering;* thus we say that the potential scatters the particle and that the *amplitude to be scattered by a potential is* $-(i/\hbar)V$ *per unit volume and per unit time.*

With this interpretation we can describe K_V in the following way. K_V is, of course, a sum over alternative ways in which the particle may move from point a to point b. The alternatives are:

1. The particle may not be scattered at all $[K_0(b,a)]$.
2. The particle may be scattered once $[K^{(1)}(b,a)]$.
3. The particle may be scattered twice $[K^{(2)}(b,a)]$.
   Etc.

In accordance with this interpretation, the various paths of the particle are diagramed in Fig. 6-2.

Each one of these alternatives is itself a sum over alternatives. Consider, for example, the kernel for a single scattering, $K^{(1)}(b,a)$.

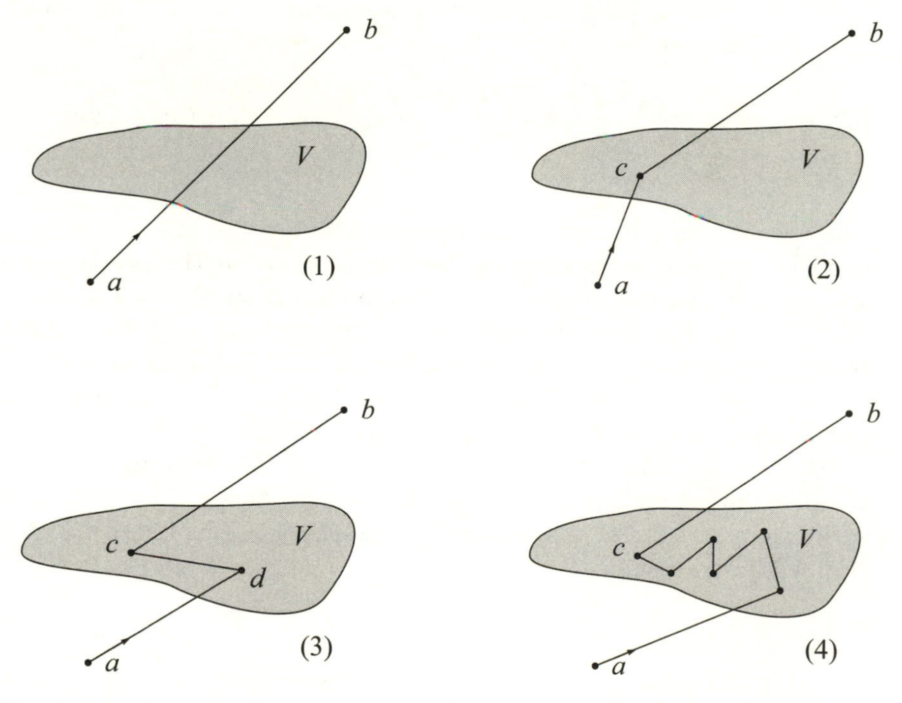

Fig. 6-2 In (1) a particle moves from a to b through the potential V without being scattered. The amplitude for this is $K_0(b,a)$. In (2) the particle is scattered once at c as it moves through the potential V. The amplitude for this is $K^{(1)}(b,a)$. In (3) the particle is scattered twice with the amplitude $K^{(2)}(b,a)$. And in (4) it is scattered n times, the last scattering taking place at c. The total amplitude for motion from a to b with any number of scatterings is $K_0 + K^{(1)} + K^{(2)} + \cdots + K^{(n)} + \cdots$.

One of the alternatives which comprise this kernel consists of the following motion. The particle starts at point a, moves as a free particle to the point $x_c, t_c = c$, is there scattered by the potential $V(c)$, after which it moves as a free particle from point c to the final point b. The amplitude for this path is

$$K_0(b,c)\left[-\frac{i}{\hbar}V(c)\,dx_c\,dt_c\right]K_0(c,a) \tag{6.12}$$

(Remember that in our convention we are using the motion of the particle is traced by reading the formulas from right to left.)

The construction of this amplitude follows the rule stated in Sec. 2-5, namely, that the amplitudes for events occurring in succession in time multiply. The completed form for the kernel $K^{(1)}$ is obtained by adding up all such alternatives by integrating over x_c and t_c, as indicated in Eq. (6.11).

Using this reasoning, we can write down the kernel $K^{(2)}$ for double scattering immediately as

$$K^{(2)}(b,a) = \left(-\frac{i}{\hbar}\right)^2 \int\!\!\int K_0(b,c)V(c)K_0(c,d)V(d)K_0(d,a)\,d\tau_c\,d\tau_d \tag{6.13}$$

where $d\tau = dx\,dt$. Reading from right to left, this formula means: The particle moves as a free particle from a to d. At d the particle gets scattered by the potential $V(d)$ at that point. It then moves as a free particle from d to c, where it is scattered by the potential $V(c)$. After that it moves from c to b, again as a free particle. We sum over all the alternatives, namely, all places and times that the scattering may take place.

Here we have tacitly assumed that $t_c > t_d$. In order to avoid the complication of having to introduce this assumption explicitly in each such example, we shall make use of the convention adopted in Chap. 4 (Eq. 4.28) and assume

$$K(b,a) = 0 \qquad \text{for } t_b < t_a \tag{6.14}$$

Then Eq. (6.13) is correct without restrictions on the range of integration of t_c and t_d.

The reader may wonder what happened to the factor $\frac{1}{2}$ which appears in Eq. (6.7) but is omitted in Eq. (6.13). Note that in Eq. (6.13) the range of integration for t_d is still from t_a to t_b; however, the range of t_c

has been restricted (by the definition of Eq. (6.14)) to lie between t_d and t_b. This restriction cuts the value of the double integral in half. To see this more clearly, suppose the double integral of Eq. (6.7) is rewritten as

$$\int_{t_a}^{t_b}\int_{t_a}^{t_b} V(x(s),s)V(x(s'),s')\,ds'\,ds = \tag{6.15}$$
$$\int_{t_a}^{t_b}\int_{s}^{t_b} V(x(s),s)V(x(s'),s')\,ds'\,ds + \int_{t_a}^{t_b}\int_{t_a}^{s} V(x(s),s)V(x(s'),s')\,ds'\,ds$$

The first term on the right-hand side of this equation satisfies the restrictions implied by Eq. (6.14). By interchanging the order of integration, the second term on the right-hand side can be rewritten as

$$\int_{t_a}^{t_b}\int_{s'}^{t_b} V(x(s),s)V(x(s'),s')\,ds\,ds' \tag{6.16}$$

If the variable names s and s' are interchanged in the last expression, the value of the double integral remains the same. (This useful result is recorded in the Appendix as Eq. A.12.) This means that the first and second terms on the right-hand side of Eq. (6.15) are equal, so each one is half the value of the original double integral. This same sort of argument accounts for a factor $1/n!$ in the expression for $K^{(n)}$.

Problem 6-1 Suppose the potential can be written as $U+V$, where V is small but U is large. Suppose further that the kernel for motion in the potential of U alone can be worked out (for example, U might be quadratic in x and independent of time). Show that the motion in the complete potential $U+V$ is described by Eqs. (6.4), (6.11), (6.13), and (6.14) with K_0 replaced by K_U, where K_U is the kernel for motion in the potential U alone. Thus we can consider V as a perturbation on the potential U. We can say that $-(i/\hbar)V$ is the amplitude to be scattered by the perturbing part of the potential (per unit volume and per unit time). K_U is the amplitude for the motion in the system in the unperturbed potential U.

Problem 6-2 Suppose a system consists of two particles which interact only through a potential $V(x,y)$, where x represents the coordinates of the first particle and y represents the coordinates of the second (see Sec. 3-8 and Eq. 3.75). Apart from this interaction, the particles are free. If V were 0, then K would be simply a product of the two free-particle kernels. Using this fact, develop a perturbation expansion for $K_V(x_b,y_b,t_b;x_a,y_a,t_a)$. By what rules of physical reasoning can the various terms in this expression be described?

6-2 AN INTEGRAL EQUATION FOR K_V

Before applying the results of the preceding paragraphs to a special example, we shall develop some mathematical relations involving the kernels and wave functions of systems moving in a potential field. Using the results so far obtained, we can write Eq. (6.4) as follows:

$$K_V(b,a) = K_0(b,a) - \frac{i}{\hbar}\int K_0(b,c)V(c)K_0(c,a)\,d\tau_c \tag{6.17}$$
$$+\left(-\frac{i}{\hbar}\right)^2 \int\int K_0(b,c)V(c)K_0(c,d)V(d)K_0(d,a)\,d\tau_c\,d\tau_d + \cdots$$

Alternatively, this expression could be written as

$$K_V(b,a) = K_0(b,a) \tag{6.18}$$
$$-\frac{i}{\hbar}\int K_0(b,c)V(c)\left[K_0(c,a) - \frac{i}{\hbar}\int K_0(c,d)V(d)K_0(d,a)\,d\tau_d + \cdots\right]d\tau_c$$

The expression in square brackets has the same form as Eq. (6.17). In both cases the sums extend over an infinite number of terms. This means that K_V can be written as

$$K_V(b,a) = K_0(b,a) - \frac{i}{\hbar}\int K_0(b,c)V(c)K_V(c,a)\,d\tau_c \tag{6.19}$$

which is an exact expression. This is an integral equation determining K_V if K_0 is known. (Note that for the situation described in Prob. 6-1, K_0 would be replaced by K_U.) Thus the path integral problem has been transformed into an integral equation.

This last result can be understood physically in the following way. The total amplitude for the transition of the system from a to b, with any number of scatterings, can be expressed as the sum of two alternatives. The first alternative is the amplitude that the transition takes place with no scatterings, which is expressed by K_0. The second alternative is the amplitude that the transition takes place with one or more scatterings, which is given by the last term of Eq. (6.19). In this last term the point c can be thought of as the point at which the *last* scattering takes place. Thus the system moves from a to c in the potential field with its motion exactly described by $K_V(c,a)$. Then at point c the final scattering takes place, after which the system moves as a free system (without scattering) to the point b, as represented by the kernel K_0. This interpretation is diagramed in Fig. 6-3.

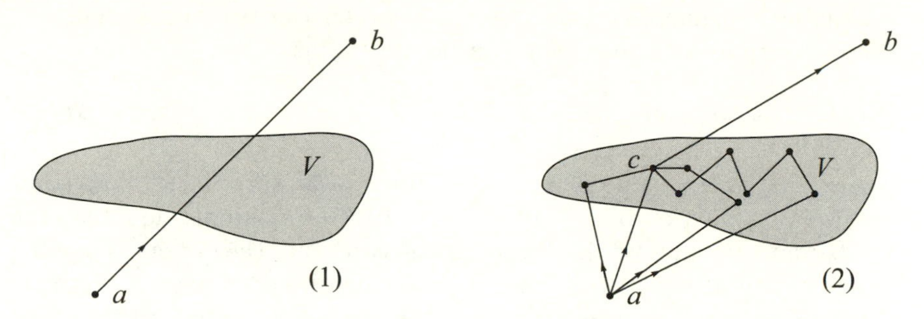

Fig. 6-3 In (1) the particle moves from a to b through the potential V as a free particle, described by the amplitude $K_0(b,a)$. In (2) the particle is scattered one or more times by V, with the last scattering taking place at c. The motion from a to c is described by $K_V(c,a)$, and that from c to b by $K_0(b,c)$. A combination of the two situations, when all positions of c are accounted for, covers all possible cases and gives $K_V(b,a)$ in the form of Eq. (6.19).

Since the last scattering could take place at any point in space and time between a and b, the amplitude for this composite motion, represented by the integrand of the last term of Eq. (6.19), must be integrated over all possible positions of the point c.

Problem 6-3 For a free particle, Eq. (4.29) reduces to

$$\frac{\partial}{\partial t_b}K_0(b,a) + \frac{i}{\hbar}\left[-\frac{\hbar^2}{2m}\frac{\partial^2}{\partial x_b^2}K_0(b,a)\right] = \delta(t_b - t_a)\delta(x_b - x_a) \tag{6.20}$$

Show, from this result and Eq. (6.19), that the kernel K_V satisfies the differential equation

$$\frac{\partial}{\partial t_b}K_V(b,a) + \frac{i}{\hbar}\left[-\frac{\hbar^2}{2m}\frac{\partial^2}{\partial x_b^2}K_V(b,a) + V(b)K_V(b,a)\right] = \delta(t_b - t_a)\delta(x_b - x_a) \tag{6.21}$$

6-3 AN EXPANSION FOR THE WAVE FUNCTION

In Sec. 3-4 we introduced the idea of a wave function and discussed some relations between wave functions and kernels. Equation (3.42) of that section shows how the wave function at time t_b can be obtained from the wave function an earlier time t_a with the help of the kernel describing the motion of the system between the two times. For our present purposes this equation can be written as

$$\psi(b) = \int K_V(b,a) f(a)\, dx_a \tag{6.22}$$

where $f(a)$ is the value of the wave function at the time $t = t_a$ (that is, $f(a)$ is a function of x_a), $\psi(b)$ is the wave function at the later time $t = t_b$,[†] and we suppose that between the two times the system is moving in the potential V, with the motion described by the kernel $K_V(b,a)$.

If the series expansion (Eq. 6.17) for K_V is substituted into this equation, the result will be a series expansion for $\psi(b)$. Thus

$$\psi(b) = \int K_0(b,a) f(a)\, dx_a - \frac{i}{\hbar} \int\int K_0(b,c) V(c) K_0(c,a)\, d\tau_c f(a)\, dx_a + \cdots \tag{6.23}$$

The first term of the series gives the wave function at the time t_b assuming the system to be free (or unperturbed, in case K_U is to be substituted for K_0) between the time t_a and t_b. Call this term ϕ. Thus

$$\phi(b) = \int K_0(b,a) f(a)\, dx_a \tag{6.24}$$

Using this definition, the series of Eq. (6.23) can be rewritten as

$$\psi(b) = \phi(b) - \frac{i}{\hbar} \int K_0(b,c) V(c) \phi(c)\, d\tau_c + \left(-\frac{i}{\hbar}\right)^2 \int\int K_0(b,c) V(c) K_0(c,d) V(d) \phi(d)\, d\tau_c\, d\tau_d + \cdots \tag{6.25}$$

In this form the series is called the *Born expansion* for ψ. If only the first two terms are included (thus only through first order in V), the result is the *first Born approximation.* It involves a single scattering by the potential V. This scattering occurs at the point c. Up to this point the system described by $\psi(c)$ is free; and after the scattering, the system moves from c to b, again free, and described by $K_0(b,c)$. An integral must be taken over all the possible points at which the scattering occurs. If three terms of the series are used (thus through terms of second order in V), the result is called the *second Born approximation*, etc.

[†]Note that our convention that $K(b,a)$ is zero for $t_b < t_a$ makes Eq. (6.22) invalid if $t_b < t_a$, but we shall not use it in this range of t values.

Problem 6-4 Using arguments similar to those leading to Eq. (6.19), show that the wave function $\psi(b)$ satisfies the integral equation

$$\psi(b) = \phi(b) - \frac{i}{\hbar} \int K_0(b,c)V(c)\psi(c)\, d\tau_c \tag{6.26}$$

This integral equation is equivalent to the Schrödinger equation

$$\frac{\partial \psi}{\partial t} = -\frac{i}{\hbar}\left[-\frac{\hbar^2}{2m}\nabla^2\psi + V\psi\right] \tag{6.27}$$

Working in one dimension only, show how the Schrödinger equation may be deduced from the integral equation.

6-4 THE SCATTERING OF AN ELECTRON BY AN ATOM

Mathematical Treatment. We have developed the concepts and formulas of the perturbation treatment in a somewhat abstract framework. Now, to develop a physical understanding of the perturbation method, we shall discuss the specific problem of the scattering of a fast electron by an atom. We envision an experiment in which a beam of electrons bombards a target, such as a thin foil of metal, and then is collected by some suitable counter, as shown in Fig. 6-4.

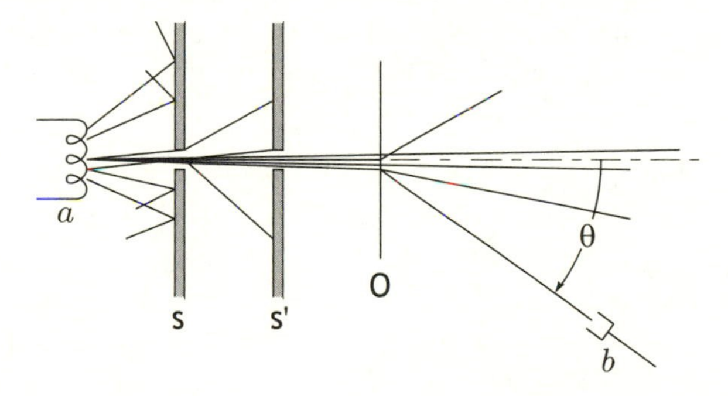

Fig. 6-4 Electrons boil off a hot filament at a, are screened into a beam by collimating holes in s and s$'$, and then strike a thin-foil target at O. Most of the electrons pass straight on without being scattered (if their energy is great enough and the target is thin enough), but some are deflected by interactions with the atoms in the target and scattered, for example, through an angle θ to b. As the counter at b is moved up and down, the relation between the relative number of scatterings and the scattering angle θ can be measured.

Suppose the energy of the scattering particles is determined by a time-of-flight method. That is, we release an electron from the source at one time, say $t = 0$, and ask for the chance that it arrives at the counter after some delay T. We can then make direct use of our result for the amplitude $K(b, a)$ to go from one place to another in a definite time.

We shall simplify the problem by assuming that either the foil is so thin or the interaction is so weak that each electron can interact with, at most, one atom. Actually, this assumption is quite realistic for many scattering experiments. Furthermore, most multiple scatterings can be analyzed in terms of the simple scattering from one atom. Thus we shall discuss the interaction between a single electron and a single atom.

The center of the atom will be taken as the center of a coordinate system in which the electrons are released at the point a, as in Fig. 6-5, at the time $t = 0$. A counter placed at the point b tells us whether or not, at the time $t = T$, the electron arrives at the point b. We shall make the following approximations:

1. The interaction can be represented by a first-order Born approximation. That is, the electron is scattered only once by the atom.
2. The atom can be represented by a potential° $V(\mathbf{r})$ fixed in space and constant in time.

Actually, the atom presents a very complicated system interacting with the electron, and the interaction between the electron and the atom is really more complicated than can be represented by a simple potential $V(\mathbf{r})$. The electron could excite or ionize the atom and lose energy in

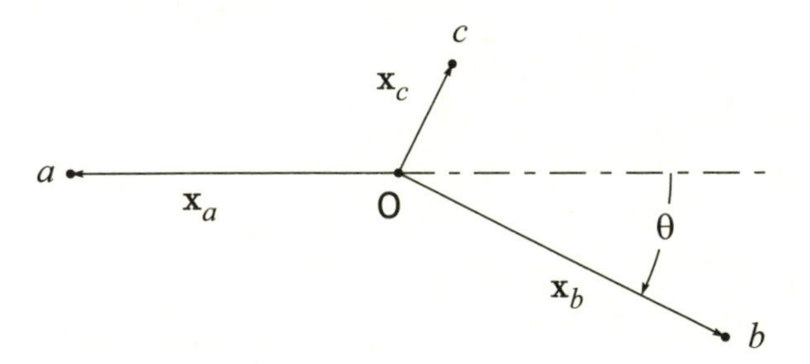

Fig. 6-5 The geometry of the scattering problem. The electron starts at a and move as a free particle to c, where it is scattered by the atomic potential $V(\mathbf{x}_c)$. Afte the scattering, it moves as a free particle to the counter at b, which is located a the end of vector $\mathbf{x}_b$ from the scattering center O. In this process, the electron ha been scattered through the angle θ, measured from the direction of the nonscattere beam. This process corresponds to the first-order Born approximation. If the amplitud for two scatterings, say, at d and c, is included, then the result is the second-orde approximation, etc.

the process. It can be shown, however, that if we consider only elastic collisions between the electron and the atom, so that the atom is in the same energy state after the collision as it is before, then when the approximation (1) is valid, approximation (2) is valid too.

Let $\mathbf{x}_a$ and $\mathbf{x}_b$ be the vectors from the center of the atom to the points at which the electron is released and detected, respectively. In the calculations we shall take $\mathbf{x}_a$ and $\mathbf{x}_b$ to have lengths much larger than the radius of the atom. That is, we shall assume that the atomic potential $V(\mathbf{r})$ becomes negligibly small at distances much smaller than $|\mathbf{x}_a|$ and $|\mathbf{x}_b|$. Thus during most of its flight, the electron will be moving as a free particle, and only in the vicinity of the origin will it be exposed to the potential.

The first-order Born approximation contains two terms, only the second of which is of interest to us here. The first term is the kernel $K_0(b,a)$ for the motion of the electron from a to b as a free particle, and it has already been studied sufficiently. The term of interest is then

$$\begin{aligned} K^{(1)}(b,a) &= -\frac{i}{\hbar}\int K_0(b,c)V(c)K_0(c,a)\,d\tau_c \qquad (6.28)\\ &= -\frac{i}{\hbar}\int\int_0^T\left(\frac{m}{2\pi i\hbar(T-t_c)}\right)^{3/2}\exp\left\{\frac{im|\mathbf{x}_b-\mathbf{x}_c|^2}{2\hbar(T-t_c)}\right\}\\ &\qquad\times V(\mathbf{x}_c)\left(\frac{m}{2\pi i\hbar t_c}\right)^{3/2}\exp\left\{\frac{im|\mathbf{x}_c-\mathbf{x}_a|^2}{2\hbar t_c}\right\}dt_c\,d^3\mathbf{x}_c \end{aligned}$$

Here we have used $\mathbf{x}_c$ as the vector from the origin to the point c, and $d^3\mathbf{x}_c$ represents the product of the differentials of all the components of the vector $\mathbf{x}_c$. The integral over t_c gives (see Appendix, Eq. A.5)

$$\begin{aligned} K^{(1)}(b,a) &= -\frac{i}{\hbar}\left(\frac{m}{2\pi i\hbar T}\right)^{5/2}T \qquad (6.29)\\ &\qquad\times\int\left(\frac{1}{R_{ca}}+\frac{1}{R_{bc}}\right)\exp\left\{\frac{im}{2\hbar T}(R_{ca}+R_{bc})^2\right\}V(\mathbf{x}_c)\,d^3\mathbf{x}_c \end{aligned}$$

where $R_{ca}=|\mathbf{x}_c-\mathbf{x}_a|$ and $R_{bc}=|\mathbf{x}_b-\mathbf{x}_c|$. Using these definitions, as well as $r_a=|\mathbf{x}_a|$ and $r_b=|\mathbf{x}_b|$, we write

$$R_{ca}=r_a\left(1-\frac{2\mathbf{x}_a\cdot\mathbf{x}_c}{r_a^2}+\frac{|\mathbf{x}_c|^2}{r_a^2}\right)^{1/2}\approx r_a+\mathbf{i}_a\cdot\mathbf{x}_c \qquad (6.30)$$

$$R_{bc}=r_b\left(1-\frac{2\mathbf{x}_b\cdot\mathbf{x}_c}{r_b^2}+\frac{|\mathbf{x}_c|^2}{r_b^2}\right)^{1/2}\approx r_b-\mathbf{i}_b\cdot\mathbf{x}_c \qquad (6.31)$$

where $\mathbf{i}_a$ and $\mathbf{i}_b$ are unit vectors in the direction of $-\mathbf{x}_a$ and $\mathbf{x}_b$, respectively (that is, $\mathbf{i}_a=-\mathbf{x}_a/r_a$), and we have made use in the approximation the fact that r_a is much larger than any value of $|\mathbf{x}_c|$ for which the potential is not negligible. It is necessary to keep the first-order terms in $|\mathbf{x}_c|$ only in the argument of the exponential, since this factor is quite sensitive to small relative changes in phase. Here we need

$$(R_{ca} + R_{bc})^2 \approx (r_a + r_b)^2 + 2(r_a + r_b)(\mathbf{i}_a \cdot \mathbf{x}_c - \mathbf{i}_b \cdot \mathbf{x}_c) \tag{6.32}$$

Using these approximations, the kernel can be written as

$$K^{(1)}(b, a) \approx -\frac{i}{\hbar} \left(\frac{m}{2\pi i \hbar T} \right)^{5/2} T \left(\frac{1}{r_a} + \frac{1}{r_b} \right) \exp\left\{ \frac{im}{2\hbar T}(r_a + r_b)^2 \right\} \tag{6.33}$$
$$\times \int \exp\left\{ \frac{im}{\hbar T}(r_a + r_b)(\mathbf{i}_a \cdot \mathbf{x}_c - \mathbf{i}_b \cdot \mathbf{x}_c) \right\} V(\mathbf{x}_c)\, d^3\mathbf{x}_c$$

Physical Interpretation. We can deduce some of the physical characteristics of the motion from a study of Eq. (6.33). In the time T the electron has traveled the total distance of $r_a + r_b$. Thus its velocity during this time is $u = (r_a + r_b)/T$ and its energy is $mu^2/2$, while its momentum has the magnitude mu. In writing these expressions we are making the assumption that the energy of the electron is not changed by the scattering process.

That these values for the velocities, energy, and momentum are consistent can be verified from an inspection of the exponential factor appearing in front of the integral of Eq. (6.33). The phase of this exponential term is $im(r_a + r_b)^2/2\hbar T$, and the derivative of this phase with respect to T gives the frequency as

$$\omega = \frac{m}{2\hbar} \frac{(r_a + r_b)^2}{T^2} \tag{6.34}$$

With u defined as above this means that the energy is $mu^2/2$ (see Eq. 3.15).

Differentiating the phase with respect to r_b yields the wave number at the point b as

$$k = \frac{m}{\hbar} \frac{r_a + r_b}{T} \tag{6.35}$$

which means that the magnitude of the momentum is mu (see Eq. 3.12).

Problem 6-5 The integral over t_c in Eq. (6.28) can be performed approximately using the method of stationary phase. By studying the application of such a method to this integral, show that most of the contribution to the integral comes from values of t_c near the region

$t_c = r_a/u$, the time at which the electron would arrive at the center of the atom if it moved in a classical manner.

With the velocity of the electron defined as $u = (r_a + r_b)/T$ define the incoming vector momentum $\mathbf{p}_a$ as

$$\mathbf{p}_a = mu\mathbf{i}_a \tag{6.36}$$

and the outgoing vector momentum $\mathbf{p}_b$ as

$$\mathbf{p}_b = mu\mathbf{i}_b \tag{6.37}$$

Then Eq. (6.33) can be written as

$$K^{(1)}(b,a) \approx -\frac{i}{\hbar}\left(\frac{m}{2\pi i\hbar}\right)^{5/2} \frac{u}{T^{1/2}r_a r_b} \exp\left\{\frac{i}{\hbar}\frac{mu^2}{2}T\right\} \times \int \exp\left\{\frac{i}{\hbar}(\mathbf{p}_a - \mathbf{p}_b)\cdot\mathbf{x}_c\right\} V(\mathbf{x}_c)\, d^3\mathbf{x}_c \tag{6.38}$$

Call the negative change in momentum, or the momentum transfer,

$$\breve{\mathbf{p}} = \mathbf{p}_a - \mathbf{p}_b$$

and define the quantity $v(\breve{\mathbf{p}})$ as

$$v(\breve{\mathbf{p}}) = \int e^{i(\breve{\mathbf{p}}/\hbar)\cdot\mathbf{r}} V(\mathbf{r})\, d^3\mathbf{r} \tag{6.39}$$

The probability that an electron arrives at the point b is given by the square of the absolute value of the kernel $K_V(b,a)$. Thus the probability will depend upon the first term in the series expansion of this kernel, namely, $K_0(b,a)$, which is likely to be so large as to completely overshadow the small perturbation term $K^{(1)}(b,a)$.

For this reason it is customary in most scattering experiments to collimate the incoming beam with suitable shields so that those electrons which are not scattered by the atoms in the target are confined to the region of a particular line (or direction), as shown in Fig. 6-6. Of course, there will be some diffraction by the collimating shields, such as that studied in Secs. 3-2 and 3-3, which means that some nonscattered electrons will appear outside this central beam. However, with suitable collimation, and for positions suitably far away from the collimated beam, the number of electrons diffracted by the collimator will be very small compared to the number scattered by the atoms in the target.

In such a region the probability of arrival for an electron is given, at least to first order, by the square of the absolute value of $K^{(1)}(b,a)$ alone. Using Eqs. (6.38) and (6.39), this probability is

$$\frac{P(b)}{\text{unit volume}} = \frac{1}{\hbar^2}\left(\frac{m}{2\pi\hbar}\right)^5 \frac{u^2}{Tr_a^2 r_b^2}|v(\breve{\mathbf{p}})|^2 \tag{6.40}$$

In this last expression the factor $v(\breve{\mathbf{p}})$ contains the characteristics of the atomic potential and the dependence of the kernel upon the relative directions of $\mathbf{x}_a$ and $\mathbf{x}_b$. It is completely independent of the dimensions of the experimental equipment. The effects of such dimensions are represented by the remaining factors of Eq. (6.40). For example, the term $1/r_a^2$ can be easily seen to result from the idea that the chance for an electron to actually hit the atom varies inversely as r_a^2. The application of such an idea might be questioned in this experiment in view of the fact that we have supposed some collimating shields are present. However, this collimation has a negligible effect over *atomic* dimensions. From the point of view of a target atom the beam of oncoming electrons appears to consist of electrons spreading in all directions from a single source.

In a similar manner, after the scattering, the electrons spread out again in all directions from the scattering atom. Thus the chance per unit volume to find an electron in the counter varies inversely as r_b^2. Since the more interesting features of the experiment are contained in the function $v(\breve{\mathbf{p}})$, we shall give special attention to this function in the next section.

The additional factors depend on the particular normalization of our kernel. We can interpret the formula more easily if we give it as a ratio. We compare the probability of finding a scattered particle at b to the probability of finding one at a point d behind the atom at the same total distance $r_a + r_b$ (and at the same time T, to keep the velocity the same) if no scattering occurred, as shown in Fig. 6-7. That is, we calculate $P(d)$ per unit volume, as if no atom were present.

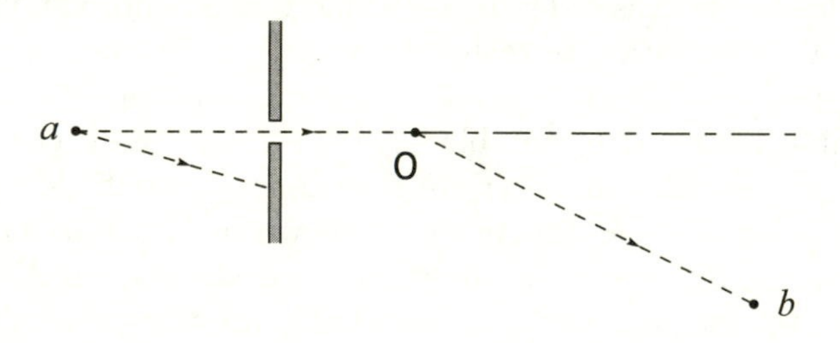

Fig. 6-6 Principle of collimation to eliminate the zero-order term at b. Only electrons which have been scattered at least once can get from a to b with any reasonable probability. Thus the zero-order term in the perturbation expansion of $K_V(b,a)$ will contribute a negligible amount and can be neglected. The first term of importance is $K^{(1)}(b,a)$.

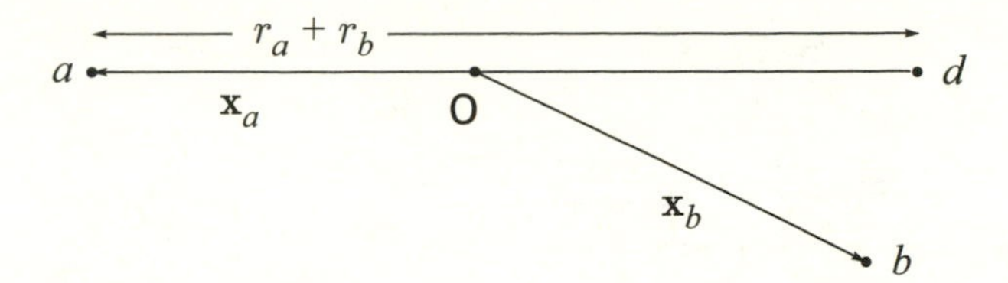

Fig. 6-7 If d and b are the same total distance from O, namely r_b, then the difference (or ratio) of numbers of electrons arriving at the two points can depend only on the scattering phenomenon. If d is in the direct line of nonscattered electrons, the ratio of the number arriving at b to that arriving at d if no scattering source were present is the probability of scattering to b.

The result is $|K_0(d,a)|^2$ or

$$\frac{P(d)}{\text{unit volume}} = \left(\frac{m}{2\pi\hbar}\right)^3 \frac{u^2}{T(r_a + r_b)^2} \tag{6.41}$$

so that

$$\frac{P(b)}{P(d)} = \left(\frac{m}{2\pi\hbar^2}\right)^2 |v(\breve{\mathbf{p}})|^2 \frac{(r_a + r_b)^2}{r_a^2 r_b^2} \tag{6.42}$$

We shall interpret the last factor geometrically in the next section, where we shall also give more detailed attention to the function $v(\breve{\mathbf{p}})$.

The Cross Section for Scattering. It is convenient to describe the characteristics of an atom in a scattering experiment by means of the concept of a *cross section.* The utility of such a concept stems from the convenience of thinking along the lines of classical physics. The cross section $d\sigma/d\Omega$ is defined as the effective target area (from a classical point of view) of the atom that must be hit by an electron in order that the electron be scattered into a unit solid angle. This solid angle is measured around a sphere whose center is at the atom. The cross section is thus a function of the scattering angle, i.e., the angle between $\mathbf{x}_a$ and $\mathbf{x}_b$. In terms of such a classical model we can determine the probability that an electron arrives at the point b.

If particles starting from the origin were to hit a small target of area $d\sigma$ at distance r_a, these particles would be removed from the region d, where they would have spread out over an area $[(r_a + r_b)/r_a]^2\, d\sigma$. Instead they are sent out in a solid angle $d\Omega$ toward b and are therefore spread out over an area $r_b^2\, d\Omega$ there, as shown in Fig. 6-8. Hence the ratio of the probability of finding them at b to that of finding them

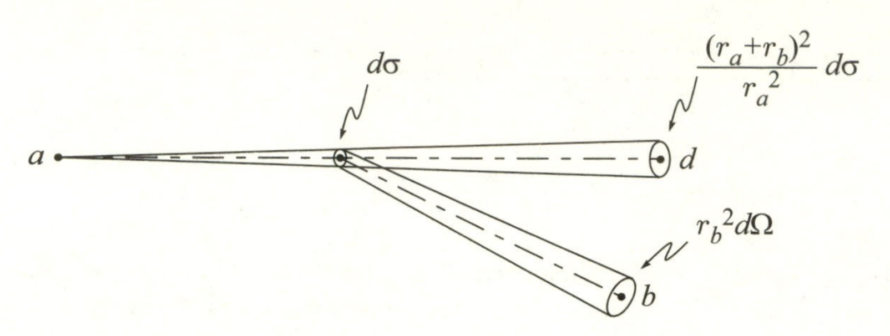

Fig. 6-8 Particles striking an area $d\sigma$ of the target are deflected through an angle θ into an area measured by the solid angle $d\Omega$. If no target had been present, those particles would have proceeded to point d. Instead, they proceed to point b, spreading out into the area $r_b^2\,d\Omega$. The probability of finding a particle at d is inversely proportional to the area over which the beam would have spread in arriving at d. Similarly, the probability of finding the particle at b is inversely proportional to the area $r_b^2\,d\Omega$ over which the beam of scattered particles spreads in traveling from the target to b. If we take the ratio of these areas, we have the inverse ratio of the associated probabilities. From this point of view we say that all of the particles which hit the target area $d\sigma$ are scattered, and through a particular angle θ. Of course, actually only a few particles which hit the target are scattered at all and only a fraction of these through the angle θ. Thus, the area element $d\sigma$ which we have used in this calculation is the *effective* cross-sectional area for scattering through the angle θ measured in terms of the element of solid angle $d\Omega$ into which the particles are scattered.

at d if there were no target is the inverse ratio of these areas,

$$\frac{P(b)}{P(d)} = \frac{[(r_a + r_b)/r_a]^2\,d\sigma}{r_b^2\,d\Omega} \tag{6.43}$$

On comparing Eqs. (6.42) and (6.43), we see that the cross section per unit solid angle is

$$\frac{d\sigma}{d\Omega} = \left(\frac{m}{2\pi\hbar^2}\right)^2 |v(\breve{\mathbf{p}})|^2 \tag{6.44}$$

The main advantage of an expression in terms of cross section instead of using Eq. (6.40) directly is this: Equation (6.44) does not depend on particular experimental conditions, so the cross sections obtained in one or another experiment can be directly compared, whereas probabilities per unit volume cannot be.

It must be emphasized that this idea of an effective target is purely classical and is convenient in recording scattering probabilities. There is no direct relation between it and the size of the atom, nor is the scattering mechanism to be thought of as localized exactly over such an area. For example, the shadow one would expect to find classically behind the target is not there in the classical sense (with sharp boundaries); for since we are dealing with a wave phenomenon, there is diffraction into the shadow.

Special Forms of the Atomic Potential. The results obtained when the atomic potential $V(\mathbf{r})$ is assumed to have various forms are shown in the following problems.

Problem 6-6 Suppose the potential is that of a central force. Thus $V(\mathbf{r}) = V(r)$. Show that $v(\breve{\mathbf{p}})$ can be written as

$$v(\breve{\mathbf{p}}) = v(\breve{p}) = \frac{4\pi\hbar}{\breve{p}} \int_0^\infty \left(\sin\frac{\breve{p}r}{\hbar}\right) V(r)\, r\, dr \tag{6.45}$$

Suppose $V(r)$ is the Coulomb potential $-Ze^2/r$. In this case the integral for $v(\breve{p})$ is oscillatory at the upper limit. But convergence of the integral can be artificially forced by introducing the factor $e^{-\epsilon r}$ and then taking the limit of the result as $\epsilon \to 0$. Following through this calculation, show that the cross section corresponds to the Rutherford cross section

$$\frac{d\sigma_{\text{Ruth}}}{d\Omega} = \frac{4m^2Z^2e^4}{\breve{p}^4} = \frac{Z^2e^4}{16(mu^2/2)^2\sin^4(\theta/2)} \tag{6.46}$$

where e = charge on a proton

$$\breve{p} = 2p\sin(\theta/2) = 2mu\sin(\theta/2) \tag{6.47}$$

θ = angle between the vectors $\mathbf{i}_a$ and $\mathbf{i}_b$

The result of Prob. 6-6 is, accidentally, exact. That is, the first-order Born approximation gives the exact value of the probability for scattering in a Coulomb potential. This does not mean the higher-order terms are zero; it means, rather, that they contribution only to the phase of the scattering amplitude. Since the probability is the absolute square of the amplitude, it is independent of the phase. Thus a first-order Born approximation, which gives the correct value for the probability, is not exact for the amplitude. This case of a Coulomb scattering is amusing, for there is also another accident. A completely classical treatment of such a scattering problem, i.e., treating the electrons as charged point masses, gives the same result.

Problem 6-7 Suppose the potential energy $V(\mathbf{r}) = -e\phi(\mathbf{r})$ is the result of a charge distribution $\rho(\mathbf{r})$ so that

$$\nabla^2\phi(\mathbf{r}) = -4\pi\rho(\mathbf{r}) \tag{6.48}$$

By assuming that $\rho(\mathbf{r})$ goes to 0 as $|\mathbf{r}| \to \infty$, multiplying Eq. (6.48) by $e^{i(\breve{\mathbf{p}}/\hbar)\cdot\mathbf{r}}$, and integrating twice over $\mathbf{r}$, show that $v(\breve{\mathbf{p}})$ can be expressed in terms of $\rho(\mathbf{r})$ as

$$v(\breve{\mathbf{p}}) = -\frac{4\pi\hbar^2 \mathrm{e}}{\breve{p}^2} \int e^{i(\breve{\mathbf{p}}/\hbar)\cdot\mathbf{r}} \rho(\mathbf{r})\, d^3\mathbf{r} \tag{6.49}$$

An atom can be represented in terms of its charge density. At the nucleus the charge density is singular, so that it can be represented as a Dirac delta function of $\mathbf{r}$ of strength $Z\mathrm{e}$, where Z is the atomic number of the nucleus. Then if $\rho_e(\mathbf{r})$ is the density of atomic electrons, $v(\breve{\mathbf{p}})$ is

$$v(\breve{\mathbf{p}}) = -\frac{4\pi\hbar^2 \mathrm{e}^2}{\breve{p}^2} \left[Z - \int e^{i(\breve{\mathbf{p}}/\hbar)\cdot\mathbf{r}} \rho_e(\mathbf{r})\, d^3\mathbf{r} \right] \tag{6.50}$$

The quantity in the brackets is called the *form factor* for electron scattering. (Incidentally, a similar form factor appears in X-ray scattering. The theory of X-ray scattering shows that only the atomic electrons, and not the nucleus, contribute to the scattering. Thus the form factor for X-ray scattering is the same but with the Z omitted.)

Problem 6-8 In an atom the potential follows the Coulomb law only for very small radii. As the radius is increased the atomic electrons gradually shield, or cancel out, the nuclear change until, for sufficiently large values of r, the potential is zero. The shielding effect of atomic electrons can be accounted for in a very rough approximate manner with the formula

$$V(r) = -\frac{Z\mathrm{e}^2}{r} e^{-r/a} \tag{6.51}$$

In this expression a is called the radius of the atom. It is not the same as the outer radius of the atom as used by chemists, but instead is given by $a_0/Z^{1/3}$, where the Bohr radius is $a_0 = \hbar^2/m\mathrm{e}^2 = 0.0529$ nm.

Show that in such a potential

$$v(\breve{p}) = -\frac{4\pi Z \mathrm{e}^2}{(\breve{p}/\hbar)^2 + (1/a)^2} \tag{6.52}$$

and hence

$$\frac{d\sigma}{d\Omega} = \frac{Z^2 \mathrm{e}^4}{(mu^2/2)^2 [4\sin^2(\theta/2) + (\hbar/pa)^2]^2} \tag{6.53}$$

The total cross section σ_T is defined as the integral of $d\sigma/d\Omega$ over the unit sphere; thus

$$\sigma_T = \int_0^{4\pi} \frac{d\sigma}{d\Omega}\, d\Omega \tag{6.54}$$

In the present example show that

$$\sigma_T = \pi a^2 \frac{(2Ze^2/u\hbar)^2}{1 + (\hbar/2pa)^2} \tag{6.55}$$

Problem 6-9 Suppose we introduce that fact that the atomic nucleus has a finite radius given by

$$r_N = 1.2 \text{ fm} \times (\text{mass number})^{1/3} \tag{6.56}$$

and assume that the nuclear change is distributed approximately uniformly in a sphere of this radius. What is the effect of this assumption on the cross section for the scattering of electrons by atoms at large values of the momentum transfer $\breve{p}$?

Show how the nuclear radius can be determined along with some of the details of the nuclear charge distribution by making use of this effect. How large must the momentum p of the incoming electrons be in order to produce an appreciable effect? Would one observe more carefully the large or small scattering angles? Why?

Note: In this type of experiment the required electron momentum is so high that actually the relativistic formula $E = \sqrt{(mc^2)^2 + (pc)^2} - mc^2$ must be used to find the kinetic energy. So, strictly, we should not be allowed to use nonrelativistic formulas to describe the interaction. However, the relations between momentum and wavelength and between energy and frequency are not changed in the relativistic region. Since it is the wavelength which determines the resolving power of this "electron microscope," the momentum calculated by nonrelativistic formulas is still correct.

Problem 6-10 Consider a diatomic molecule containing two atoms, A and B, arranged with their centers at the points given by the vectors **a** and **b**. Using the Born approximation, show that the amplitude for an electron to be scattered from such a molecule is

$$K^{(1)} = e^{i(\breve{\mathbf{p}}/\hbar)\cdot\mathbf{a}} f_A(\breve{\mathbf{p}}) + e^{i(\breve{\mathbf{p}}/\hbar)\cdot\mathbf{b}} f_B(\breve{\mathbf{p}}) \tag{6.57}$$

where f_A and f_B are the amplitudes for scattering by the two atoms individually when each atom is located at the center of a coordinate system. (Within the Born approximation, these f values are real for spherically symmetric potentials.) The atomic binding does not change the charge distributions around the nuclei very much (except for very light nuclei such as hydrogen) because the binding forces affect only a few of the outermost electrons.

Using Eq. (6.57), show that the probability of scattering at a particular value of $\breve{\mathbf{p}}$ is proportional to $f_A^2 + f_B^2 + 2f_Af_B\cos(\breve{\mathbf{p}}\cdot\mathbf{d}/\hbar)$, where $\mathbf{d}$ is $\mathbf{a}-\mathbf{b}$.

Problem 6-11 Suppose the diatomic molecules are oriented in a random fashion. Show that the electron scattering averaged over a group of such molecules is proportional to

$$f_A^2 + f_B^2 + 2f_Af_B\frac{\sin(|\breve{\mathbf{p}}||\mathbf{d}|/\hbar)}{|\breve{\mathbf{p}}||\mathbf{d}|/\hbar}$$

How can this result be generalized to the case of polyatomic molecules?

These results form the basis of electron diffraction techniques which make possible the determination of the form of molecules. The values of f computed through the Born approximation are real and the result is valid for electron energies usually used in diffraction experiments on molecules (the order of 1 keV). However, if the molecule includes the very heaviest atoms, such as uranium, the atomic potential is too large for the results to be adequately described by the Born approximation, and small corrections are necessary.

Problem 6-12 Assume that $V(\mathbf{r})$ is independent of time and show that the time integral of the second-order scattering term $K^{(2)}(b,a)$ gives

$$K^{(2)}(b,a) = \frac{1}{2}\left(\frac{m}{2\pi\hbar^2}\right)^2\left(\frac{m}{2\pi i\hbar T}\right)^{3/2}\iint\frac{R_{bc}+R_{cd}+R_{da}}{R_{bc}R_{cd}R_{da}} \times\exp\left\{\frac{im}{2\hbar T}(R_{bc}+R_{cd}+R_{da})^2\right\}V(\mathbf{x}_d)V(\mathbf{x}_c)\,d^3\mathbf{x}_c\,d^3\mathbf{x}_d \tag{6.58}$$

where the points a, d, c, and b are arranged as shown in Fig. 6-9. The term R_{cd} stands for the distance from point d to point c, etc.

Assume that $V(\mathbf{r})$ becomes negligibly small at distances which are short compared to r_a or r_b. Show that the cross section is given by $d\sigma/d\Omega = |f|^2$, where the scattering amplitude f, including the first-order term, is

$$f = \frac{-m}{2\pi\hbar^2}\int e^{-i(\mathbf{p}_b/\hbar)\cdot\mathbf{x}_c}V(\mathbf{x}_c)e^{i(\mathbf{p}_a/\hbar)\cdot\mathbf{x}_c}\,d^3\mathbf{x}_c + \frac{1}{2}\left(\frac{-m}{2\pi\hbar^2}\right)^2\iint e^{-i(\mathbf{p}_b/\hbar)\cdot\mathbf{x}_c}V(\mathbf{x}_c)\frac{e^{i(p/\hbar)R_{cd}}}{R_{cd}}V(\mathbf{x}_d)e^{i(\mathbf{p}_a/\hbar)\cdot\mathbf{x}_d}\,d^3\mathbf{x}_c\,d^3\mathbf{x}_d + \text{higher-order terms} \tag{6.59}$$

Here $\mathbf{p}_b$ is the momentum of the electron traveling in the direction of $\mathbf{x}_b$ and $\mathbf{p}_a$ is the momentum of the electron traveling in the direction

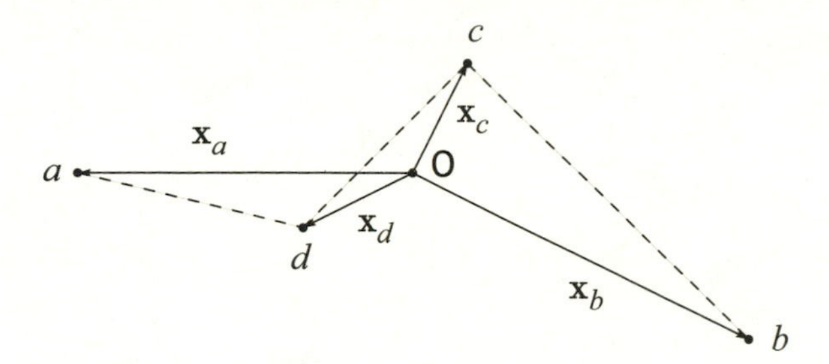

Fig. 6-9 To increase the accuracy of scattering calculations, we can take account of second-order terms in the perturbation expansion. Here, as in Fig. 6-2 (3), we picture the electron as being scattered at two separate points in the atomic potential. Thus the electron starts at a; moves as a free particle to d, where it is scattered; then moves as a free particle to c, where it is scattered again; and finally moves as a free particle to b, where it is collected by the counter. The points d and c can lie at any position in space. The atomic potential at these positions depends upon the radius vectors $\mathbf{x}_d$ and $\mathbf{x}_c$, measured from the center of the atom 0.

of $-\mathbf{x}_a$. The magnitude of the momentum is p, and it is approximately unchanged by an elastic scattering of the electron from the (relatively massive) atom.

One might expect that in a situation in which the Born approximation is not adequate it would be worthwhile to compute the second-order term as a correction. But in practice it seems that in this application Eq. (6.59) is a kind of asymptotic series. If the second term makes an appreciable correction (say 10 per cent or more) the higher terms are not much smaller and the true correction cannot be gotten easily by this method. Of course, if it is a problem in which the errors of the Born approximation are small (say less than 1 per cent), the second term will be adequate to find the corrections.

The Wave Function Treatment of Scattering. In the scattering experiment which we have described we have assumed that the initial state of the incoming electron was that of a free particle with momentum $\mathbf{p}_a$. We have assumed that the value of the momentum is determined by a time-of-flight technique (i.e., the total time required to travel the distance $r_a + r_b$ is T).

It is not necessary to use such a technique. Any device which enables us to determine the momentum is equally satisfactory. So suppose we generalize our picture of scattering phenomena with the help of the wave function method.

Suppose the incoming electrons have momentum $\mathbf{p}_a$ and energy $E_a = p_a^2/2m$. Then the wave function for the incoming electrons is

$$\phi_a(\mathbf{x}, t) = e^{i(\mathbf{p}_a/\hbar)\cdot\mathbf{x}} e^{-(i/\hbar)E_a t} \tag{6.60}$$

Then, using the first two terms of Eq. (6.25), the wave function for the outgoing electrons is, to first order,

$$\psi(\mathbf{x}_b, t_b) = e^{i(\mathbf{p}_a/\hbar)\cdot\mathbf{x}_b} e^{-(i/\hbar)E_a t_b} - \frac{i}{\hbar}\int_0^{t_b}\int K_0(\mathbf{x}_b, t_b; \mathbf{x}_c, t_c)V(\mathbf{x}_c, t_c)e^{i(\mathbf{p}_a/\hbar)\cdot\mathbf{x}_c} e^{-(i/\hbar)E_a t_c}\, d^3\mathbf{x}_c\, dt_c \tag{6.61}$$

The first term represents the alternative of the particle passing through the potential region without scattering. The second term represents the alternative of the particle scattering, summed over all possible scattering locations. This second term is called $\psi_s(\mathbf{x}_b, t_b)$, the scattered wave.

Problem 6-13 Assume that $V(\mathbf{r}, t)$ is independent of time. Substitute the free-particle kernel K_0 into Eq. (6.61) and integrate over t_c to show° that

$$\psi(\mathbf{x}_b, t_b) = e^{-(i/\hbar)E_a t_b}\left[e^{i(\mathbf{p}_a/\hbar)\cdot\mathbf{x}_b} - \frac{m}{2\pi\hbar^2}\int \frac{e^{i(p/\hbar)R_{bc}}}{R_{bc}} V(\mathbf{x}_c)e^{i(\mathbf{p}_a/\hbar)\cdot\mathbf{x}_c}\, d^3\mathbf{x}_c\right] \tag{6.62}$$

where R_{bc} is the distance from the variable point of integration $\mathbf{x}_c$ to the final point $\mathbf{x}_b$ and p is the magnitude of the momentum of the electron.

Once again suppose that the potential drops to 0 for distances which are short compared to either r_a or r_b. Show that Eq. (6.62) can be written as

$$\psi(\mathbf{x}_b, t_b) = e^{-(i/\hbar)E_a t_b}\left[e^{i(\mathbf{p}_a/\hbar)\cdot\mathbf{x}_b} + f(\theta)\frac{e^{i(p/\hbar)r_b}}{r_b}\right] \tag{6.63}$$

where the scattering amplitude $f(\theta)$ is defined in terms of $v(\breve{\mathbf{p}})$ (see Eq. 6.39) as

$$f(\theta) = -\frac{m}{2\pi\hbar^2}v(\breve{\mathbf{p}}) \tag{6.64}$$

The last term of Eq. (6.63), $f(\theta)e^{i(p/\hbar)r_b}/r_b$, can be thought of as the spatial part of the scattered wave function. It has the form of a spherical wave radiating outward from the center of the scattering atom. The amplitude of this spherical wave at some particular scattering angle depends upon that angle through the function $f(\theta)$ which, by Eq. (6.64), varies with the momentum transfer $\breve{\mathbf{p}}$. Thus the complete wave function for the

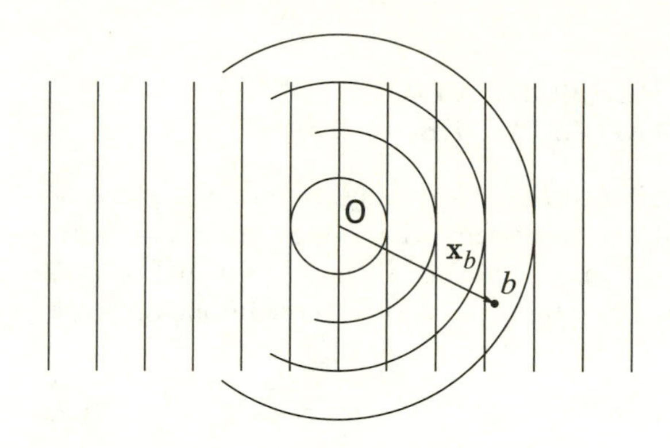

Fig. 6-10 An electron, represented by its equivalent wave, moves toward the atom at O. The strongest amplitude is for the electron to continue on undisturbed as a plane wave with momentum $\mathbf{p}_a$. A weaker amplitude is for the electron to be scattered and move away from O as a spherical wave. The resulting amplitude of ending up at some point b, located at $\mathbf{x}_b$ relative to the atom at O, is then made up of two parts. The first is the nonscattered amplitude given by the plane wave $e^{i(\mathbf{p}_a/\hbar)\cdot\mathbf{x}_b}$. To this is added the scattered wave of spherical form $e^{i(p/\hbar)r_b}/r_b$ times the scattering amplitude $f(\theta)$. The combination of these two waves gives the spatial part of the scattered wave function.

electron after scattering can be thought of as the sum of two terms. The first term is the plane wave of the nonscattered alternative, $e^{i(\mathbf{p}_a/\hbar)\cdot\mathbf{x}_b}$, and the second term is the spherical wave of the scattered alternative, as indicated in Fig. 6-10. Use this point of view to derive the formula for the cross section $d\sigma/d\Omega$.

Problem 6-14 Use the wave function approach to discuss the scattering of an electron from a sinusoidally oscillating field whose potential is given by

$$V(\mathbf{x},t) = U(\mathbf{x})\cos\omega t \tag{6.65}$$

Show that in the first-order Born approximation the energy of the outgoing wave is changed by either $+\hbar\omega$ or $-\hbar\omega$. What happens in the higher-order terms?

6-5 TIME-DEPENDENT PERTURBATIONS AND TRANSITION AMPLITUDES

The Transition Amplitude. An especially useful form of the perturbation theory occurs if the unperturbed problem corresponds to a potential U independent of time, for then we have seen in Eq. (4.59) that the unperturbed kernel can be expanded as (now in one dimension for convenience)

$$K_U(b,a) = \sum_n \phi_n(x_b)\phi_n^*(x_a)e^{-(i/\hbar)E_n(t_b-t_a)} \qquad \text{for } t_b > t_a \tag{6.66}$$

in terms of the eigenfunctions $\phi_n(x)$ and eigenvalues E_n of the unperturbed hamiltonian. Let us look at our series for $K_V(b,a)$ after substituting this expression for K_U. Writing out the first two terms, it is (compare Eq. 6.10)

$$\begin{aligned} K_V(b,a) &= \sum_n \phi_n(x_b)\phi_n^*(x_a)e^{-(i/\hbar)E_n(t_b-t_a)} \\ &\quad - \frac{i}{\hbar}\sum_m\sum_n \int_{t_a}^{t_b}\int_{-\infty}^{\infty} \phi_m(x_b)\phi_m^*(x_c)e^{-(i/\hbar)E_m(t_b-t_c)} \\ &\qquad\qquad \times V(x_c,t_c)\phi_n(x_c)\phi_n^*(x_a)e^{-(i/\hbar)E_n(t_c-t_a)}\,dx_c\,dt_c \\ &\quad + \cdots \end{aligned} \tag{6.67}$$

It is clear that within each term the variable x_a will appear in some energy eigenfunction, like $\phi_n^*(x_a)$, and the x_b likewise, so we can always write K_V in the form

$$K_V(b,a) = \sum_m\sum_n \lambda_{mn}(t_b,t_a)\phi_m(x_b)\phi_n^*(x_a) \tag{6.68}$$

where the λ's are coefficients depending on t_b, t_a. We shall call these coefficients *transition amplitudes.* To zero order in V, this must reduce to K_U, so to this order $\lambda_{mn} = \delta_{mn}e^{-(i/\hbar)E_n(t_b-t_a)}$. If we expand λ in a series in increasing orders of V, we obtain

$$\lambda_{mn} = \delta_{mn}e^{-(i/\hbar)E_n(t_b-t_a)} + \lambda_{mn}^{(1)} + \lambda_{mn}^{(2)} + \cdots \tag{6.69}$$

and comparison to Eq. (6.67) shows

$$\begin{aligned} \lambda_{mn}^{(1)} &= -\frac{i}{\hbar}\int_{t_a}^{t_b}\int_{-\infty}^{\infty} \phi_m^*(x_c) \\ &\qquad \times V(x_c,t_c)\phi_n(x_c)e^{-(i/\hbar)[E_m(t_b-t_c)+E_n(t_c-t_a)]}\,dx_c\,dt_c \end{aligned} \tag{6.70}$$

Problem 6-15 Recall that in Prob. 5-4 we defined a particular integral as the transition amplitude to go from state $\psi(x)$ to state $\chi(x)$. Show that the function λ_{mn} satisfies this definition when the initial state is the eigenfunction $\phi_n(x)$ and the final state is the eigenfunction $\phi_m(x)$.

Define, for brevity,

$$V_{mn}(t_c) = \int_{-\infty}^{\infty} \phi_m^*(x_c) V(x_c, t_c) \phi_n(x_c)\, dx_c \tag{6.71}$$

(This is called the *matrix element* of V between states m and n.) Then Eq. (6.70) can be written

$$\lambda_{mn}^{(1)} = -\frac{i}{\hbar} e^{-(i/\hbar)(E_m t_b - E_n t_a)} \int_{t_a}^{t_b} V_{mn}(t_c) e^{+(i/\hbar)(E_m - E_n)t_c}\, dt_c \tag{6.72}$$

This is an important result of the time-dependent perturbation theory.

The coefficient λ_{mn} is the amplitude for the system to be found in state m at time t_b if it was in state n at time t_a. Suppose the wave function at t_a was $\phi_n(x_a)$. What is it at t_b? Using Eq. (3.42), we can express the wave function at t_b as

$$\begin{aligned}\int_{-\infty}^{\infty} K_V(b,a)\phi_n(x_a)\, dx_a &= \sum_j \sum_k \lambda_{jk}\phi_j(x_b) \int_{-\infty}^{\infty} \phi_k^*(x_a)\phi_n(x_a)\, dx_a \\ &= \sum_j \lambda_{jn}\phi_j(x_b)\end{aligned} \tag{6.73}$$

That is, the wave function at t_b is in the form $\sum_m C_m \phi_m(x_b)$.

This expansion in terms of eigenfunctions was first introduced in Eq. (4.48). Now we can assign a deeper meaning to the constants C_m. We can interpret C_m as the amplitude that the system is in state $\phi_m(x)$. In this particular case, $C_m = \lambda_{mn}$ is the amplitude for the system to be in state $\phi_m(x)$ at time t_b if it was in $\phi_n(x)$ at time t_a.

With no perturbation acting, a system once in state n is always in state n, with an amplitude varying in time. So, to zero order,

$$\lambda_{mn} = \delta_{mn} e^{-(i/\hbar)E_n(t_b - t_a)}$$

We can interpret the first-order term by the rule (see Fig. 6-11) that: *The amplitude to be scattered from state n to state m within a time dt is* $-(i/\hbar)V_{mn}(t)\, dt$.

Problem 6-16 Interpret Eq. (6.71) as a sum over alternatives; i.e., identify the alternatives.

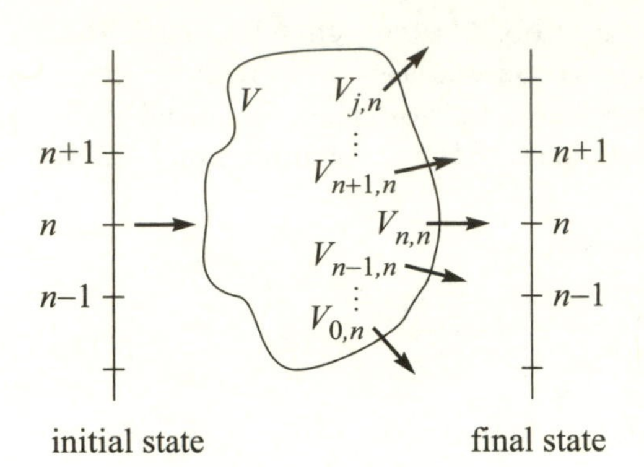

Fig. 6-11 A system initially in the nth energy state is subjected to a potential V which "scatters" the system into all of the states available to it. The amplitude for scattering into the mth state is proportional to V_{mn}. In particular, the amplitude to be scattered from the state n to the state m in the time interval dt is $-(i/\hbar)V_{mn}(t)\,dt$.

Problem 6-17 Interpret Eq. (6.72) by explaining the meaning of each term. Then explain and verify the equation for the second-order coefficient

$$\lambda_{mn}^{(2)} = \left(-\frac{i}{\hbar}\right)^2 \int_{t_a}^{t_b} \left[\int_{t_a}^{t_c} \sum_j e^{-(i/\hbar)E_m(t_b-t_c)} V_{mj}(t_c) \times e^{-(i/\hbar)E_j(t_c-t_d)} V_{jn}(t_d) e^{-(i/\hbar)E_n(t_d-t_a)}\,dt_d\right] dt_c \tag{6.74}$$

Problem 6-18 Derive and interpret the integral equation

$$\lambda_{mn}(t_b,t_a) = \delta_{mn} e^{-(i/\hbar)E_m(t_b-t_a)} - \frac{i}{\hbar}\int_{t_a}^{t_b} e^{-(i/\hbar)E_m(t_b-t_c)} \sum_j V_{mj}(t_c)\lambda_{jn}(t_c,t_a)\,dt_c \tag{6.75}$$

Problem 6-19 Consider $\lambda_{mn}(t_b)$ as a function of the final time t_b. Show, using either Eq. (6.75) or (6.69), that

$$\frac{d}{dt_b}\lambda_{mn}(t_b) = -\frac{i}{\hbar}\left[E_m\lambda_{mn}(t_b) + \sum_j V_{mj}(t_b)\lambda_{jn}(t_b)\right] \tag{6.76}$$

Give a direct physical interpretation of this result. Next, deduce this result directly from the Schrödinger equation. *Hint:* Use Eq. (6.73) and substitute into the Schrödinger equation. Note that Eq. (6.76), with the initial condition $\lambda_{mn}(t_a) = \delta_{mn}$, could be used to determine the λ's directly.

We can interpret all of the terms in Eq. (6.69) using the rule that $-(i/\hbar)V_{mn}(t)\,dt$ is the amplitude that the potential V will scatter (or induce a transition) from a state n to a state m during the time interval dt. We can go from the state n to the state m by 0, 1, 2, or more scatterings. We can go directly between the two states, i.e., with no scatterings, only if $m = n$. Thus the first term in the expansion is proportional to δ_{mn}.

The second term, given by Eq. (6.72), gives the amplitude that the transition will take place as the result of a single scattering. The amplitude for the particle to be found in the initial state n is $e^{-(i/\hbar)E_n(t_c-t_a)}$ at the time t_c. (In this case the phrase "to be found in the state n" should be interpreted as "available for scattering from the state n by the potential V.") The amplitude to be scattered by the potential $V(t_c)$ from the state n to the state m is $-(i/\hbar)V_{mn}(t_c)$. Finally, the amplitude to be found in the state m (which in this case means "the amplitude that the state m shall be available to the particle at the time the scattering takes place") at the time t_b is proportional to $e^{-(i/\hbar)E_m(t_b-t_c)}$. This scattering (at time t_c) can take place at any time between t_a and t_b. Therefore, an integration over the time t_c is carried out between these two end points.

The third term, given by Eq. (6.74), is the amplitude for a transition as the result of a double, or second-order, scattering. The first scattering takes the system from its initial state n to the intermediate state j at time t_d. The system then stays in this state until the time t_c, when its availability for scattering is again measured by an exponential function, $e^{-(i/\hbar)E_j(t_c-t_d)}$. Another scattering takes place at the time t_c and carries the system from the state j to the state m. We integrate over all of the possible alternate times for the scatterings at t_d and t_c, requiring only that t_d be earlier than t_c. Next, we add over all the possible states j into which the system may have been scattered in the intermediate interval.

The terms of Eq. (6.69), which we have just interpreted, give the results of the general time-dependent perturbation theory. It is applicable when the unperturbed system has a *constant hamiltonian* and thus *definite energy values.* Next, we shall study some special cases of this theory in more detail.

First-order transitions. First, let us take the case that the final state m is different from the initial state n and let us consider only the first Born approximation, i.e., the second term in Eq. (6.69). The result will be applicable for small values of V. The amplitude that we make the transition from n to m is

$$\lambda_{mn}^{(1)} = -\frac{i}{\hbar} e^{-(i/\hbar)(E_m t_b - E_n t_a)} \int_{t_a}^{t_b} V_{mn}(t) e^{+(i/\hbar)(E_m - E_n)t}\, dt \tag{6.77}$$

This is a very important special formula in the time-dependent perturbation theory. Suppose as a first example that $V(x,t) = V(x)$ is not an explicit function of time. If we take the interval of time from 0 to T, then since V_{mn} is constant, we have

$$\begin{aligned} \lambda_{mn}^{(1)} e^{+(i/\hbar)(E_m t_b - E_n t_a)} &= -\frac{i}{\hbar} V_{mn} \int_0^T e^{+(i/\hbar)(E_m - E_n)t}\, dt \\ &= -V_{mn} \frac{e^{+(i/\hbar)(E_m - E_n)T} - 1}{E_m - E_n} \end{aligned} \tag{6.78}$$

The probability of a transition during the time interval T is then

$$P(n \to m) = |\lambda_{mn}^{(1)}|^2 = \frac{|V_{mn}|^2}{(E_m - E_n)^2} \left[4 \sin^2 \frac{(E_m - E_n)T}{2\hbar} \right] \tag{6.79}$$

We see that for at least a long interval T this probability is a rapidly oscillating function of the energy difference $E_m - E_n$. If E_m and E_n differ appreciably, i.e., if $|E_m - E_n| \gg |V_{mn}|$, this probability is very small. This means that the probability that the energy in the final state will be modified appreciably from that in the initial state by a very weak constant perturbation is very small. One might ask: How can the energy be expected to change at all by the large amount $E_m - E_n$ as a result of the small disturbance V_{mn}? The answer is that we have considered V to start suddenly at the time $t = 0$, and the definiteness of this time permits, by the uncertainty principle, a large uncertainty in the energy (see Eq. (5.19) and associated discussion).

Problem 6-20 Suppose V is turned on and off slowly. For example, let $V(x,t) = V(x)g(t)$, where $g(t)$ is smooth, as shown in Fig. 6-12.

$$g(t) = \begin{cases} \frac{1}{2} e^{+\gamma t} & \text{for } t < 0 \\ 1 - \frac{1}{2} e^{-\gamma t} & \text{for } 0 < t < T/2 \\ 1 - \frac{1}{2} e^{+\gamma(t-T)} & \text{for } T/2 < t < T \\ \frac{1}{2} e^{-\gamma(t-T)} & \text{for } T < t \end{cases} \tag{6.80}$$

The rise time of the function $g(t)$ is $1/\gamma$. Supposing that $1/\gamma \ll T$, show that the probability given by Eq. (6.79) is reduced by a factor $\{1 + [(E_m - E_n)/\hbar\gamma]^2\}^{-2}$. In this definition of $g(t)$ we still have a discontinuity in the second derivative with respect to time. Smoother functions make still further reductions.

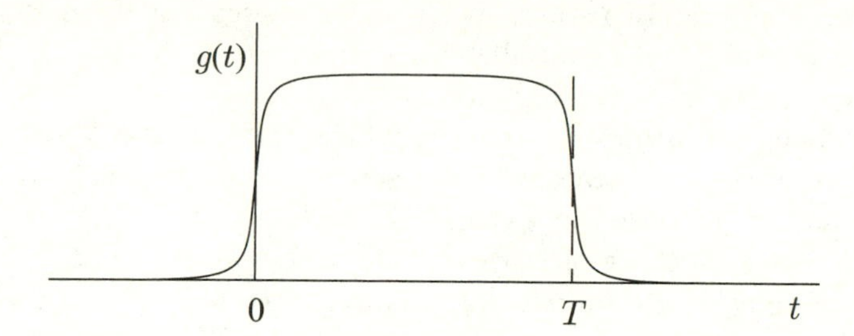

Fig. 6-12 The potential effecting the transition from n to m is turned on and off slowly with the time variation $g(t)$, shown here. As this time factor becomes smoother (e.g., as discontinuities appear in successively higher derivatives) the probability of a transition becomes smaller.

If it should happen that E_m and E_n are exactly the same energy, we find $P(n \to m) = |V_{mn}|^2T^2/\hbar^2$. This grows as the square of the time. It means that a concept of "transition probability per unit time" is not meaningful in this case. This formula holds only for T short enough that $|V_{mn}|T \ll \hbar$. It turns out that if only two states of exactly the same perturbed energy are involved, the probability of being found in the first goes as $\cos^2(|V_{mn}|T/\hbar)$ and of being found in the second as $\sin^2(|V_{mn}|T/\hbar)$, while our formula is only a first approximation to this.

Problem 6-21 Consider the special case that the perturbing potential V has no matrix elements except between the two states 1 and 2; and further, suppose these states are degenerate, that is, suppose $E_1 = E_2$. Let $V_{12} = V_{21} = v$ and let V_{11}, V_{22}, and all other V_{mn} be zero. Show that

$$\lambda_{11} = 1 - \frac{v^2T^2}{2\hbar^2} + \frac{v^4T^4}{24\hbar^4} - \cdots = \cos\frac{vT}{\hbar} \tag{6.81}$$

$$\lambda_{12} = -i\frac{vT}{\hbar} + i\frac{v^3T^3}{6\hbar^3} - \cdots \quad = -i\sin\frac{vT}{\hbar} \tag{6.82}$$

Problem 6-22 In Prob. 6-21 we have $V_{12} = V_{21}$, so that V_{12} is real. Show that even if V_{12} is complex, the physical results are the same (let $v = |V_{12}|$).

Such systems swing back and forth from one state to the other. A further conclusion can be drawn from this result. Suppose the perturbation acts for an extremely long time so that $|V_{mn}|T \gg \hbar$. Then if the system is investigated at an arbitrary time T, which is somewhat indefinite, the probabilities of being in either the first or second state are, on the average, equal. That is, a small indefinite perturbation acting for a very long time between two states at the same energy makes these states have equal probability. This will be useful when we discuss the theory of statistical mechanics in Chap. 10.

The case of great importance is that in which the values allowed for E_m, the energy of the final state, are not separate and discrete but lie in a continuum, or at least are extremely closely spaced. Let us say that $\rho(E)\,dE$ is the number of states in the range of energy E to $E+dE$. Then we can ask for the probability to go to some state in this continuum. First we see that to go to any state for which $|E_m - E_n|$ is large is very unlikely. It is most likely that the final state will be one of nearly the same energy as the initial E_n (within an error $\pm|V_{mn}|$). The total chance to go into any state is

$$\begin{aligned}\sum_{m=1}^{\infty} P(n \to m) &= \sum_{m=1}^{\infty} |V_{mn}|^2 \frac{4\sin^2[(E_m - E_n)T/2\hbar]}{(E_m - E_n)^2} \\ &\approx \int |V_{mn}|^2 \frac{4\sin^2[(E_m - E_n)T/2\hbar]}{(E_m - E_n)^2} \rho(E_m)\,dE_m\end{aligned} \tag{6.83}$$

The quantity $\{4\sin^2[(E_m - E_n)T/2\hbar]\}/(E_m - E_n)^2$ is very large if $E_m \approx E_n$, reaching a maximum of $T^2/\hbar^2$, whereas it is much smaller if E_m and E_n differ appreciably (relative to $\hbar/T$), as shown in Fig. 6-13. Thus almost all the contribution to the integral over E_m comes when E_m is in the neighborhood of the value E_n.

If $|V_{mn}|$ varies slowly enough with m that we can replace it with a typical value, and furthermore if $\rho(E_m)$ likewise does not vary rapidly,

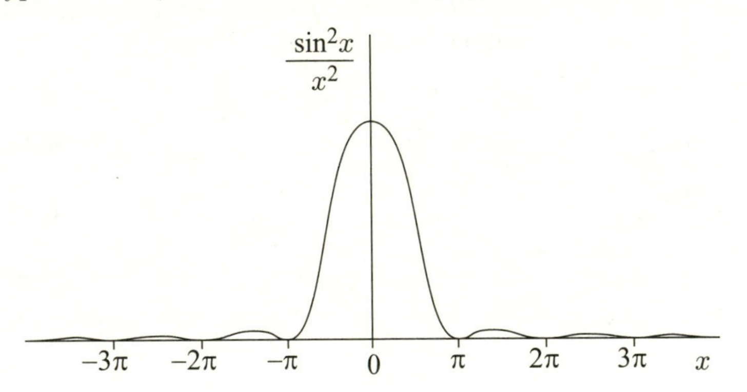

Fig. 6-13 In this figure the energy difference $E_m - E_n$ is replaced by the variable x. When these two energies are approximately equal (thus x is very small) the function $(\sin^2 x)/x^2$ approaches its maximum value. For large values of the difference the function becomes very small. Thus, in expressions involving this function, the most important contributions come from the central region, that is, the region where the two energies are approximately equal.

then to a good approximation we can replace the integral of Eq. (6.83) by the expression

$$4|V_{mn}|^2\rho(E_n)\int\frac{\sin^2[(E_m-E_n)T/2\hbar]}{(E_m-E_n)^2}\,dE_m \tag{6.84}$$

Since $\int_{-\infty}^{\infty}[(\sin^2 x)/x^2]\,dx=\pi$, the integral of Eq. (6.84) has the value $\pi T/2\hbar$ and we obtain the result that the probability for a transition to some state in the continuum is

$$P(n\to m)=\frac{2\pi}{\hbar}|V_{mn}|^2\rho(E_n)T \tag{6.85}$$

and that the energy in the final state is the same as the energy in the initial state.

From these results we can write the probability of a transition per unit time in the form

$$\frac{dP(n\to m)}{dt}=\frac{2\pi}{\hbar}|M_{n\to m}|^2\rho(E) \tag{6.86}$$

where $M_{n\to m}$ is called the *matrix element for the transition* and $\rho(E)$ is the density of levels in the final state. In our case $M_{n\to m}$ is V_{mn}. If we went to a higher-order expansion of λ_{mn}, it would be more complicated. Another way to write this expression is that the probability of a transition per unit time from state n to some particular state m is

$$\frac{dP(n\to m)}{dt}=\frac{2\pi}{\hbar}|M_{n\to m}|^2\delta(E_m-E_n) \tag{6.87}$$

Then when we sum over a group of final states, only those with energies $E_m=E_n$ survive. Since $\sum_m(\quad)\to\int(\quad)\rho(E_m)\,dE_m$, we get as a result Eq. (6.86).

We may illustrate Eq. (6.86) by an example which we have previously discussed from a different point of view, namely, the scattering of an electron in a potential (see Sec. 6-4). Suppose an otherwise free particle has an interaction with a potential $V(\mathbf{r})$ and we wish to discuss the scattering of this particle from an initial state of a definite momentum to a final state with another definite momentum in a new direction. We suppose that the state n, the initial state, is a plane wave of momentum $\mathbf{p}_a$ so that the wave function $\phi_n(\mathbf{x})$ is $e^{i(\mathbf{p}_a/\hbar)\cdot\mathbf{x}}/\sqrt{\text{Vol}}$ (where "Vol" is the volume of the enclosing box as described in Sec. 4-3). Likewise, suppose the final state is a plane wave of momentum $\mathbf{p}_b$ so that the wave function $\phi_m(\mathbf{x})$ is $e^{i(\mathbf{p}_b/\hbar)\cdot\mathbf{x}}/\sqrt{\text{Vol}}$. The matrix element V_{mn} is

$$V_{mn}=\frac{1}{\text{Vol}}\int e^{-i(\mathbf{p}_b/\hbar)\cdot\mathbf{x}}V(\mathbf{x})e^{i(\mathbf{p}_a/\hbar)\cdot\mathbf{x}}\,d^3\mathbf{x}=\frac{1}{\text{Vol}}v(\breve{\mathbf{p}}) \tag{6.88}$$

where $\breve{\mathbf{p}} = \mathbf{p}_a - \mathbf{p}_b$. In the scattering, the energy will be conserved so that $p_a^2/2m = p_b^2/2m$. This means that the magnitudes of momenta $\mathbf{p}_a$ and $\mathbf{p}_b$ are the same. Let us call that magnitude p, so that

$$|\mathbf{p}_b| = |\mathbf{p}_a| = p$$

By our usual convention for writing differential elements of momentum, the number of states which have their momenta in the volume element of momentum space $d^3\mathbf{p}_b$ is $\text{Vol}\, d^3\mathbf{p}_b/(2\pi\hbar)^3 = \text{Vol}\, p^2\, dp\, d\Omega/(2\pi\hbar)^3$, where $d\Omega$ is the element of solid angle which contains the momentum vector $\mathbf{p}_b$. An element dE of the energy range is connected to the element of momentum space by

$$dE = d\left(\frac{p^2}{2m}\right) = \frac{p\, dp}{m} \tag{6.89}$$

Thus the density of momentum states for particles traveling into the solid angle $d\Omega$ is

$$\rho(E) = mp\, d\Omega \frac{\text{Vol}}{(2\pi\hbar)^3} \tag{6.90}$$

Substituting these relations into Eq. (6.86), we find the probability of transition per second into the element of solid angle $d\Omega$ to be

$$\frac{dP}{dt} = \frac{mp}{(2\pi\hbar^2)^2} \frac{d\Omega}{\text{Vol}} |v(\breve{\mathbf{p}})|^2 \tag{6.91}$$

We define an effective target area or cross section for scattering into $d\Omega$ as $d\sigma$ (see Sec. 6-4). The number of particles that will hit this area in time dt is the cross-sectional area times the velocity of the particles coming in, $u_a = p_a/m$, times dt, times the density of incoming particles. Thus

$$\frac{dP}{dt} = \frac{d\sigma\, u_a}{\text{Vol}} \tag{6.92}$$

Therefore the cross section is

$$\frac{d\sigma}{d\Omega} = \left(\frac{m}{2\pi\hbar^2}\right)^2 |v(\breve{\mathbf{p}})|^2 \tag{6.93}$$

which is exactly what we obtained in Eq. (6.44).

Problem 6-23 Show that the same result is obtained for $d\sigma/d\Omega$ if the wave functions $\phi(\mathbf{x})$ have the specific normalization of unity for a box of unit volume, e.g., $\phi_n(\mathbf{x}) = e^{i(\mathbf{p}_a/\hbar)\cdot\mathbf{x}}$.

Problem 6-24 Suppose that the potential V is periodic in time. For example, suppose $V(x,t) = V(x)(e^{i\omega t} + e^{-i\omega t})$. Show that the probability for a transition to take place is small unless the final state is one of the two values (1) $E_{\text{final}} = E_{\text{initial}} + \hbar\omega$ (corresponding to an absorption of energy) or (2) $E_{\text{final}} = E_{\text{initial}} - \hbar\omega$ (corresponding to an emission of energy). This means that Eq. (6.86) is unchanged, but the density of states $\rho(E)$ must be calculated at these new values of E. Or, in analogy with Eq. (6.87), we have

$$\frac{dP(n \to m)}{dt} = \frac{2\pi}{\hbar}|M_{n\to m}|^2[\delta(E_m - E_n - \hbar\omega) + \delta(E_m - E_n + \hbar\omega)] \tag{6.94}$$

Problem 6-25 It has been argued that the equations of the electrodynamics must, like those of mechanics, be converted to a quantized form on the basis of the photoelectric effect. Here an electron of energy $\hbar\omega$ is occasionally emitted from a thin layer of metal under the influence of light of frequency ω. Is this impossible if matter obeys the quantum laws but light is still represented as a continuous wave? What arguments can you adduce for the necessity of giving up a classical description of electrodynamics, in view of the results of Prob. 6-24?

Problem 6-26 Suppose we have two discrete energy levels E_1 and E_2, neither of which is in the continuum. Let a transition be induced by a potential of the form $V(x,t) = V(x)g(t)$. Show that the probability of transition is

$$P(1 \to 2) = |V_{12}|^2|\phi(\omega_0)|^2/\hbar^2 \tag{6.95}$$

if $g(t)$ is representable by the Fourier transform

$$g(t) = \int_{-\infty}^{\infty} \phi(\omega)e^{i\omega t}\,\frac{d\omega}{2\pi} \tag{6.96}$$

and $\omega_0 = (E_2 - E_1)/\hbar$.

If $g(t)$ is a statistically irregular function familiar from the theory of noise (called filtered white noise), the value of $\phi(\omega)$ given by the inverse transform

$$\phi(\omega) = \int_{-T}^{T} g(t)e^{-i\omega t}\,dt \tag{6.97}$$

depends on the integration range T. If T is very large, $|\phi(\omega_0)|^2$ can be shown to be proportional to T. Thus we get a transition probability proportional to the time and to the "intensity" or "power" (mean-square value of g per second) at frequency ω_0 per unit frequency range. In virtue of this, the probability for the transition of an atom in a continuous spectrum of light is proportional to (1) the exposure time and (2) the intensity of light at the frequency of absorption $(E_2 - E_1)/\hbar$.

The Higher-order Terms. It is interesting to look at the second-order term in the perturbation expansion. This term is of special importance in problems where $V_{mn} = 0$ for those particular states n and m of interest. Let us suppose that we have such a problem, and suppose further that there are other states $j \neq n$ for which $V_{jn} \neq 0$. The first-order term is 0, and so long as $m \neq n$ the zero-order term is likewise zero. Thus the lowest-order term which enters into the calculation of the transition amplitude is the second.

Suppose that the potential $V(\mathbf{x})$ is independent of t. Then the second-order term in the transition element is $\lambda_{mn}^{(2)}$; and if $T = t_b - t_a$, we have from Eq. (6.74)

$$\begin{aligned}
&\lambda_{mn}^{(2)} e^{+(i/\hbar)(E_m t_b - E_n t_a)} \\
&= \left(-\frac{i}{\hbar}\right)^2 \sum_j V_{mj} V_{jn} \int_0^T \int_0^{t_c} e^{+(i/\hbar)(E_m - E_j)t_c} e^{+(i/\hbar)(E_j - E_n)t_d}\, dt_d\, dt_c \\
&= \frac{i}{\hbar} \sum_j V_{mj} V_{jn} \int_0^T e^{+(i/\hbar)(E_m - E_j)t_c} \frac{e^{+(i/\hbar)(E_j - E_n)t_c} - 1}{E_j - E_n}\, dt_c \\
&= \sum_j \frac{V_{mj} V_{jn}}{E_j - E_n} \left[\frac{e^{+(i/\hbar)(E_m - E_n)T} - 1}{E_m - E_n} - \frac{e^{+(i/\hbar)(E_m - E_j)T} - 1}{E_m - E_j} \right]
\end{aligned} \tag{6.98}$$

The first of the two terms in brackets has the same time dependence as we have seen in our first-order result. Therefore if the second term is neglected for a moment we see that the net result would again be to make transitions to states where $E_m = E_n$, with a probability proportional to T. The probability per unit time has the same form as Eq. (6.86) but with $M_{n \to m}$ now given by

$$M_{n \to m} = \sum_j \frac{V_{mj} V_{jn}}{E_j - E_n} \tag{6.99}$$

If the states lie in a continuum, the sum becomes an integral.

Equation (6.99) is correct in the circumstance that it is impossible to go by a first-order transition from state n to state m or to any state with the same energy as the initial state. Under these circumstances $V_{jn} = 0$ for states such that $E_j = E_n$. Then the second term in brackets in Eq. (6.98) is never large; for it cannot be large unless $E_n - E_j$ is nearly zero, and then V_{jn} in the numerator is zero. All the effects come from the first term, and Eq. (6.99) is correct. Furthermore, in the sum over j in Eq. (6.98) there is no ambiguity at the pole where $E_j = E_m$; for the numerator vanishes at this same value of E_j.

On the other hand, in some situations it may be true that a first-order transition is possible to some other continuum state (e.g., a nucleus may decay in more than one way). In such a case the sum in Eq. (6.99) is meaningless; for we must define what to do near the pole. It is the neglected second term in Eq. (6.98) which comes to our rescue here and shows that the correct expression for $M_{n\to m}$ (now including the first-order term for generality) is

$$M_{n\to m} = V_{mn} + \sum_j \frac{V_{mj}V_{jn}}{E_j - E_n - i\epsilon} \tag{6.100}$$

in the limit $\epsilon \to 0$. How this comes about we shall now analyze.

First we may notice that for large T we cannot get a large probability of transition (proportional to T, that is) unless E_n and E_m are practically equal (within about $\hbar/T$). This is evident for the first term in Eq. (6.98). For the second term large amplitudes can arise only if $E_j \approx E_m$; but if E_m is not very close to E_n, the factor in front is a smooth function of E_j for E_j near E_m. Taking it as nearly constant for a small range near $E_j = E_m$, we see that the second term can be approximated as some constant times

$$\frac{e^{(i/\hbar)\epsilon T} - 1}{\epsilon}\, d\epsilon$$

where $\epsilon = E_m - E_j$ is to be integrated over a small range, say $-\delta$ to $+\delta$. But

$$\begin{aligned}\int_{-\delta}^{\delta} \frac{e^{(i/\hbar)\epsilon T} - 1}{\epsilon}\, d\epsilon &= \int_{-T\delta/\hbar}^{T\delta/\hbar} \frac{e^{iy} - 1}{y}\, dy \\ &= \int_{-T\delta/\hbar}^{T\delta/\hbar} \left(\frac{\cos y - 1}{y} + i\frac{\sin y}{y}\right) dy\end{aligned} \tag{6.101}$$

The first integral is that of an odd function and vanishes. The second approaches a finite limit as $T \to \infty$ (and therefore as $T\delta/\hbar \to \infty$). That is,

$$2i \int_0^{\infty} \frac{\sin y}{y}\, dy = 2\pi i$$

so no large transition probability occurs. A large effect can arise only in case E_n and E_m are essentially equal, for then the double coincidence of the two poles from $(E_j - E_n)^{-1}$ and $(E_m - E_j)^{-1}$ can make the second term important. Therefore, we continue the analysis, assuming E_m and E_n are nearly equal.

The sum over j in Eq. (6.98) can be divided into two regions by choosing a very small energy Δ and breaking the sum up into a part A for which $|E_j - E_n| \geq \Delta$ and a part B for which $|E_j - E_n| < \Delta$.

We choose Δ to be small enough that the factor $V_{mj}V_{jn}$ does not vary appreciably when E_j varies about E_n over the energy range 2Δ. This is some finite energy, and we shall take T so long that $\hbar/T \ll \Delta$, which means that $|E_n - E_m| \ll \Delta$.

First, for part A, $|E_j - E_m| \geq \Delta$. Then the second term cannot become large, for its poles are avoided. Only the first contributes, and the contribution is

$$a\frac{e^{ix}-1}{x}\frac{T}{\hbar} \tag{6.102}$$

where $x = (1/\hbar)(E_m - E_n)T$ and

$$a = \sum_j^{(A)} \frac{V_{mj}V_{jn}}{E_j - E_n}$$

The sum extends over all E_j except for those within $\pm\Delta$ of E_m. This sum is nearly independent of Δ, and as $\Delta \to 0$ it is the definition of a principal-value integral. That is, in the limit $\Delta \to 0$ we can write

$$a = V_{mn} + \sum_j V_{mj}V_{jn}\text{P.P.}\frac{1}{E_j - E_n} \tag{6.103}$$

where P.P. is the principal part and we have reinstated the first-order term, in case it does not vanish.

For the region B we take $V_{mj}V_{jn}$ to be constant at its value for $E_j - E_m = 0$. That is, we replace

$$\sum_j^{(B)} V_{mj}V_{jn}F(E_j) \quad \text{by} \quad \left[\sum_j V_{mj}V_{jn}\delta(E_j - E_m)\right]\int_{E_m-\Delta}^{E_m+\Delta} F(E_j)\,dE_j \tag{6.104}$$

We can write this as bI, where

$$b = \sum_j V_{mj}V_{jn}\delta(E_j - E_m) \tag{6.105}$$

and

$$I = \int_{E_m-\Delta}^{E_m+\Delta} \frac{1}{E_j - E_n}\left[\frac{e^{(i/\hbar)(E_m-E_n)T}-1}{E_m - E_n} - \frac{e^{(i/\hbar)(E_m-E_j)T}-1}{E_m - E_j}\right] dE_j \tag{6.106}$$

Now we put $(1/\hbar)(E_m - E_n)T = x$ and $(1/\hbar)(E_j - E_n)T = y$, so that $(1/\hbar)(E_m - E_j)T = x - y$, to get

$$I = \frac{T}{\hbar}\int_{-T\Delta/\hbar}^{T\Delta/\hbar} \frac{1}{y}\left[\frac{e^{ix}-1}{x} - \frac{e^{i(x-y)}-1}{x-y}\right] dy \tag{6.107}$$

This integral is most easily evaluated by contour integration, imagining y as a complex variable and changing the contour. Instead of integrating on the straight line from $-T\Delta/\hbar$ to $T\Delta/\hbar$, we go on the semicircle of radius $T\Delta/\hbar$ below the real axis. Since $T\Delta/\hbar$ is very large, the second term contributes negligibly; and since

$$\int_{-T\Delta/\hbar}^{T\Delta/\hbar} \frac{dy}{y} = i\pi$$

on this contour, we get $I = i\pi(T/\hbar)(e^{ix} - 1)/x$.

Putting the A and B parts together, we get

$$(a + i\pi b)\frac{(e^{ix} - 1)}{x}\frac{T}{\hbar} \tag{6.108}$$

for the amplitude. This gives a probability for transition of the form Eq. (6.86) with

$$\begin{aligned} M_{n\to m} &= a + i\pi b \\ &= V_{mn} + \sum_j V_{mj}V_{jn}\left[\text{P.P.}\frac{1}{E_j - E_n} + i\pi\delta(E_j - E_m)\right] \end{aligned} \tag{6.109}$$

In light of Eq. (A.10), the last bracket can be written $(E_j - E_m - i\epsilon)^{-1}$ in the limit as $\epsilon \to 0$, as we have written in Eq. (6.100).

From Eq. (6.100) we learn then that even if no direct transition is possible from n to m, nevertheless the transition can occur, as we say, through a *virtual* state. That is, we can imagine that the system goes from n to j, then from j to m. The amplitude for an indirect transition process is given by Eq. (6.99). We note that it is not right to say that it actually goes through one or another intermediate state j, but rather that in characteristic quantum-mechanical fashion there is a certain amplitude to go via the various intermediate states j, and the contributions interfere.

The intermediate states are not of the same energy as the initial and final states. The conservation of energy is not violated, for the virtual state is not permanently occupied. The strength of contribution to the sum varies inversely with this energy discrepancy.

There is nothing absolute about these intermediate states. They come from considering V as a perturbation to a system H and from speaking about the true states of $H + V$ in terms of those of H alone. If other separations are made as to what is the "unperturbed" problem and what is the "perturbation," different formulas and intermediate states will arise in the description.

When the potential depends upon time (e.g., periodically), many interesting effects result. Most of these have been observed in microwave experiments, where the perturbation $V(\mathbf{x}, t)$ is a weak electric or magnetic field with a periodic variation in time.

Problem 6-27 Derive the perturbation expansion up through the terms of the second order for potentials periodic in time.

Sometimes a transition cannot take place except by the use of two or more intermediate virtual states. Analysis of such transitions requires the calculation of third- and higher-order terms in the perturbation expansion.

Problem 6-28 Show that when a transition is impossible either directly or through a single intermediate state, but requires the use of two intermediate states, it is determined through the matrix element

$$M_{n\to m} = -\sum_j \sum_k \frac{V_{mj}V_{jk}V_{kn}}{(E_j - E_n)(E_k - E_n)} \tag{6.110}$$

This corresponds to the third-order term in the perturbation expansion.

Problem 6-29 Suppose two perturbations, $V(x, t)$ and $U(x, t)$, are acting. (Examples include a combination of DC and AC electric fields or a combination of electric and magnetic fields.) Suppose further that a certain transition cannot occur with either V or U alone, but can occur only when both act together. Under the special assumption that both V and U are constant in time, show that the matrix element determining the transition element is given by

$$M_{n\to m} = \sum_j \frac{V_{mj}U_{jn} + U_{mj}V_{jn}}{E_n - E_j} \tag{6.111}$$

Next, suppose both potentials are periodic in time but have different frequencies, ω_V and ω_U. What then is the matrix element?

Calculation of the Change in Energy of the State. In computing transition amplitudes we have considered only those states $m \neq n$. Suppose we turn our attention to the term $m = n$. Considering the zero- and first-order terms in the perturbation expansion, we have

$$e^{+(i/\hbar)E_nT}\lambda_{nn} = 1 - \frac{i}{\hbar}\int_0^T V_{nn}(t)\,dt \tag{6.112}$$

If V is constant in time, this gives $1 - (i/\hbar)V_{nn}T$. What is the meaning of this result? As a consequence of the introduction of the additional potential V into the original hamiltonian we can expect the energies of all the states of the system to be slightly altered. We can write the new energy of the state n as $E_n + \Delta E_n$. The time-dependent portion of the wave function describing this state will be $e^{-(i/\hbar)(E_n+\Delta E_n)t}$ instead of the previous $e^{-(i/\hbar)E_nt}$.

Over the period of time T during which the perturbing potential acts this relative difference in phase introduces the factor $e^{-(i/\hbar)\,\Delta E_n\,T}$. Expanding this factor to first order in time gives $1 - (i/\hbar)\,\Delta E_n\,T$. Thus we see that a first-order calculation of the energy shift in a state n due to a perturbation V is

$$\Delta E_n = V_{nn} \tag{6.113}$$

This derivation of the first-order energy shift is not correct if the system is degenerate, i.e., if there are initially several states of exactly the same energy. It turns out that in such a case terms of second order in V give equally large effects.

Adding in the second-order term in the perturbation expansion for the transition element gives

$$e^{+(i/\hbar)E_nT}\lambda_{nn} = 1 - \frac{i}{\hbar}V_{nn}T + \left(-\frac{i}{\hbar}\right)^2 \sum_j V_{nj}V_{jn}\int_0^T\int_0^{t_c} e^{-(i/\hbar)(E_j-E_n)(t_c-t_d)}\,dt_d\,dt_c \tag{6.114}$$

For the present let us assume that there is no degeneracy. Consider first the term $j = n$ in the series which is the second-order term. The integral over this particular term is just $T^2/2$. Integrals for the terms $j \neq n$ can also be performed easily to give the result

$$e^{+(i/\hbar)E_nT}\lambda_{nn} = 1 - \frac{i}{\hbar}V_{nn}T - \frac{1}{2\hbar^2}V_{nn}^2T^2 \\ + \frac{i}{\hbar}\sum_{j\neq n}\frac{|V_{jn}|^2T}{E_j - E_n}\left[1 + \frac{1 - \exp\{-(i/\hbar)(E_j - E_n)T\}}{-(i/\hbar)(E_j - E_n)T}\right] \tag{6.115}$$

The first three terms on the right-hand side of this equation represent an expansion through second order of $e^{-(i/\hbar)V_{nn}T}$. The first of the summation terms, the one corresponding to the 1 in brackets, can be interpreted as a second-order energy change. That is, the incremental energy is not just V_{nn}, but contains higher-order corrections. Writing out the energy correction through second order in the perturbation energy, we get

$$\Delta E_n = V_{nn} - \sum_{j\neq n}\frac{V_{nj}V_{jn}}{E_j - E_n} \tag{6.116}$$

This last equation gives the correct expression, through second order, for the shift in energy of nondegenerate states. This result is much more easily obtained by conventional methods, i.e., by finding solutions of

$$(H + V)\phi = E\phi \tag{6.117}$$

Furthermore, the conventional approach based on Eq. (6.117) permits simpler handling of degenerate states. However, it has been our purpose here to give an example of the use of transition amplitudes, rather than to give the simplest formulas for the computation of energy shifts.

Actually, there are more complex problems involving energy shsifts in which the method of transition amplitudes is the simplest to apply. In such applications the scheme, as we have attempted to show above, is to identify terms in a series proportional to T, T^2, etc. Then, if we remember that the amplitude to stay in the initial state is proportional to $e^{-(i/\hbar)\Delta E_nT}$ and that the series expansion is equivalent to a series expansion of this exponential, the correct expression for ΔE_n can be written down.

We have not yet discussed the last term in Eq. (6.115). If the states E_j lie in a continuum, we must also define the character of the reciprocal in the sum of Eq. (6.116). If we take it to mean the principal value, just as we found when analyzing the problem in second order for $m \neq n$, this extra term can be shown to produce an effect proportional to T and to lead to an additional correction to Eq. (6.116) of

$$\Delta' E_n = -i\pi\sum_j V_{nj}V_{jn}\delta(E_j - E_n) \tag{6.118}$$

But this cannot represent a further correction to the energy for it is purely imaginary, and the energy must be real. Let us call it $-i\hbar\gamma/2$ (the $\hbar/2$ is for convenience later) and write

$$\Delta E_n - \frac{i\hbar\gamma}{2} = V_{nn} - \sum_j \frac{V_{nj}V_{jn}}{E_j - E_n - i\epsilon} \tag{6.119}$$

This implies that the transition amplitude λ_{nn} to be in the nth state after a long time is proportional to

$$\exp\left\{-\frac{i}{\hbar}\left(\Delta E_n - \frac{i\hbar\gamma}{2}\right)T\right\} = \exp\left\{-\frac{i}{\hbar}\Delta E_n T\right\}\exp\left\{-\frac{\gamma T}{2}\right\}$$

The first factor is the energy shift. The second is easily interpreted; for the probability to be in state n after time T is $|\lambda_{nn}|^2 = e^{-\gamma T}$. It falls with time because at each instant there is a probability that a transition is made from n to some other state. That is, if all is consistent, γ must be the total probability per second of a transition from n to any state in the continuum of the same energy. This it is, because from Eq. (6.118) our γ is

$$\gamma = \frac{2\pi}{\hbar}\sum_j |V_{jn}|^2\delta(E_j - E_n) \tag{6.120}$$

So we see that the total probability per second is just the sum of Eq. (6.87) over all possible final states as required (i.e., up to the required order in V).

The reciprocal of γ is called the mean lifetime of the state. Strictly speaking, a state with a finite lifetime has no definite energy; the energy uncertainty by the Heisenberg relation is $\hbar$/lifetime, or $\hbar\gamma$.

If resonance experiments are performed to find the energy difference between two levels, each of which has a decay rate γ, the resonance is not sharp but has a definite shape. The center of the resonance determines the energy difference, and the width of the resonance gives the sum of the γ's of each level.

7

Transition Elements

IN the preceding chapter we developed the concept of a perturbation treatment for changes of state in a quantum-mechanical system. We carried out an investigation of this method as it is applied to systems whose unperturbed hamiltonians are constant in time. In this chapter we shall continue the development of the perturbation concept and generalize the treatment to cover systems where the unperturbed state may have a hamiltonian varying with time. We shall introduce a more general type of notation and attempt to broaden and deepen our understanding of the ways in which changes of state take place in a quantum-mechanical system. The notation to be introduced applies to a type of function which will be defined in the first portion of this chapter. The function is called a *transition element.*

The chapter is divided into four parts. The first part, consisting of Sec. 7-1, defines "transition amplitude" and "transition element," with the help of examples based upon the perturbation theory of Chap. 6. The second part, consisting of Secs. 7-2 to 7-4, gives some interesting general relations among transition elements. The third part, consisting of Sec. 7-5, shows the connection between transition elements defined with the help of path integrals and the treatment of quantum-mechanical transitions defined in terms of the more usual operator notation of quantum mechanics. In the last part, consisting of Sec. 7-6 and 7-7, the results learned in the preceding sections are applied to two interesting problems of quantum mechanics.

7-1 DEFINITION OF THE TRANSTION ELEMENT

The time development of a quantum-mechanical system can be pictured as follows. At an initial time t_a the state is described by the wave function $\psi(x_a, t_a)$. At a later time t_b the original state will develop into the state $\phi(x_b, t_b)$.

At this later time suppose we ask the question: What is the probability of finding the system in the specific state $\chi(x_b, t_b)$? We know from the general principles developed in Chap. 5 that the probability of finding the system in this specified state is proportional to the square of the amplitude defined by

$$\int_{-\infty}^{\infty} \chi^*(x_b, t_b)\phi(x_b, t_b)\, dx_b$$

We also know from Chap. 3 that the function $\phi(x_b, t_b)$ can be expressed in terms of the original wave function with the help of the kernel $K(x_b, t_b; x_a, t_a)$ describing the propagation of the system between the times t_a and t_b. Thus, in determining the probability of finding the system in a specified state we can start with the original wave function $\psi(x_a, t_a)$ and bridge the time gap with the propagation kernel $K(b, a)$.

The resulting amplitude, whose absolute square gives the probability desired, we shall call the *transition amplitude*, and we shall write it in the following notation:

$$\langle \chi | 1 | \psi \rangle = \int\int \chi^*(x_b, t_b) K(b, a) \psi(x_a, t_a)\, dx_a\, dx_b \tag{7.1}$$

We wish to return to an even more basic description of the transition phenomena, and we reintroduce the action $S[x(t)]$ describing the behavior of the system between the two time limits. Thus we write the transition amplitude as

$$\langle \chi | 1 | \psi \rangle_S = \int\int\int_{x_a}^{x_b} \chi^*(x_b, t_b) e^{iS/\hbar} \psi(x_a, t_a)\, \mathcal{D}x(t)\, dx_a\, dx_b \tag{7.2}$$

Here we have made the notation a bit more explicit by attaching the subscript S to the transition amplitude to indicate the action for which the integral was calculated. The path integral is to be taken over all paths that go from x_a to x_b and the result of this path integral is multiplied by the two wave functions, then integrated over the space variables at the two limits.

Before proceeding further, we shall define the notation more completely to cover a more general situation. We introduce the functional $F[x(t)]$ without (for the present) describing its physical nature. With this functional we define a *transition element* as

$$\langle \chi | F | \psi \rangle_S = \int\int\int_{x_a}^{x_b} \chi^*(x_b, t_b) F[x(t)] e^{iS/\hbar} \psi(x_a, t_a)\, \mathcal{D}x(t)\, dx_a\, dx_b \tag{7.3}$$

Here F is any functional of $x(t)$ which does not involve $x(t)$ at the end points x_a or x_b or beyond the end points. In the special case that $F = 1$, the integral of Eq. (7.3) is a transition amplitude.

It is difficult to understand transition elements at an intuitive level. One approach toward such understanding involves a classical analogy. Picture a small particle moving with brownian motion. At some initial time t_a the particle is at x_a. We wish to determine the probability that the particle arrives at the point x_b at the time t_b. For quantum-mechanical particles, we talk about starting from an initial state and arriving at some final state. Thus, the point x_a for the brownian particle is analogous to the initial wave function $\psi(x_a)$ in Eq. (7.2), and the point x_b to $\chi(x_b)$. Furthermore, the solution of the quantum-mechanical problem requires integration over the variables x_a and x_b of the initial and final states — a step unnecessary in our classical problem.

We would solve the classical problem by considering all possible paths for the particle's motion. We would weigh each path with the function defining the probability that the particle actually follows such a path and then integrate the weighed contributions for all such paths. The weighing function is analogous to the term $e^{iS/\hbar}$ appearing in the integral of Eq. (7.2).

The final position in such a problem would not be a single point, but rather a small interval, x_b to $x_b + dx_b$. The result, when properly normalized, would be the distribution function $P(x_b)$ giving the relative probability of arriving in the (differential) vicinity of x_b. This function is analogous to the transition amplitude of Eq. (7.2) in the case that $\psi(x_a)$ and $\chi(x_b)$ are Dirac delta functions of position.

Now suppose we wish to know more about the motion than simply the relative probability to arrive at x_b. For example, we may wish to find the acceleration experienced by the particle at some particular instant, say 2 seconds after it starts. But now we need the weighed average of the acceleration, i.e., the acceleration for each possible path with each path weighted by the function defining the probability of the path. Such a weighted average is analogous to the transition element of Eq. (7.3). The property of interest, such as the acceleration at some time t_c, replaces the functional $F[x(t)]$ in the integral of Eq. (7.3). The classical problem could be solved by a path integral very similar in form to Eq. (7.3).

In the remainder of this chapter we shall make use of this analogy, and we shall occasionally refer to transition elements as "weighted averages." However, it must be kept in mind that the weighting function in quantum mechanics is a complex function. Thus the result is not an "average" in the ordinary sense.

The path integral method of solving brownian-motion problems as described in this classical analogy is actually a very powerful method. It will be developed in detail in Sec. 12-6. For now, we attempt to further clarify the notion of a transition element with the help of the perturbation theory developed in Chap. 6.

Perturbations. Suppose the action describing the development of the system can be separated into two parts, so that $S = S_0 + \sigma$. We suppose that the first part S_0 leads to simple path integrals, whereas the remaining part σ is small enough that we can apply a perturbation scheme. We write the exponential function of Eq. (7.2) as

$$e^{iS/\hbar} = e^{iS_0/\hbar} e^{i\sigma/\hbar} \tag{7.4}$$

Using Eq. (7.3), the transition element of Eq. (7.2) becomes

$$\langle \chi | 1 | \psi \rangle_{S_0+\sigma} = \langle \chi | e^{i\sigma/\hbar} | \psi \rangle_{S_0} \tag{7.5}$$

The exponential function can be expanded to give

$$\langle\chi|1|\psi\rangle_{S_0+\sigma} = \langle\chi|1|\psi\rangle_{S_0} + \frac{i}{\hbar}\langle\chi|\sigma|\psi\rangle_{S_0} - \frac{1}{2\hbar^2}\langle\chi|\sigma^2|\psi\rangle_{S_0} + \cdots \tag{7.6}$$

This expansion is a generalized version of Eq. (6.3) and forms the basis of the perturbation theory. The transition elements which arise in most quantum-mechanical problems result from this expansion.

Suppose the perturbation action σ results from a perturbation potential, so that

$$\sigma = -\int_{t_a}^{t_b} V(x(t),t)\,dt \tag{7.7}$$

Then the first-order perturbation is given by the transition element

$$\langle\chi|\sigma|\psi\rangle_{S_0} = -\int_{t_a}^{t_b} \langle\chi|V[x(t),t]|\psi\rangle_{S_0}\,dt \tag{7.8}$$

To evaluate this element, we need to solve the integral

$$\langle\chi|V[x(t),t]|\psi\rangle_{S_0} = \int\!\!\int\!\!\int_{x_a}^{x_b} \chi^*(x_b)V[x(t),t]e^{iS_0/\hbar}\psi(x_a)\,\mathcal{D}x(t)\,dx_a\,dx_b \tag{7.9}$$

The first step in the solution of this integral is the same as the solution for the perturbation kernel $K^{(1)}$ described in Eqs. (6.8) to (6.11). This solution for the path integral is followed by integration over both end points, x_a and x_b, as well as an integral over the midpoint x_c. That is,

$$\langle\chi|V[x(t),t]|\psi\rangle_{S_0} = \int\!\!\int\!\!\int \chi^*(x_b)K_0(b,c)V(c)K_0(c,a)\psi(x_a)\,dx_c\,dx_a\,dx_b \tag{7.10}$$

We have now arrived at an expression which combines three concepts previously introduced. First, we have made use of the propagation rule for a wave function as defined in Eq. (3.42). Next, we have made use of the amplitude function as defined in Eq. (5.31), which gives the amplitude that a system known to be in one state will be found in another state. Lastly, we have made use of the first-order perturbation theory given in Eq. (6.11) for the kernel describing the propagation in time. All of these ideas combined give the transition element of Eq. (7.10). The absolute square of this element is the probability that a system starting in state ψ and acted upon by the small potential $V(x,t)$ will be found at a later time in state χ (if state χ would not be reached for $V=0$, that is, if $\langle\chi|1|\psi\rangle_{S_0}=0$).

We can use Eq. (3.42) to shorten our notation, in the same way that the notation of Eq. (6.23) was shortened into the form of Eq. (6.25). We define

$$\psi(x_c, t_c) = \int_{-\infty}^{\infty} K_0(c, a)\psi(x_a)\, dx_a \tag{7.11}$$

which is the wave function that would result at time t_c from the initial wave function if there were no perturbation. In a similar way we define

$$\chi^*(x_c, t_c) = \int_{-\infty}^{\infty} \chi^*(x_b) K_0(b, c)\, dx_b \tag{7.12}$$

as the complex conjugate of the wave function which, at $t = t_c$, would result in the function $\chi(x_b)$ at time t_b if there were no perturbation. (See Eq. (4.38) and the following discussion, including Prob. 4-7.)

In terms of these new wave functions, the first-order term in the perturbation expansion can be simplified to read

$$\left\langle \chi \left| \int_{t_a}^{t_b} V[x(t), t]\, dt \right| \psi \right\rangle_{S_0} = \int_{t_a}^{t_b} \int_{-\infty}^{\infty} \chi^*(x_c) V(c) \psi(x_c)\, dx_c\, dt_c \tag{7.13}$$

We see here that the transition amplitude written in this form is a generalization of the transition amplitude λ_{mn} which was introduced in Sec. 6-5. If the wave functions on the right-hand side of Eq. (7.13) are eigenfunctions, then the resulting transition amplitude is identical with $\lambda_{mn}^{(1)}$, as defined by Eq. (6.70).

Thus the evaluation of a transition element of a functional $F[x(t)]$, which depends only on x at a particular time t (that is, an ordinary function of $x(t)$), or of a time integral of such a functional presents no problem. The evaluation of a transition element for functionals involving the values of x at two separate times is also easy. This occurs, for example, in the second-order perturbation term. This can be written as

$$\frac{1}{2\hbar^2}\langle \chi | \sigma^2 | \psi \rangle_{S_0} = \frac{1}{2\hbar^2} \int_{t_a}^{t_b} \int_{t_a}^{t_b} \langle \chi | V[x(t), t] V[x(s), s] | \psi \rangle\, dt\, ds \tag{7.14}$$

The integrand of this last equation is itself a transition element, and it is written as

$$\langle \chi | V[x(t), t] V[x(s), s] | \psi \rangle = \iint \chi^*(c) V(c) K_0(c, d) V(d) \psi(d)\, dx_d\, dx_c \tag{7.15}$$

where we have substituted $t_d = s$ and $t_c = t$ if $s < t$ or $t_d = t$ and $t_c = s$ if $t < s$.

Thus the second-order term in the perturbation expansion becomes

$$\frac{1}{2\hbar^2}\left\langle \chi \left| \int_{t_a}^{t_b} V[x(t),t]\,dt \int_{t_a}^{t_b} V[x(s),s]\,ds \right| \psi \right\rangle_{S_0} = \frac{1}{2\hbar^2}\iint \chi^*(c)V(c)K_0(c,d)V(d)\psi(d)\,dx_d\,dt_d\,dx_c\,dt_c \tag{7.16}$$

This can be recognized as a generalization of the transition amplitude defined in Eq. (6.74). Expressions involving three or more functionals are also readily written down.

Equation (7.6) corresponds also to a more general type of perturbation theory. For example, consider the case of the particle interacting with an oscillator. After integrals have been carried out over the coordinates describing the oscillator, the resulting action can be written as $S_0 + \sigma$, where (see Sec. 3-10)

$$\sigma = \frac{-1}{M\omega\sin\omega T}\int_{t_a}^{t_b}\int_{t_a}^{t} g(x(t),t)g(x(s),s)\sin\omega(t_b-t)\sin\omega(s-t_a)\,ds\,dt \tag{7.17}$$

with $g(x(t),t)$ characterizing the interaction of the particle and oscillator, and $T = t_b - t_a$.

We have noted that path integrals involving such complicated actions are very hard to evaluate indeed; but if the effect of the complicated term σ is expected to be small, we can obtain useful results with less effort with the help of the perturbation expansion of Eq. (7.6). To illustrate, we find the first-order term in such an expansion (i.e., the first Born approximation). Using Eq. (7.17) for σ, we must evaluate the term $(i/\hbar)\langle\chi|\sigma|\psi\rangle_{S_0}$. This term can be written as

$$\frac{i}{\hbar}\langle\chi|\sigma|\psi\rangle_{S_0} = \frac{-i}{\hbar M\omega\sin\omega T} \times \int_{t_a}^{t_b}\int_{t_a}^{t} \langle\chi|g[x(t),t]g[x(s),s]|\psi\rangle_{S_0}\sin\omega(t_b-t)\sin\omega(s-t_a)\,ds\,dt \tag{7.18}$$

so that the difficult part of the problem is reduced to finding

$$\langle\chi|g[x(t),t]g[x(s),s]|\psi\rangle_{S_0}$$

But this we have already done in Eq. (7.15), except that g replaces V. Therefore, we write

$$\langle\chi|g[x(t),t]g[x(s),s]|\psi\rangle = \iint \chi^*(c)g(x(t_c),t_c)K_0(c,d)g(x(t_d),t_d)\psi(d)\,dx_d\,dx_c \tag{7.19}$$

This expression can be substituted into Eq. (7.18) to obtain the final result for the first Born approximation, $(i/\hbar)\langle\chi|\sigma|\psi\rangle_{S_0}$

Transition elements will come up more frequently in succeeding chapters. In each example they can be evaluated in the straightforward manner which we have illustrated here. For that reason, very little of the material in the remainder of this chapter is really essential to the work that follows. Nevertheless, there are two reasons for the inclusion of this material in this book. First, it is possible to obtain a very general relation between transition elements. This relation might well serve as an alternative starting point for the foundations of quantum mechanics. Second, for many people already familiar with the more conventional operator notation of quantum mechanics, it is helpful to have examples of the translation from the more customary representation into that which is used in this book, such as expressions in the form of Eq. (7.3).

With the rules for translation available, the subject matter of the later chapters, developed as it is from the path integral approach, can be appreciated in terms of more familiar symbolic concepts.

The relations discussed in the remainder of this chapter are independent of the form of the wave functions which describe either the initial or final state of the system, and which are used in defining the transition element. For this reason we shall abbreviate our notation by omitting any specific reference to these wave functions. Thus a transition element will be written as $\langle F\rangle_S$ instead of $\langle\chi|F|\psi\rangle_S$.

7-2 FUNCTIONAL DERIVATIVES

We are embarking on a mathematical development which leads to an interesting relation between transition elements. This relation finds its most elegant expression in terms of a mathematical idea, the functional derivative. Since this idea may not be familiar, we describe it in this section.

The functional $F[x(t)]$ gives a number for each function $x(t)$ that we may choose. We may ask: How much does this number change if we make a very small change in the argument function $x(t)$? Thus, for small $\eta(t)$, how much is $F[x(t)+\eta(t)]-F[x(t)]$? The effect to first order in η (assuming it exists, etc.) is some linear expression in η, say, $\int K(s)\eta(s)\,ds$. Then $K(s)$ is called the functional derivative of $F[x(t)]$ with respect to variation of the function $x(t)$ at s. It is written $\delta F/\delta x(s)$. That is, to first order,

$$F[x+\eta]=F[x]+\int\frac{\delta F}{\delta x(s)}\eta(s)\,ds+\cdots \tag{7.20}$$

This $\delta F/\delta x(s)$ depends on the function $x(t)$, of course, and also on the value of s. Thus it is a functional of $x(t)$ and a function of time s.

We may look at it another way. Suppose time is divided into very many steps of small interval ϵ, the values of the time being t_i ($t_{i+1} = t_i + \epsilon$). The function $x(t)$ can now be specified approximately by giving the value x_i that it takes on at each of the times t_i. The functional $F[x(t)]$ now is a number depending on all the x_i; that is, it becomes an ordinary function of the variables x_i,

$$F[x(t)] \to F(\ldots, x_i, x_{i+1}, \ldots) \tag{7.21}$$

Now we can consider $\partial F/\partial x_i$, the derivative of F with respect to just one of these several variables. Our functional derivative is just this partial derivative, taken at the point $t_i = s$, and then divided by ϵ. That is

$$\frac{\delta F}{\delta x(s)} \to \frac{1}{\epsilon}\frac{\partial F}{\partial x_i} \tag{7.22}$$

This we can see as follows. If we alter the path from $x(t)$ to $x(t)+\eta(t)$, we change all the x_i from x_i to $x_i + \eta_i$ (where $\eta_i = \eta(t_i)$), so that the first-order change in our function is

$$F(\ldots, x_i + \eta_i, x_{i+1} + \eta_{i+1}, \ldots) - F(\ldots, x_i, x_{i+1}, \ldots) = \sum_i \frac{\partial F}{\partial x_i}\eta_i \tag{7.23}$$

from the ordinary rules of partial differentiation. If now we call $(1/\epsilon)(\partial F/\partial x_i) = K_i$, the last sum is $\sum_i K_i \eta_i \epsilon$, which in the limit becomes $\int K(s)\eta(s)\,ds$. So if this limit exists as $\epsilon \to 0$, then it is equal to $\delta F/\delta x(s)$.

One can also use the ideas of differentials. Just as we can write

$$df = \sum_i \frac{\partial f}{\partial x_i}\,dx_i$$

so we can write for the first variation of any functional

$$\delta F = \int \frac{\delta F}{\delta x(s)}\delta x(s)\,ds \tag{7.24}$$

where $\delta x(s)$ is the differential change in path at $x(s)$.

Problem 7-1 If $S[x(t)] = \int_{t_a}^{t_b} L(\dot{x}, x, t)\,dt$, show that, for any s inside the range t_a to t_b,

$$\frac{\delta S}{\delta x(s)} = -\frac{d}{ds}\left(\frac{\partial L}{\partial \dot{x}}\right) + \frac{\partial L}{\partial x} \tag{7.25}$$

where the partial derivatives are evaluated at $t = s$.

Problem 7-2 If $F[x(t)] = x(t)$, show that

$$\frac{\delta F}{\delta x(s)} = \delta(t - s) \tag{7.26}$$

Problem 7-3 If

$$F[j(\mathbf{r},t)] = \exp\left\{\tfrac{1}{2}\iiiint j(\mathbf{r}_1,t_1)j(\mathbf{r}_2,t_2)R(\mathbf{r}_2 - \mathbf{r}_1, t_2 - t_1)\, d^3\mathbf{r}_2\, dt_2\, d^3\mathbf{r}_1\, dt_1\right\}$$

where the integrals extend over all space and all time, show that

$$\frac{\delta F}{\delta j(\mathbf{x},s)} = F\iint j(\mathbf{r},t)\tfrac{1}{2}[R(\mathbf{r}-\mathbf{x},t-s) + R(\mathbf{x}-\mathbf{r},s-t)]\, d^3\mathbf{r}\, dt \tag{7.27}$$

Note that the function $j(\mathbf{r},t)$ is a function of the four variables (r_x, r_y, r_z, t). Thus the single coordinate s, as used in Eq. (7.24), for example, must be replaced by the set of coordinates (x, y, z, s) in specifying the point at which the functional derivative is evaluated.

The general relation between functionals which we mentioned at the end of the preceding section may be obtained by trying to develop a formula for the transition element of $\delta F/\delta x(s)$. This we can do most easily in this way. Consider, using an abbreviated notation,

$$\langle F\rangle_S = \int F[x(t)]e^{(i/\hbar)S[x(t)]}\, \mathcal{D}x(t) \tag{7.28}$$

Now in the integral over paths substitute $x(t) + \eta(t)$ for the variable $x(t)$. For fixed $\eta(t)$, $\mathcal{D}[x(t) + \eta(t)] = \mathcal{D}x(t)$, because $d[x_i + \eta_i] = dx_i$. But the integral is unchanged by a substitution of its variable. Hence

$$\begin{aligned}\langle F\rangle_S &= \int F[x(t) + \eta(t)]e^{(i/\hbar)S[x(t)+\eta(t)]}\, \mathcal{D}x(t) \qquad (7.29)\\ &= \int F[x(t)]e^{(i/\hbar)S[x(t)]}\, \mathcal{D}x(t) + \int\left[\int \frac{\delta F}{\delta x(s)}\eta(s)\, ds\right] e^{(i/\hbar)S[x(t)]}\, \mathcal{D}x(t)\\ &\quad + \frac{i}{\hbar}\int F[x(t)]\left[\int \frac{\delta S}{\delta x(s)}\eta(s)\, ds\right] e^{(i/\hbar)S[x(t)]}\, \mathcal{D}x(t) + \cdots\end{aligned}$$

expanding the exponential and displaying only to first order. The zero-order term is exactly $\langle F\rangle_S$ again, so the remaining terms must all vanish. In particular, the first-order term must vanish for any $\eta(s)$, so that we conclude the relation

$$\left\langle \frac{\delta F}{\delta x(s)}\right\rangle_S = -\frac{i}{\hbar}\left\langle F\frac{\delta S}{\delta x(s)}\right\rangle_S \tag{7.30}$$

This general relation has many important consequences.

It would be possible to use Eq. (7.30) as a starting point to define the laws of quantum mechanics. One could work backwards to reproduce, for example, Eq. (7.6). If some generalization of quantum mechanics is desired, one might suppose such a generalization is included in the action S appearing in the term $e^{iS/\hbar}$, or perhaps start with a form like Eq. (7.30) and introduce modifications with the help of the differential notation. Julian Schwinger has been investigating the formulation of quantum mechanics suggested by Eq. (7.30).

We can see how the relation of Eq. (7.30) comes about in another way by imagining our time split into intervals ϵ and functionals replaced by functions of the points x_i corresponding to t_i. Then consider the path integral

$$\int \frac{\partial F}{\partial x_k} e^{(i/\hbar)S[x(t)]}\,\mathcal{D}x(t) \tag{7.31}$$

where t_k is some intermediate time not at either end point. The path integral is simply an integral over all the points x_i. So we integrate by parts to get

$$\int \frac{\partial F}{\partial x_k} e^{(i/\hbar)S[x(t)]}\,\mathcal{D}x(t) = -\frac{i}{\hbar}\int F\frac{\partial S}{\partial x_k} e^{(i/\hbar)S[x(t)]}\,\mathcal{D}x(t) \tag{7.32}$$

dropping the integrated part.

Problem 7-4 Discuss why the integrated part vanishes.

The result is

$$\left\langle \frac{\partial F}{\partial x_k}\right\rangle_S = -\frac{i}{\hbar}\left\langle F\frac{\partial S}{\partial x_k}\right\rangle_S \tag{7.33}$$

which has the same content as Eq. (7.30).

It is better to write these relations as differentials,

$$\langle \delta F\rangle_S = -\frac{i}{\hbar}\langle F\,\delta S\rangle_S \tag{7.34}$$

for then the specific variables on which F and S depend need not be indicated

Problem 7-5 Argue that Eq. (7.34) may be misleading, for Eq. (7.33) applies only to rectangular coordinates. Do this by studying the corresponding relation where spherical coordinates, for example, are used and we wish to find $\langle \partial F/\partial r_k\rangle_S$.

7-3 TRANSITION ELEMENTS OF SOME SPECIAL FUNCTIONALS

The relation of Eq. (7.34) has many interesting implications. In this section we shall investigate some of them. We shall take the special case of a one-dimensional particle moving in a potential $V(x)$.

Suppose the action over the path of the particle is given by

$$S = \int_{t_a}^{t_b} \left[\frac{m}{2}\dot{x}^2(t) - V(x(t))\right] dt \tag{7.35}$$

Upon application of the small variation $\delta x(t)$ to each path there results (to first order)

$$\delta S = -\int_{t_a}^{t_b} [m\ddot{x} + V'(x)]\, \delta x(t)\, dt \tag{7.36}$$

Using Eq. (7.34), we have

$$\langle \delta F \rangle = \frac{i}{\hbar}\left\langle F \int_{t_a}^{t_b} [m\ddot{x} + V'(x)]\, \delta x(t)\, dt \right\rangle \tag{7.37}$$

Alternatively, we could return to the point of view used in developing Eq. (7.33). That is, we imagine time divided into small slices of length ϵ. In this case the action S can be written as

$$S = \sum_{i=1}^{N-1} \left[\frac{m}{2}\frac{(x_{i+1} - x_i)^2}{\epsilon} - V(x_i)\epsilon\right] \tag{7.38}$$

If we select a particular time t_k and, as before, let x_k be the associated position of a path, then

$$\frac{\partial S}{\partial x_k} = -m\left(\frac{x_{k+1} - x_k}{\epsilon} - \frac{x_k - x_{k-1}}{\epsilon}\right) - V'(x_k)\epsilon \tag{7.39}$$

Upon application of Eq. (7.33) there results

$$\left\langle \frac{\partial F}{\partial x_k} \right\rangle = \frac{i}{\hbar}\epsilon\left\langle F\left[m\left(\frac{x_{k+1} - 2x_k + x_{k-1}}{\epsilon^2}\right) + V'(x_k)\right]\right\rangle \tag{7.40}$$

In this last expression the factor involving an ϵ^2 in the denominator is actually the acceleration $\ddot{x}$ evaluated at the time t_k. Thus Eq. (7.40) is just a special example of Eq. (7.37). In particular, it corresponds to Eq. (7.37) if $\delta x(t)$ is zero for all $t \neq t_k$. If $\delta x(t)$ is assigned the value $\epsilon \cdot \delta x_k \cdot \delta(t - t_k)$, then Eq. (7.40) results. Actually Eq. (7.40), since it holds for all k, is completely equivalent to Eq. (7.37) in a more detailed notation.

In Eq. (7.37) suppose we choose the special functional $F = 1$. Then $\delta F = 0$ and we have

$$\frac{i}{\hbar}\left\langle \int [m\ddot{x} + V'(x)]\, \delta x(t)\, dt \right\rangle = 0 \tag{7.41}$$

Since this result holds for any arbitrary choice of $\delta x(t)$, it must be that

$$\langle m\ddot{x}\rangle = -\langle V'(x)\rangle \tag{7.42}$$

at all values of time. This is the quantum-mechanical analogue of Newton's law. Making use of the classical analogue for a transition element, described in Sec. 7-1, this result says that the weighted "average" of the mass times acceleration at any time (where "averaged" means "averaged over all paths with the weight $e^{iS/\hbar}$") is equal to the weighted "average" of the force (negative gradient of the potential) at the same time.

As another example, suppose F is some arbitrary nonzero functional of all position variables *except* x_k. Then the left-hand side of Eq. (7.40) is zero (since $\partial F/\partial x_k = 0$) and there results

$$\left\langle F(x_1, \ldots, x_{k-1}, x_{k+1}, \ldots, x_N)\left[m\frac{x_{k+1} - 2x_k + x_{k-1}}{\epsilon^2} + V'(x_k)\right]\right\rangle = 0 \tag{7.43}$$

This equation says that the transition element of $m\ddot{x} + V'(x)$, averaged over all paths, is zero at t_k even if these paths are weighted with an arbitrary functional, so long as the functional is independent of the position of the path at the time t_k of interest.

Suppose, however, the functional does depend upon the position of the path at the moment of interest. In particular, suppose simply that the functional F is x_k. Applying Eq. (7.40), we have

$$\begin{aligned}\langle 1\rangle &= \frac{i}{\hbar}\epsilon\left\langle mx_k\left(\frac{x_{k+1} - 2x_k + x_{k-1}}{\epsilon^2}\right) + x_k V'(x_k)\right\rangle \\ &= \frac{i}{\hbar}\left\langle mx_k\left(\frac{x_{k+1} - x_k}{\epsilon} - \frac{x_k - x_{k-1}}{\epsilon}\right) + \epsilon x_k V'(x_k)\right\rangle\end{aligned} \tag{7.44}$$

If we suppose that the potential $V(x)$ is a smooth function, then in the limit as $\epsilon \to 0$ we find that $\epsilon x_k V'(x_k)$ becomes negligible in comparison with the remaining terms. The result is

$$\left\langle m\frac{x_{k+1} - x_k}{\epsilon}x_k\right\rangle - \left\langle x_k m\frac{x_k - x_{k-1}}{\epsilon}\right\rangle = \frac{\hbar}{i}\langle 1\rangle \tag{7.45}$$

This last equation involves the product of position variables x and momentum variables $m\dot{x}$. In the first term the momentum is evaluated first as a linear average corresponding to the time $t_k + \epsilon/2$, and the position is taken at t_k. In the second term the position is again taken at t_k, but the momentum corresponds to the time $t_k - \epsilon/2$. Thus this equation says that the transition element of a product of position and momentum depends upon the order in time of these two quantities.

Later on, when we make a translation into the more usual operator notation, we shall see (Sec. 7-5) that both the operator equation of motion, corresponding to Eq. (7.42), and the operator commutation laws of Eq. (7.45) have been derived from the same fundamental relation, Eq. (7.34).

We can derive a further result from Eq. (7.45) which will give us a better idea of the characteristics of the paths which are important in quantum mechanics. Consider the two terms

$$\left\langle x_k m \frac{x_k - x_{k-1}}{\epsilon} \right\rangle \tag{7.46}$$

and

$$\left\langle x_{k+1} m \frac{x_{k+1} - x_k}{\epsilon} \right\rangle \tag{7.47}$$

These two terms differ from each other only in order ϵ, since they are the same quantity calculated at two times differing by the interval ϵ. Thus we are justified in substituting Eq. (7.47) for the second term in Eq. (7.45). The result is

$$\left\langle m \frac{x_{k+1} - x_k}{\epsilon} (x_k - x_{k+1}) \right\rangle = \frac{\hbar}{i} \langle 1 \rangle \tag{7.48}$$

Alternatively, we can write this as

$$\left\langle \left(\frac{x_{k+1} - x_k}{\epsilon} \right)^2 \right\rangle = -\frac{\hbar}{im\epsilon} \langle 1 \rangle \tag{7.49}$$

This equation says that the transition element of the square of the velocity is of the order $1/\epsilon$, and thus becomes infinite as ϵ approaches zero. This result implies that the important paths for a quantum-mechanical particle are not those which have a definite slope (or velocity) everywhere, but are instead quite irregular on a very fine scale, as suggested by the sketch of Fig. 7-1. In fact, these irregularities are such that the "average" square velocity does not exist, where we have used the classical analogue in referring to an "average."

If some average velocity is defined for a short time interval Δt, as, for example, $[x(t + \Delta t) - x(t)]/\Delta t$, the "mean" square value of this is $-\hbar/(im\,\Delta t)$. That is, the "mean" square value of a velocity averaged over a short time interval is finite, but its value becomes larger as the interval becomes shorter.

It appears that quantum-mechanical paths are very irregular. However, these irregularities average out over a reasonable length of time to produce a reasonable drift, or "average" velocity, although for short intervals of time the "average" value of the velocity is very high.

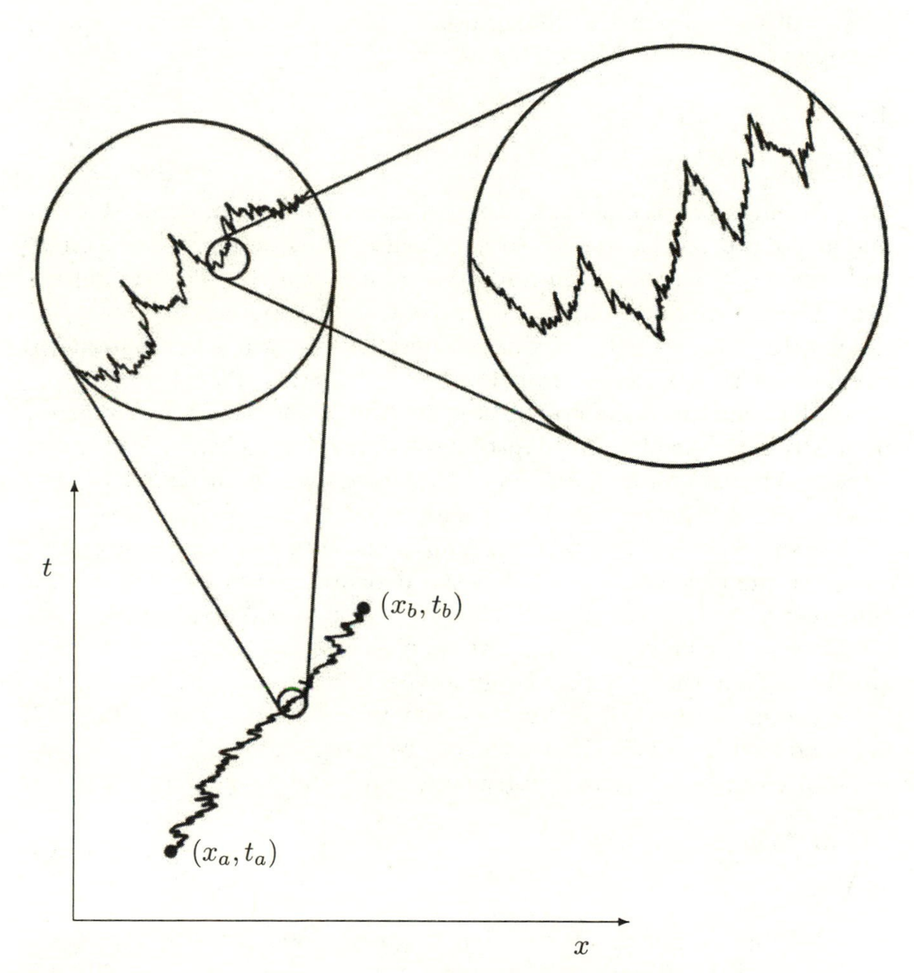

Fig. 7-1 Typical paths of a quantum-mechanical particle are highly irregular on a fine scale, as shown in the sketch. Thus, although a mean velocity can be defined, no mean-square velocity exists at any point. In other words, the paths are nondifferentiable.

Problem 7-6 Show, for a particle moving in three-dimensional space x, y, z,

$$\langle (x_{k+1} - x_k)^2 \rangle = \langle (y_{k+1} - y_k)^2 \rangle = \langle (z_{k+1} - z_k)^2 \rangle = -\frac{\hbar\epsilon}{im}\langle 1 \rangle \tag{7.50}$$

$$\begin{aligned}\langle (x_{k+1} - x_k)(y_{k+1} - y_k) \rangle &= \langle (x_{k+1} - x_k)(z_{k+1} - z_k) \rangle \\ &= \langle (y_{k+1} - y_k)(z_{k+1} - z_k) \rangle = 0\end{aligned} \tag{7.51}$$

It will not do to write the transition element of the kinetic energy simply as

$$\frac{1}{2}\left\langle m\left(\frac{x_{k+1} - x_k}{\epsilon}\right)^2\right\rangle \tag{7.52}$$

for this quantity becomes infinite as ϵ approaches zero. How shall we find an appropriate expression to represent the kinetic energy? We might make the heuristic guess that only those functionals F which might appear in some kind of a physical perturbation problem may be of importance. How can we get the kinetic energy through a perturbation? If the mass of the particle were perturbed by a factor $1 + \eta$ (with η very small) for some short interval of time Δt, the action would be perturbed by $\eta\,\Delta t(m/2)\dot{x}^2$, which is proportional to the kinetic energy. We are led to ask: What would be the form of the first-order perturbation $\langle \sigma \rangle_{S_0}$ if m were changed to $m(1 + \eta)$ for a short time?

For simplicity we can take the short time to be just ϵ, the step used to define the time spacing, so that the first-order term divided by $\epsilon\eta$ is the kinetic energy. The perturbation in S of Eq. (7.38) (if the m in the $i = k$ term is changed to $m + \eta m$) is clearly $\epsilon\eta(m/2)(x_{k+1} - x_k)^2/\epsilon^2$. But this is not the only change in the path integral if m changes. The normalizing factors $1/A$ for each m vary as $m^{1/2}$, so a factor $(1 + \eta/2)$ is introduced from this. Hence the entire first-order change in the path integral when m is so changed becomes, after dividing by $\eta\epsilon$,

$$\frac{i}{\hbar}\left\langle \frac{m}{2}\left(\frac{x_{k+1} - x_k}{\epsilon}\right)^2 + \frac{\hbar}{2i\epsilon}\right\rangle \tag{7.53}$$

which should be satisfactory for $i/\hbar$ times the kinetic energy.

Using Eq. (7.49), one might expect this to vanish; but Eq. (7.49) is valid only as $\epsilon \to 0$ to the order $1/\epsilon$. The quantity in Eq. (7.53) is, in fact, finite as $\epsilon \to 0$. The expression can be rewritten by expanding the quadratic term. In Eq. (7.40) let F be $x_{k+1} - x_k$. If terms of lowest order in ϵ are kept, the result is

$$\left\langle \frac{m}{2}\left(\frac{x_{k+1}-x_k}{\epsilon}\right)\left(\frac{x_k - x_{k-1}}{\epsilon}\right)\right\rangle = \left\langle \frac{m}{2}\left(\frac{x_{k+1}-x_k}{\epsilon}\right)^2\right\rangle + \frac{\hbar}{2i\epsilon}\langle 1\rangle \tag{7.54}$$

Thus we can define the left-hand side of Eq. (7.54) as the transition element of the kinetic energy.

We see from this result that the easiest way to produce satisfactory transition elements involving powers of the velocities is to replace these powers by a product of velocities, each factor of which is taken at a slightly different time.

In simple problems the transition elements can sometimes be evaluated directly. For such problems the same results can also be obtained by using the relations among transition elements which we derived in Sec. 7-2. These relations may supply us with soluble differential equations for the transition elements. We shall give a few illustrations, but it will be readily seen that the examples for which the method works must be so simple that a direct evaluation would not really be much more difficult.

For our first example, consider the case of a free particle going from x_a to x_b in the total time interval T. Let us find the transition element of the position at the time t, that is, $x(t)$. Of course, this is some function of t and it is clear that

$$\langle x(0)\rangle = x_a\langle 1\rangle \qquad \langle x(T)\rangle = x_b\langle 1\rangle \tag{7.55}$$

Since any potentials acting on the particle are constant in space (i.e., no forces act), the second derivative of the transition element of position is zero in accordance with Eq. (7.42). Thus an integration gives

$$\langle x(t)\rangle = \left[x_a + \frac{t}{T}(x_b - x_a)\right]\langle 1\rangle \tag{7.56}$$

Note that the expression in the brackets is just the value of $x(t)$ along the classical path $\bar{x}(t)$.

Problem 7-7 Show that for any quadratic action

$$\langle x(t)\rangle = \bar{x}(t)\langle 1\rangle \tag{7.57}$$

As a somewhat less trivial example, let us try to evaluate the transition element $\langle x(t)x(s)\rangle$ for the same free-particle conditions. Since this is a function of two times, we can write it as $g(t,s)$. The second derivative with respect to t is

$$\frac{\partial^2 g(t,s)}{\partial t^2} = \langle \ddot{x}(t)x(s)\rangle \tag{7.58}$$

This transition element can be worked out by substituting $F = x(s)$ into Eq. (7.40). For $s \neq t$, following the arguments leading to Eq. (7.42), the result is $-(1/m)\langle V'(x(t))x(s)\rangle$; while for $s = t$, following the arguments leading to Eq. (7.44), we find that the transition element of Eq. (7.58) is of order $1/\epsilon$. In the limit of small ϵ we have

$$\frac{\partial^2 g(t,s)}{\partial t^2} = \langle \ddot{x}(t)x(s)\rangle = \frac{\hbar}{im}\delta(t-s)\langle 1\rangle - \frac{1}{m}\langle V'(x(t))x(s)\rangle \tag{7.59}$$

Since for our free particle the potential is independent of position, the second term on the right of Eq. (7.59) vanishes. The resulting equation may be solved by dividing the region of interest into two parts. For $t < s$

$$g(t,s) = a(s)t + b(s) \tag{7.60}$$

while for $t > s$

$$g(t,s) = A(s)t + B(s) \tag{7.61}$$

Thus the first derivative of the function g with respect to t jumps by the quantity $A(s) - a(s)$ as t goes from just below to just above s, and in accordance with Eq. (7.59), $A(s) - a(s) = (\hbar/im)\langle 1\rangle$.

The boundary conditions state that

$$\begin{aligned} \langle x(0)x(s)\rangle &= x_a\langle x(s)\rangle = x_a\bar{x}(s)\langle 1\rangle \\ \langle x(T)x(s)\rangle &= x_b\bar{x}(s)\langle 1\rangle \end{aligned} \tag{7.62}$$

This is not enough information to determine all of the four functions $a(s)$, $b(s)$, $A(s)$, and $B(s)$, but we can either make use of the relation

$$\frac{\partial^2 g}{\partial s^2} = \frac{\hbar}{im}\delta(s-t)\langle 1\rangle \tag{7.63}$$

obtained by differentiating $g(t,s)$ with respect to s, or else notice that $g(t,s)$ must be symmetric in t and s. One can conclude that the functions $a(s)$, $b(s)$, $A(s)$, and $B(s)$ must all be linear in s. The boundary conditions are now sufficient to determine the solution. The result is

$$\langle x(t)x(s)\rangle = \begin{cases} \left[\bar{x}(t)\bar{x}(s) + \dfrac{\hbar}{im}\dfrac{t(T-s)}{T}\right]\langle 1\rangle & \text{for } t < s \\[2ex] \left[\bar{x}(t)\bar{x}(s) + \dfrac{\hbar}{im}\dfrac{s(T-t)}{T}\right]\langle 1\rangle & \text{for } t > s \end{cases} \tag{7.64}$$

That this result is right can be verified by inspection. The product of two classical paths taken at different times, $\bar{x}(t)\bar{x}(s)$, is the solution of the homogeneous equations obtained by setting the right-hand sides of Eqs. (7.59) and (7.63) equal to zero, which satisfies the necessary boundary conditions. The last terms on the right of the pair of equations (7.64) are the special solutions of the inhomogeneous equations (7.59) and (7.63), which are zero at the end points.

The transition element of the product of two positions taken at two different times contains more than just the product of the two corresponding positions along the classical path. There is a small additional term which is purely quantum-mechanical in nature. This additional term is consistent with our picture of quantum-mechanical motion. Even though the particle moving between fixed end points will be found on the average along the classical path, it has an amplitude for motion along all alternative paths. This fact must be remembered when considering the transition element of the product of positions at two different times. All the possible positions among all the various alternatives must be accounted for in the transition element, and this accounting introduces the extra term. Only at the specified end points are no other alternatives possible.

We can better understand the significance of this result if we make use once again of the terminology from our classical analogue. Suppose the path of the particle goes through a particularly large value of x at some time s. Then the "average" value of x at a later time t is not just the ordinary average $\bar{x}(t)$. There is a correlation with the previous large deflection. Therefore, the "average" product is not just the product of "averages."

In this and other applications of the classical analogue, we remember that the "average" referred to is defined with the help of the weighting function $e^{iS/\hbar}$. This weighting function is not positive definite, and is in fact complex. Thus we develop such purely quantum-mechanical results as that of Eqs. (7.64), wherein the extra correlation term is pure imaginary!

Problem 7-8 Find the transition element $\langle x(t)x(s)\rangle = g(t,s)$ when the potential is not constant but, rather, corresponds to that of a forced harmonic oscillator. Do this by obtaining differential equations for $g(t,s)$ and trying the solution

$$\langle x(t)x(s)\rangle = g(t,s) = [\bar{x}(t)\bar{x}(s) + G(t,s)]\langle 1\rangle \tag{7.65}$$

Obtain an equation for $G(t,s)$ showing that G is independent of the end-point values x_a and x_b and of the forcing function $f(t)$. Show in general that, with $T = t_b - t_a$,

$$G(t,s) = \begin{cases} \dfrac{\hbar}{im} \dfrac{\sin \omega t \, \sin \omega(T-s)}{\omega \sin \omega T} & \text{for } t < s \\ \dfrac{\hbar}{im} \dfrac{\sin \omega s \, \sin \omega(T-t)}{\omega \sin \omega T} & \text{for } t > s \end{cases} \tag{7.66}$$

7-4 GENERAL RESULTS FOR QUADRATIC ACTIONS

Evidently if the action S is a quadratic form, transition elements of many functionals can be determined readily. This suggests that we extend our consideration into a somewhat more general class of functionals. The technique to be used is the same as that described in Sec. 3-5. For example, we note that with a quadratic action S we can easily evaluate the transition element of $\exp\{(i/\hbar)\int f(t)x(t)\,dt\}$, where $f(t)$ is any arbitrary function of time. The transition element of such a functional can be written as

$$\left\langle e^{(i/\hbar)\int f(t)x(t)\,dt} \right\rangle = \tag{7.67}$$

$$\iiint_{x_a}^{x_b} \chi^*(x_b) e^{(i/\hbar)[S+\int f(t)x(t)\,dt]} \psi(x_a)\,\mathcal{D}x(t)\,dx_a\,dx_b$$

If the original action S is gaussian, then so is the action

$$S' = S + \int_{t_a}^{t_b} f(t)x(t)\,dt$$

Thus the path integrals on the right of Eq. (7.67) can be carried out by the methods of Sec. 3-5. If S'_{cl} is the extremum of the action S', then the factor $\exp\{iS'_{cl}/\hbar\}$ can be extracted as a factor for the path integral of Eq. (7.67). The remaining factor is a path integral over the paths $y(t)$, which run from zero to zero during the allowed time interval. (We set $x(t) = \bar{x}(t) + y(t)$, where $\bar{x}(t)$ is the classical path corresponding to the extremum of the action.)

The integral over the paths $y(t)$ does not depend upon the function $f(t)$, since this function appears in the action S' multiplying only a linear term in $x(t)$, and we have seen (Eq. 3.49) that the remaining path integral involves only the quadratic parts of S' which are not more than the quadratic parts of S. This means that the path integral on the right-hand side of Eq. (7.67) can be reduced to an exponential function multiplied by the transition element $\langle 1 \rangle$. The result is

$$\left\langle \exp\left\{\frac{i}{\hbar}\int_{t_a}^{t_b} f(t)x(t)\,dt\right\}\right\rangle = \langle 1 \rangle \exp\left\{\frac{i}{\hbar}(S'_{cl} - S_{cl})\right\} \tag{7.68}$$

Once the extremum S'_{cl} has been evaluated, the extremum S_{cl} can be obtained from it by setting $f(t)$ identically equal to zero. The action of the forced harmonic oscillator, described by Eq. (3.66), is a special case of this action S'_{cl}.

Problem 7-9 Use this result to show that if S corresponds to a harmonic oscillator

$$S = \frac{m}{2}\int_{t_a}^{t_b} [\dot{x}^2 - \omega^2 x^2]\,dt$$

then

$$\begin{aligned}\left\langle \exp\left\{\frac{i}{\hbar}\int_{t_a}^{t_b} f(t)x(t)\,dt\right\}\right\rangle &= \langle 1\rangle \exp\left\{\frac{i}{\hbar}\frac{m\omega}{2\sin\omega(t_b-t_a)}\right. \\ &\times\left[\frac{2x_b}{m\omega}\int_{t_a}^{t_b} f(t)\sin\omega(t-t_a)\,dt + \frac{2x_a}{m\omega}\int_{t_a}^{t_b} f(t)\sin\omega(t_b-t)\,dt\right. \\ &\left.\left. -\frac{2}{m^2\omega^2}\int_{t_a}^{t_b}\int_{t_a}^{t} f(t)f(s)\sin\omega(t_b-t)\sin\omega(s-t_a)\,ds\,dt\right]\right\}\end{aligned}$$

where x_a and x_b are the initial and final coordinates of the oscillator.

From the transition element given by Eq. (7.68) we can obtain the transition element of $x(t)$ itself by another method. Suppose we differentiate Eq. (7.68) with respect to $f(t)$. The result is

$$\begin{aligned}\left\langle x(t)\exp\left\{\frac{i}{\hbar}\int_{t_a}^{t_b} f(t)x(t)\,dt\right\}\right\rangle &= \langle 1\rangle \frac{\hbar}{i}\frac{\delta}{\delta f(t)}\exp\left\{\frac{i}{\hbar}(S'_{cl}-S_{cl})\right\} \\ &= \langle 1\rangle \frac{\delta S'_{cl}}{\delta f(t)}\exp\left\{\frac{i}{\hbar}(S'_{cl}-S_{cl})\right\} \qquad (7.69)\end{aligned}$$

Therefore, by evaluating both sides when $f(t) = 0$, we obtain

$$\langle x(t)\rangle = \langle 1\rangle \left.\frac{\delta S'_{cl}}{\delta f(t)}\right|_{f=0} \qquad (7.70)$$

We can continue this process to get the second derivative as

$$\begin{aligned}\langle x(t)x(s)\rangle &= \langle 1\rangle \left(\frac{\hbar}{i}\right)^2 \frac{\delta^2}{\delta f(t)\,\delta f(s)}\exp\left.\left\{\frac{i}{\hbar}(S'_{cl}-S_{cl})\right\}\right|_{f=0} \\ &= \langle 1\rangle \left[\frac{\hbar}{i}\frac{\delta^2 S'_{cl}}{\delta f(t)\,\delta f(s)} + \frac{\delta S'_{cl}}{\delta f(t)}\frac{\delta S'_{cl}}{\delta f(s)}\right]_{f=0} \qquad (7.71)\end{aligned}$$

Actually, since S'_{cl} is quadratic only in f (see Eq. 3.66), the transition element of a factor of any number of x's can be directly evaluated in terms of $\delta S'_{cl}/\delta f(t)$ and the quantity $\delta^2 S'_{cl}/\delta f(t)\,\delta f(s)$, which is independent of f. This explains the form of Eqs. (7.64) and (7.65) and permits the transition element of a factor of three x's to be written down.

Problem 7-10 Show, for any quadratic functional, if we write $\langle x(t)\rangle = \bar{x}(t)\langle 1\rangle$ and $\langle x(t)x(s)\rangle = [\bar{x}(t)\bar{x}(s) + G(t,s)]\langle 1\rangle$, that

$$\langle x(t)x(s)x(u)\rangle = [\bar{x}(t)\bar{x}(s)\bar{x}(u) + \bar{x}(t)G(s,u) + \bar{x}(s)G(t,u) + \bar{x}(u)G(t,s)]\langle 1\rangle$$

Find the transition element of the product of four x's. [*Suggestion:* Since $S'_{cl} - S_{cl}$ is quadratic in f and zero for $f = 0$, it must have the mathematical form $S'_{cl} - S_{cl} = \frac{1}{2}\iint f(t)f(s)G(t,s)\,dt\,ds + \int \bar{x}(t)f(t)\,dt$, where G and $\bar{x}$ are some functions.]

7-5 TRANSITION ELEMENTS AND THE OPERATOR NOTATION

In this and the following sections we shall see how transition elements look in the conventional notation of wave functions and operators. This will help the reader who is familiar with that form of expression to relate the results of path integral calculations to other results that he already knows.

If F is a functional only of x at a single time, say, the function $V(x_k)$ at time t_k, we know from Eq. (7.10) how to evaluate its transition element. Similarly, if F depends on the value of $x(t)$ at two different times, Eq. (7.15) tells us what to do.

Let us consider next the case that F represents the momentum at time t_k and make use of the approximation that the time axis is cut up into slices of length ϵ. Thus

$$F = m\frac{x_{k+1} - x_k}{\epsilon} \tag{7.72}$$

Then we have

$$\left\langle \chi \left| m\frac{x_{k+1} - x_k}{\epsilon} \right| \psi \right\rangle_S = \frac{m}{\epsilon}\left[\langle \chi | x_{k+1} | \psi\rangle_S - \langle \chi | x_k | \psi\rangle_S\right] \tag{7.73}$$

The right-hand side of Eq. (7.73) can be written as

$$\frac{m}{\epsilon}\left[\int \chi^*(x,t+\epsilon)x\psi(x,t+\epsilon)\,dx - \int \chi^*(x,t)x\psi(x,t)\,dx\right] \tag{7.74}$$

Now making use of the wave equation

$$\psi(x,t+\epsilon) = \psi(x,t) + \epsilon\frac{\partial\psi}{\partial t} = \psi - \frac{i\epsilon}{\hbar}H\psi \tag{7.75}$$

$$\chi^*(x,t+\epsilon) = \chi^*(x,t) + \epsilon\frac{\partial\chi^*}{\partial t} = \chi^* + \frac{i\epsilon}{\hbar}[H\chi]^* \tag{7.76}$$

from Prob. 4-3, where H is the hamiltonian belonging to the S. Therefore, to first order in ϵ

$$\int \chi^*(x,t+\epsilon)x\psi(x,t+\epsilon)\,dx = \int \chi^*(x,t)x\psi(x,t)\,dx \tag{7.77}$$
$$-\frac{i\epsilon}{\hbar}\left[\int \chi^*(x,t)x[H\psi(x,t)]\,dx - \int [H\chi(x,t)]^*x\psi(x,t)\,dx\right]$$

By Eq. (4.30) this last integral can be written as $\int \chi^*(x,t)[Hx\psi(x,t)]\,dx$, or more simply we have

$$\langle\chi|m\dot{x}|\psi\rangle = -\frac{im}{\hbar}\int \chi^*(xH - Hx)\psi\,dx \tag{7.78}$$

using the operator notation. This is the same as

$$-\frac{im}{\hbar}\int \chi^*\frac{\hbar^2}{m}\frac{\partial\psi}{\partial x}\,dx = \int \chi^*\frac{\hbar}{i}\frac{\partial\psi}{\partial x}\,dx \tag{7.79}$$

where we have used the result of Prob. 4-4. The operator $(\hbar/i)\partial/\partial x$ is called the momentum operator or, more specifically, the operator representing momentum in the x direction. We already see why. Constructing the transition element of $m\dot{x}$ is equivalent to putting the operator $(\hbar/i)\partial/\partial x$ between χ^* and ψ, just as constructing the transition element of x is equivalent to putting x between χ^* and ψ. These relations can be understood, perhaps with greater clarity, if we go over to the momentum representation. If

$$\chi(p) = \int_{-\infty}^{\infty} \chi(x)e^{-i(p/\hbar)x}\,dx$$
$$\psi(p) = \int_{-\infty}^{\infty} \psi(x)e^{-i(p/\hbar)x}\,dx \tag{7.80}$$

are the momentum representations of χ and ψ, one can show that

$$\int_{-\infty}^{\infty} \chi^*(x)\frac{\hbar}{i}\frac{\partial\psi(x)}{\partial x}\,dx = \int_{-\infty}^{\infty} \chi^*(p)p\psi(p)\,\frac{dp}{2\pi\hbar} \tag{7.81}$$

Problem 7-11 Show this.

Another way to see this relation is the following. Consider the transition amplitude given by

$$\langle\chi|1|\psi\rangle = \int_{-\infty}^{\infty}\int_{-\infty}^{\infty} \chi^*(x_b,t_b)K(x_b,t_b;x_a,t_a)\psi(x_a,t_a)\,dx_a\,dx_b \tag{7.82}$$

Now suppose the x_a origin is shifted left by a small amount Δ. Calling the new variable x'_a, we have

$$x_a = x'_a - \Delta \tag{7.83}$$

Using this new variable rather than the old x_a will not alter the transition amplitude of Eq. (7.82). It becomes

$$\begin{aligned}\langle\chi|1|\psi\rangle = &\int_{-\infty}^{\infty}\int_{-\infty}^{\infty}\int_{x'_a}^{x_b} \chi^*(x_b, t_b) \\ &\times \exp\left\{\frac{i}{\hbar}\sum_{i=2}^{N-1} S[x_{i+1}, t_{i+1}; x_i, t_i] + \frac{i}{\hbar}S[x_2, t_2; x'_a - \Delta, t_a]\right\} \\ &\times \psi(x'_a - \Delta, t_a)\, \mathcal{D}x(t)\, dx'_a\, dx_b\end{aligned} \tag{7.84}$$

where the path integral for the kernel has been written out explicitly, using the methods of Eq. (2.22).

Next, we expand $S[x_2, t_2; x'_a - \Delta, t_a]$ and $\psi(x'_a - \Delta, t_a)$ in Taylor series and keep only the first-order terms. The exponential function becomes

$$\begin{aligned}&\exp\left\{\frac{i}{\hbar}\sum_{i=2}^{N-1} S[x_{i+1}, t_{i+1}; x_i, t_i] + \frac{i}{\hbar}S[x_2, t_2; x'_a, t_a]\right\} \\ &\quad\times\left[1 - \frac{i}{\hbar}\Delta\frac{\partial S[x_2, t_2; x'_a, t_a]}{\partial x'_a}\right]\end{aligned} \tag{7.85}$$

We may drop the prime notation in the integral defining the transition amplitude, since x'_a is a variable of integration. The form of Eq. (7.84) now becomes

$$\begin{aligned}\langle\chi|1|\psi\rangle = &\iint \chi^*(b)K(b,a)\psi(a)\, dx_a\, dx_b - \frac{i}{\hbar}\Delta\iint \chi^*(b)K(b,a) \\ &\times\left[\frac{\partial S[x_2, t_2; x_a, t_a]}{\partial x_a}\psi(x_a, t_a) + \frac{\hbar}{i}\frac{\partial\psi(x_a, t_a)}{\partial x_a}\right] dx_a\, dx_b\end{aligned} \tag{7.86}$$

where we retain the notation that point x_2 is spaced along the path $x(t)$ only by the short time interval ϵ from the point $x_a = x_1$ and $t_2 = t_a + \epsilon$.

The first term on the right of Eq. (7.86) is identical to the transition amplitude on the left. This means that the remaining term must be zero. But this remaining term is a combination of two transition elements. Thus

$$\left\langle\chi\left|-\frac{\partial S[x_2, t_a + \epsilon; x_a, t_a]}{\partial x_a}\right|\psi\right\rangle = \left\langle\chi\left|1\right|\frac{\hbar}{i}\frac{\partial\psi(x_a, t_a)}{\partial x_a}\right\rangle \tag{7.87}$$

In the convention of Eq. (2.22) we use the classical action along each of the short segments of the path. Thus the action $S[b, a]$ appearing in

Eq. (7.87) is the classical action for the initial path element. Its negative derivative with respect to x_a is the classical definition of the momentum at x_a (see Eq. 2.11). So we can write

$$\langle \chi | p_a | \psi \rangle = \left\langle \chi \left| 1 \right| \frac{\hbar}{i} \frac{\partial \psi}{\partial x_a} \right\rangle \tag{7.88}$$

which is the same result as that obtained in Eqs. (7.78) and (7.79).

Sometimes working with a complicated S that results perhaps from the partial elimination of interacting parts, we would like to identify the functional $p(t)$ which corresponds to the momentum at time t. The work of the preceding paragraph suggests a general definition. The first-order change in the transition amplitude $\langle \chi | 1 | \psi \rangle$, if all coordinates corresponding to times previous to t are shifted by $-\Delta$, is this Δ times $\langle \chi | p(t) | \psi \rangle$. From this principle the momentum functional may be found for an arbitrarily complicated S. In a like manner, the hamiltonian or energy functional can be defined by shifting the time variables, as we shall describe in Sec. 7-7.

Problem 7-12 Show, if g is any function of position only, that

$$\left\langle \chi \left| \frac{dg}{dt} \right| \psi \right\rangle = \left\langle \chi \left| \frac{g(x_{k+1}) - g(x_k)}{\epsilon} \right| \psi \right\rangle$$
$$= -\frac{i}{\hbar} \int_{-\infty}^{\infty} \chi^* (gH - Hg) \psi \, dx \tag{7.89}$$

Consider the case that g is a function of the time as well. Show that the transition element of dg/dt is equivalent to the transition element of the operator $-(i/\hbar)(Hg - gH) + \partial g/\partial t$.

Problem 7-13 Show that

$$\langle \chi | m\ddot{x} | \psi \rangle = -\frac{i}{\hbar} \int_{-\infty}^{\infty} \chi^* (pH - Hp) \psi \, dx \tag{7.90}$$

and argue for any quantity A, given in terms of an operator or otherwise, that dA/dt is equivalent to $-(i/\hbar)(AH - HA) + \partial A/\partial t$.

Next we consider an expression F involving two quantities evaluated in rapid succession, such as

$$F = m \frac{x_{k+1} - x_k}{\epsilon} x_k \tag{7.91}$$

This evidently gives

$$\langle \chi | F | \psi \rangle = \frac{m}{\epsilon} \int_{-\infty}^{\infty} \int_{-\infty}^{\infty} \chi^*(x, t+\epsilon) x K(x, t+\epsilon; y, t) y \psi(y, t) \, dy \, dx$$
$$- \frac{m}{\epsilon} \int_{-\infty}^{\infty} \chi^*(x, t) x^2 \psi(x, t) \, dx \tag{7.92}$$

where $t = t_k$. In developing Eq. (4.12) from Eq. (4.2) we saw

$$\int_{-\infty}^{\infty} K(x,t+\epsilon;y,t)f(y)\,dy = f(x) - \frac{i\epsilon}{\hbar}Hf(x) \tag{7.93}$$

so that the first integral in Eq. (7.92) is

$$\frac{m}{\epsilon}\int_{-\infty}^{\infty} \chi^*(x,t+\epsilon)x\left(1-\frac{i\epsilon}{\hbar}H\right)x\psi(x,t)\,dx \tag{7.94}$$

Expressing χ^* by Eq. (7.76) and using the hermitian property of H, we find that this integral is

$$\frac{m}{\epsilon}\int_{-\infty}^{\infty} \chi^*(x,t)\left(1+\frac{i\epsilon}{\hbar}H\right)x\left(1-\frac{i\epsilon}{\hbar}H\right)x\psi(x,t)\,dx \tag{7.95}$$

$$= \frac{m}{\epsilon}\int_{-\infty}^{\infty} \chi^*(x)x^2\psi(x)\,dx + \frac{im}{\hbar}\int_{-\infty}^{\infty} \chi^*(x)(Hx-xH)x\psi(x)\,dx$$

Thus finally

$$\left\langle \chi \left| m\frac{x_{k+1}-x_k}{\epsilon}x_k \right| \psi \right\rangle = \frac{im}{\hbar}\int_{-\infty}^{\infty} \chi^*(x,t)(Hx-xH)x\psi(x,t)\,dx$$

$$= \int_{-\infty}^{\infty} \chi^*(x,t)px\psi(x,t)\,dx \tag{7.96}$$

the last step following from Eq. (7.78).

This is an example of the general rule: In writing the integral definition of the transition element for a set of quantities corresponding to a succession of times, the corresponding operators are written in order from right to left, according to the order in time of the original transition element. If there is a finite time interval Δt between them, a K, or alternatively the operator $e^{+(i/\hbar)L\,\Delta t}$, must be inserted. (For an example, see Prob. 7-16). As the time interval ϵ between two successive quantities approaches zero the K approaches a Dirac delta function and the rule results.

Problem 7-14 Show that the transition amplitude of $(m/\epsilon)(x_{k+1}-x_k)f(x_{k+1})$ is equivalent to that of $(f \cdot p)$.

Problem 7-15 Show that the rule works for two successive momenta, that is,

$$\left\langle \chi \left| m\frac{x_{k+1}-x_k}{\epsilon}m\frac{x_k-x_{k-1}}{\epsilon} \right| \psi \right\rangle = \iint \chi^*(y,t)pp\psi(x,t)\,dx\,dy \tag{7.97}$$

$$= -\hbar^2 \iint \chi^*(y,t)\frac{\partial^2}{\partial x^2}\psi(x,t)\,dx\,dy$$

Problem 7-16 Show that

$$\left\langle \chi \left| x_j m \frac{x_{k+1} - x_k}{\epsilon} \right| \psi \right\rangle = \iint \chi^*(x,t) x K(x,t;y,s) \frac{\hbar}{i} \frac{\partial}{\partial y} \psi(y,s)\, dy\, dx \tag{7.98}$$

if $t_j = t$ and $t_k = s$, provided $t_j > t_k$. What happens if $t_j < t_k$?

Notice that the square of the momentum p^2 corresponds to pp, or two *successive* velocities times mass multiplied together (as in Prob. 7-15). It does *not* correspond to the simple square of velocity at one time, $\langle \chi | m^2 (x_{k+1} - x_k)^2/\epsilon^2 | \psi \rangle$; for that goes to infinity as $m\hbar/i\epsilon$ when $\epsilon \to 0$, as we have seen in Sec. 7-3, particularly in Eq. (7.49). The difference between this expression $m\hbar/i\epsilon$ and the left-hand side of Eq. (7.97) is in fact p^2 in the limit. That is,

$$\left\langle \chi \left| m^2 \frac{(x_{k+1} - x_k)^2}{\epsilon^2} \right| \psi \right\rangle = \frac{m\hbar}{i\epsilon} \langle \chi | 1 | \psi \rangle + \left\langle \chi \left| m \frac{x_{k+1} - x_k}{\epsilon} m \frac{x_k - x_{k-1}}{\epsilon} \right| \psi \right\rangle \tag{7.99}$$

Problem 7-17 Prove this, using Eq. (7.40) with

$$F = m \frac{x_{k+1} - x_k}{\epsilon}$$

7-6 THE PERTURBATION SERIES FOR A VECTOR POTENTIAL

The singular behavior of the transition element of the square of the velocity, as shown in Eq. (7.49), has as a consequence the fact that many expressions involving velocity must be translated with care. For example, the lagrangian for a particle of charge e in an electromagnetic field is

$$L(\dot{\mathbf{x}}, \mathbf{x}, t) = \frac{m}{2} |\dot{\mathbf{x}}|^2 - \mathrm{e}\phi(\mathbf{x}, t) + \frac{\mathrm{e}}{c} \dot{\mathbf{x}} \cdot \mathbf{A}(\mathbf{x}, t) \tag{7.100}$$

Let us take $\phi = 0$ and ask for the effect of the vector potential $\mathbf{A}$ considered as a perturbation. That is, with $S_0 = (m/2) \int |\dot{\mathbf{x}}|^2\, dt$, $\sigma = (\mathrm{e}/c) \int \dot{\mathbf{x}} \cdot \mathbf{A}(\mathbf{x}, t)\, dt$, we develop a series for use in a perturbation treatment and solve for the resulting transition elements. Thus

$$\langle e^{i\sigma/\hbar} \rangle_{S_0} = \langle 1 \rangle_{S_0} + \frac{i}{\hbar} \langle \sigma \rangle_{S_0} - \frac{1}{2\hbar^2} \langle \sigma^2 \rangle_{S_0} + \cdots \tag{7.101}$$

The first-order term is $ie/\hbar c$ times the expression

$$\left\langle \int_{t_a}^{t_b} \dot{\mathbf{x}}\cdot\mathbf{A}(\mathbf{x},t)\,dt \right\rangle \tag{7.102}$$

We wish to translate this to operator notation. In defining σ for a discontinuous path (a series of steps of time length ϵ) we might at first expect to write either

$$\sigma = \frac{e}{c}\sum_k (\mathbf{x}_{k+1}-\mathbf{x}_k)\cdot\mathbf{A}(\mathbf{x}_k,t_k) \tag{7.103}$$

or

$$\sigma = \frac{e}{c}\sum_k (\mathbf{x}_{k+1}-\mathbf{x}_k)\cdot\mathbf{A}(\mathbf{x}_{k+1},t_{k+1}) \tag{7.104}$$

Either one, in the limit of a continuous path, gives the integral for σ. But if we look at a particular component of $\mathbf{A}$, say, A_x, we find that $A_x(\mathbf{x}_{k+1},t_{k+1})$ differs from $A_x(\mathbf{x}_k,t_k)$ by approximately

$$(\mathbf{x}_{k+1}-\mathbf{x}_k)\cdot\boldsymbol{\nabla}A_x + \epsilon\frac{\partial A_x}{\partial t} \tag{7.105}$$

which, when multiplied by $\mathbf{x}_{k+1}-\mathbf{x}_k$ again, might be expected to be of second order in ϵ for each k, thus leading to a term of first order in ϵ when the sum over k is performed. But our paths are not continuous and the transition element of the mean square of $x_{k+1}-x_k$ is of first order. In fact (see Prob. 7-6)

$$\begin{aligned}
(x_{k+1}-x_k)^2 &\approx -\frac{\hbar\epsilon}{im}\\
(x_{k+1}-x_k)(y_{k+1}-y_k) &\approx 0\\
(y_{k+1}-y_k)^2 &\approx -\frac{\hbar\epsilon}{im}
\end{aligned}$$

etc., to first order in ϵ. Hence Eq. (7.103) differs from Eq. (7.104) by approximately

$$\frac{e}{c}\sum_k \frac{\hbar\epsilon}{im}\boldsymbol{\nabla}\cdot\mathbf{A}(\mathbf{x}_k,t_k) = \frac{e}{c}\frac{\hbar}{im}\int \boldsymbol{\nabla}\cdot\mathbf{A}\,dt \tag{7.106}$$

a zero-order term. So it is imperative to decide which form is correct.

The general answer to such a question was given in Chap. 2. There the rule given was that S is replaced by $\sum_k S_{cl}[x_{k+1},t_{k+1};x_k,t_k]$, where S_{cl} is the classical action to go from one point to a neighboring point. It is not necessary to calculate this action exactly, but only sufficiently closely to resolve ambiguities. Equations (7.103) and (7.104) are not sufficiently closely calculated for this purpose, but the classical action for a short interval is very close to

$$S_{cl}[k+1,k] = \frac{m}{2}\frac{|\mathbf{x}_{k+1}-\mathbf{x}_k|^2}{\epsilon} + \frac{\mathrm{e}}{c}(\mathbf{x}_{k+1}-\mathbf{x}_k)\cdot\left(\frac{\mathbf{A}(k+1)+\mathbf{A}(k)}{2}\right) \tag{7.107}$$

Therefore, the correct expression for σ is the average of Eqs. (7.103) and (7.104), so that the transition element of Eq. (7.102) is

$$\left\langle \sum_k (\mathbf{x}_{k+1}-\mathbf{x}_k)\cdot\tfrac{1}{2}[\mathbf{A}(\mathbf{x}_{k+1},t_{k+1})+\mathbf{A}(\mathbf{x}_k,t_k)] \right\rangle \tag{7.108}$$

Leaving the sum over k for later evaluation (as an integral over time) the result is the operator $(1/2m)(\mathbf{p}\cdot\mathbf{A}+\mathbf{A}\cdot\mathbf{p})$ (see Prob. 7-12).

That is, in an electromagnetic potential, the first-order term in the perturbation expansion Eq. (7.101) has the same form as the first-order term given in Eq. (6.11), but with the quantity V replaced by the operator $-(\mathrm{e}/2mc)(\mathbf{p}\cdot\mathbf{A}+\mathbf{A}\cdot\mathbf{p})$.

This conclusion is not true for the second-order term in Eq. (7.101). The second-order term requires our finding

$$-\frac{1}{2}\left(\frac{\mathrm{e}}{\hbar c}\right)^2\left\langle\left[\int \dot{\mathbf{x}}\cdot\mathbf{A}(\mathbf{x},t)\,dt\right]^2\right\rangle = \tag{7.109}$$

$$-\frac{1}{2}\left(\frac{\mathrm{e}}{\hbar c}\right)^2\sum_k\sum_j\left\langle\left[(\mathbf{x}_{k+1}-\mathbf{x}_k)\cdot\left(\frac{\mathbf{A}(k+1)+\mathbf{A}(k)}{2}\right)\right]\right.$$
$$\left.\times\left[(\mathbf{x}_{j+1}-\mathbf{x}_j)\cdot\left(\frac{\mathbf{A}(j+1)+\mathbf{A}(j)}{2}\right)\right]\right\rangle$$

Nothing special happens for the terms with $j \neq k$, and we obtain in fact precisely the second-order term expected by comparison to Eq. (6.13) with V replaced by the operator $-(\mathrm{e}/2mc)(\mathbf{p}\cdot\mathbf{A}+\mathbf{A}\cdot\mathbf{p})$. But when $j=k$, the coincidence of the two velocities gives a new term. In view of Eq. (7.49) and Prob. 7-6 we get an additional quantity

$$-\frac{1}{2}\left(\frac{\mathrm{e}}{\hbar c}\right)^2\left\langle\frac{i\hbar\epsilon}{m}\sum_k\left(\frac{\mathbf{A}(k+1)+\mathbf{A}(k)}{2}\right)^2\right\rangle \tag{7.110}$$

which is equivalent to $-i(\mathrm{e}^2/2\hbar mc^2)\int[\mathbf{A}(\mathbf{x},t)\cdot\mathbf{A}(\mathbf{x},t)]\,dt$ and has the same effect as the first-order action of a potential $(\mathrm{e}^2/2mc^2)\mathbf{A}\cdot\mathbf{A}$.

Thus the perturbation expansion for the action of a vector potential has the same form as Eq. (6.17). The potential V is replaced by the operator $-(\mathrm{e}/2mc)(\mathbf{p}\cdot\mathbf{A}+\mathbf{A}\cdot\mathbf{p})+(\mathrm{e}^2/2mc^2)\mathbf{A}\cdot\mathbf{A}$. We have shown it to second order in $\mathbf{A}$, but a little consideration shows it is true to any order.

The hamiltonian for a particle in a vector potential $\mathbf{A}$ is

$$H = \frac{1}{2m}\left(\mathbf{p} - \frac{\mathrm{e}}{c}\mathbf{A}\right)\cdot\left(\mathbf{p} - \frac{\mathrm{e}}{c}\mathbf{A}\right) \tag{7.111}$$

It differs from that of a free particle, namely $(1/2m)\mathbf{p}\cdot\mathbf{p}$, by just this operator $-(\mathrm{e}/2mc)(\mathbf{p}\cdot\mathbf{A} + \mathbf{A}\cdot\mathbf{p}) + (\mathrm{e}^2/2mc^2)\mathbf{A}\cdot\mathbf{A}$. This is a much easier way to arrive at the result we have just obtained.

7-7 THE HAMILTONIAN

Using what we have so far derived, it would be very easy to write down the transition amplitude for the hamiltonian. We take the transition amplitude for the square of the momentum, divide it by $2m$, and add the transition amplitude for the potential. In this way the hamiltonian itself at the time t_k could be written as

$$H_k = \frac{m}{2}\left(\frac{x_{k+1} - x_k}{\epsilon}\right)\left(\frac{x_k - x_{k-1}}{\epsilon}\right) + V(x_k) \tag{7.112}$$

while in operator form we have the transition element of the hamiltonian as

$$\langle\chi|H|\psi\rangle = \int_{-\infty}^{\infty} \chi^* \left[\frac{p^2}{2m} + V(x)\right]\psi\,dx = \int_{-\infty}^{\infty} \chi^* H\psi\,dx \tag{7.113}$$

Although this method for defining the transition amplitude for the hamiltonian gives a perfectly correct result, it is somewhat artificial, since it does not exhibit the important relationship between the hamiltonian and time. Therefore, we shall next consider an alternative definition of this transition element based upon an investigation of the changes made in a state when it is displaced in time. This approach will also enable us to define H_k given only the form of S, no matter how complicated.

To carry out this investigation, we break up the time axis into infinitesimal intervals, just as we did in defining path integrals. Now, however, it is important to point out that the subdivision of time into equal intervals is not necessary. Clearly, any subdivision of time into equal intervals is not necessary. Any subdivision into instants t_i will be satisfactory; the process of taking limits is characterized by having the largest spacing $t_{i+1} - t_i$ approach zero.

For simplicity, our system will consist of a single particle moving in one dimension. The action is represented by the sum

$$S = \sum_i S[x_{i+1}, t_{i+1}; x_i, t_i] \tag{7.114}$$

where

$$S[x_{i+1}, t_{i+1}; x_i, t_i] = \int_{t_i}^{t_{i+1}} L(\dot{x}, x, t)\, dt \tag{7.115}$$

The integral in this expression is taken along the classical path between x_i at t_i and x_{i+1} at t_{i+1}. For our one-dimensional example we can write, with sufficient accuracy,

$$S[x_{i+1}, t_{i+1}; x_i, t_i] = \left[\frac{m}{2}\left(\frac{x_{i+1} - x_i}{t_{i+1} - t_i}\right)^2 - V(x_i)\right](t_{i+1} - t_i) \tag{7.116}$$

The normalizing factor associated with an integral over x_i at the time t_i is the same one we have used before, namely

$$A = \left(\frac{2\pi i\hbar(t_{i+1} - t_i)}{m}\right)^{1/2} \tag{7.117}$$

The relation of H to the change in a state with displacement in time can now be studied. Consider a state $\psi(t)$ specified within a space-time region R. Now imagine that at the same time t we consider another state $\psi_\delta(t)$, specified within another region R_δ. Suppose the region R_δ is exactly the same as R except that it is earlier by a time δ, that is, displaced bodily towards the past by a time δ. All the apparatus required to prepare the system for R_δ is identical to that for R but is operated a time δ sooner. If the lagrangian L depends explicitly on time, it too must be displaced; i.e., the state ψ_δ is obtained from the L used for the state ψ except that in writing L_δ we use the time variable $t + \delta$.

Now we ask: How does the state ψ_δ differ from ψ? In any measurement the chance of finding the system in some fixed region R' is different depending on whether the region of origin was R or R_δ. Consider the change in transition amplitude $\langle\chi|1|\psi_\delta\rangle$ produced by the shift in time δ. We can consider this shift as effected by decreasing all values of t_i by δ for $i \leq k$ and leaving all t_i fixed for $i > k$.

If the reader looks ahead at this point, it may appear to him that we are headed for trouble. Clearly, it is our intention eventually to take a limit as all infinitesimal time intervals are decreased to zero. However, with the present setup, at least one time interval $t_{k+1} - t_k$ has a lower bound, so that it cannot be indefinitely decreased. This difficulty could be straightened out by assuming the time shift δ to be itself a function of time. We can imagine that it is turned on smoothly before $t = t_k$ and turned off smoothly after $t = t_k$. Then keeping the time variation of δ fixed, we can let all time intervals proceed smoothly to zero, including $t_{k+1} - t_k$. We would then investigate the first-order effect of the time shift by letting the magnitude of δ approach zero. The result obtained by this more rigorous process is essentially the same as that of the procedure we are using in our present example.

Returning now to our investigation of the effect of the time shift we see that the action $S[x_{i+1}, t_{i+1}; x_i, t_i]$ as defined by Eq. (7.115) will not change so long as both t_{i+1} and t_i change by the same amount. On the other hand, $S[x_{k+1}, t_{k+1}; x_k, t_k]$ changes to $S[x_{k+1}, t_{k+1}; x_k, t_k - \delta]$. Furthermore, the factor A associated with the integration over x_k is also altered and becomes

$$A = \left(\frac{2\pi i\hbar(t_{k+1} - t_k + \delta)}{m} \right)^{1/2} \tag{7.118}$$

We use Eq. (7.2) to define the transition amplitude. Keeping in mind that the path integral depends on both the action S and the normalizing factor A (both of which are altered by our time shift) we can write the change in the transition amplitude to first order in δ as

$$\langle \chi | 1 | \psi \rangle - \langle \chi | 1 | \psi_\delta \rangle = \left\langle \chi \left| \frac{\partial S[x_{k+1}, t_{k+1}; x_k, t_k]}{\partial t_k} + \frac{\hbar}{2i(t_{k+1} - t_k)} \right| \psi \right\rangle \frac{i\delta}{\hbar} \tag{7.119}$$

the second term coming from the change in A. We wish to define the functional corresponding to the hamiltonian in quantum mechanics as

$$H_k = \frac{\partial S[x_{k+1}, t_{k+1}; x_k, t_k]}{\partial t_k} + \frac{\hbar}{2i(t_{k+1} - t_k)} \tag{7.120}$$

The first term on the right-hand side of this last equation is the definition of the classical hamiltonian. The second term is necessary in the quantum-mechanical definition in order to keep H_k finite as the time interval $t_{k+1} - t_k$ goes to zero. This last term is a consequence of the change in the normalizing factor A due to the time shift δ.

Applying this result to the specific one-dimensional example indicated by Eq. (7.116), we can write the operator H_k as

$$\begin{aligned} H_k &= \frac{m}{2} \left(\frac{x_{k+1} - x_k}{t_{k+1} - t_k} \right)^2 + \frac{\hbar}{2i(t_{k+1} - t_k)} + V(x_k) \\ &= \frac{m}{2} \left(\frac{x_{k+1} - x_k}{t_{k+1} - t_k} \right) \left(\frac{x_k - x_{k-1}}{t_k - t_{k-1}} \right) + V(x_k) \end{aligned} \tag{7.121}$$

The second of these equations is based upon the results obtained in Eq. (7.54). By writing the product of velocities as the product of two successive velocities, we can do away with the apparently extraneous term $\hbar/(2i(t_{k+1} - t_k))$.

Using the relation $t_\delta = t - \delta$ for all values of $t < t_k$, we have

$$\psi(t) = \psi(t_\delta) + \delta \frac{\partial \psi}{\partial t} = \psi_\delta + \delta \frac{\partial \psi}{\partial t} \tag{7.122}$$

connection the function ψ defined in the two regions R and R_δ. Thus the cycle of relations connecting operators to the Schrödinger equation and to path integrals can be closed with the result obtained by combining Eqs. (7.119), (7.120), and (7.122):

$$-\delta \left\langle \chi \middle| 1 \middle| \frac{\partial \psi}{\partial t} \right\rangle = \langle \chi | H_k | \psi \rangle \frac{i\delta}{\hbar} \tag{7.123}$$

which leads us back again to the Schrödinger equation

$$\frac{\partial \psi}{\partial t} = -\frac{i}{\hbar} H \psi \tag{7.124}$$

For arbitrarily complicated actions we can find an expression for the hamiltonian (i.e., a functional corresponding to the energy) by asking for the first-order change in the transition amplitude $\langle \chi | 1 | \psi \rangle$ when all times previous to t are shifted by $-\delta$ and writing this change as $\langle \chi | H(t) | \psi \rangle \delta$.

8

Harmonic Oscillators

THE problem of the harmonic oscillator is perhaps the simplest in quantum mechanics. We solved it completely in Prob. 3-8 when we found that the kernel for the motion of a harmonic oscillator is

$$K(x_b, T; x_a, 0) = \left(\frac{m\omega}{2\pi i\hbar \sin \omega T}\right)^{1/2} \times \exp\left\{\frac{im\omega}{2\hbar \sin \omega T}[(x_b^2 + x_a^2)\cos \omega T - 2x_b x_a]\right\} \tag{8.1}$$

If we are to make full use of this, we should look at all sorts of problems which involve harmonic oscillators, either exactly or approximately. It is the purpose of this chapter to describe several such problems, both those involving single oscillators and those involving systems of interacting harmonic oscillators. We could carry this program to extremes and include all kinds of classical vibration problems (plates, rods, etc.), but such systems are so large that it would be a waste of time to analyze the quantum-mechanical corrections. Instead, it would be better to look at systems on the atomic scale. For example, we might analyze the oscillations of the molecule CO. In so doing, we find that the potential energy between the carbon and oxygen atoms is not exactly quadratic. Nevertheless, for the lower-energy states the potential is so close to quadratic that a pure harmonic oscillator treatment is a good approximation for many purposes.

In a much more complicated polyatomic molecule, when the excitation energy is not too high, the travel of the atoms is small compared with their spacing. In this case again the potential energy is very nearly a quadratic function of the coordinates. Thus the system is approximately equivalent to a set of coupled harmonic oscillators. A solid crystal is, from one point of view, a polyatomic molecule of great size. As such it is a vast array of interacting harmonic oscillators.

As another example we can consider the electromagnetic field in a cavity. Classically, there are several patterns of standing waves, or modes, in which the field can vibrate harmonically with a definite frequency. In quantum mechanics, each of these modes constitutes a quantum oscillator.

8-1 THE SIMPLE HARMONIC OSCILLATOR

Solution from the Schrödinger Equation. In this section we shall develop a number of relations describing the simple one-dimensional harmonic oscillator. We shall begin with the language of the Schrödinger equation. Problem 2-2 gave the lagrangian describing a one-dimensional harmonic oscillator as

$$L = \frac{m}{2}(\dot{x}^2 - \omega^2 x^2) \tag{8.2}$$

The corresponding hamiltonian, which we use in the present treatment, is

$$H = \frac{p^2}{2m} + \frac{m}{2}\omega^2 x^2 \tag{8.3}$$

The wave equation is then

$$\frac{\partial \psi}{\partial t} = -\frac{i}{\hbar} H\psi = -\frac{i}{\hbar}\left(\frac{p^2}{2m} + \frac{m}{2}\omega^2 x^2\right)\psi \tag{8.4}$$

Since the hamiltonian is independent of time, the wave equation is easily separated, and it yields wave functions of steady states of definite energy E_n. The time-dependent part is proportional to $e^{-(i/\hbar)E_n t}$.

Recalling that the momentum operator p corresponds to differentiation with respect to x (see Sec. 7-5), we can write the Schrödinger equation for the spatial part of the wave function as

$$H\phi_n(x) = -\frac{\hbar^2}{2m}\frac{d^2}{dx^2}\phi_n(x) + \frac{m\omega^2}{2}x^2\phi_n(x) = E_n\phi_n(x) \tag{8.5}$$

This equation is easily solved. The result is given in many books on quantum mechanics.[1] The eigenvalues for the energy are

$$E_n = \hbar\omega(n + \tfrac{1}{2}) \tag{8.6}$$

where n is an integer: 0, 1, 2, ... The eigenfunctions $\phi_n(x)$ are

$$\phi_n(x) = \frac{1}{(2^n n!)^{1/2}}\left(\frac{m\omega}{\pi\hbar}\right)^{1/4} H_n\left(x\sqrt{\frac{m\omega}{\hbar}}\right) e^{-(m\omega/2\hbar)x^2} \tag{8.7}$$

where the functions H_n are the Hermite polynomials

$$\begin{aligned}
H_0(y) &= 1\\
H_1(y) &= 2y\\
H_2(y) &= 4y^2 - 2\\
&\vdots\\
H_n(y) &= (-1)^n e^{y^2}\frac{d^n}{dy^n}e^{-y^2}
\end{aligned} \tag{8.8}$$

[1] L.I. Schiff, "Quantum Mechanics," 2nd ed., McGraw-Hill Book Company, New York, 1955.

The Hermite polynomials are best defined by their generating function

$$e^{-t^2+2ty} = \sum_{n=0}^{\infty} H_n(y)\frac{t^n}{n!} \tag{8.9}$$

We can obtain these results in another manner. The functions $\phi_n(x)$ have been obtained by solving a differential equation, namely, the time-independent case. However, we already have a solution for the time-dependent case. From this solution we should be able to derive these functions directly. It is instructive to carry out this derivation to illustrate some of the formulas which have been derived in earlier chapters.

Solution from the Kernel. We have worked out the kernel describing the motion of an oscillator in Prob. 3-8. And we know from Eq. (4.59) that this kernel can be expanded in terms of energy eigenfunctions. That is

$$\begin{aligned}\left(\frac{m\omega}{2\pi i\hbar\sin\omega T}\right)^{1/2} &\exp\left\{\frac{im\omega}{2\hbar\sin\omega T}[(x_b^2+x_a^2)\cos\omega T - 2x_bx_a]\right\} \\ &= \sum_{n=0}^{\infty} e^{-(i/\hbar)E_nT}\phi_n(x_b)\phi_n^*(x_a)\end{aligned} \tag{8.10}$$

Using the relations

$$\begin{aligned} i\sin\omega T &= \tfrac{1}{2}e^{i\omega T}(1-e^{-2i\omega T}) \\ \cos\omega T &= \tfrac{1}{2}e^{i\omega T}(1+e^{-2i\omega T}) \end{aligned} \tag{8.11}$$

we can write the left-hand side of Eq. (8.10) as

$$\begin{aligned}&\left(\frac{m\omega}{\pi\hbar}\right)^{1/2} e^{-i\omega T/2}(1-e^{-2i\omega T})^{-1/2} \\ &\quad\times \exp\left\{-\frac{m\omega}{2\hbar}\left[(x_b^2+x_a^2)\left(\frac{1+e^{-2i\omega T}}{1-e^{-2i\omega T}}\right) - \frac{4x_bx_ae^{-i\omega T}}{1-e^{-2i\omega T}}\right]\right\}\end{aligned} \tag{8.12}$$

We can obtain a series with the form of the right-hand side of Eq. (8.10) by expanding Eq. (8.12) in powers of $e^{-i\omega T}$. Because of the initial factor $e^{-i\omega T/2}$, the terms in the expansion will be of the form $e^{-i\omega T/2}e^{-in\omega T}$ for $n = 0, 1, 2, \ldots$. This means the energy levels are given by

$$E_n = \hbar\omega(n + \tfrac{1}{2}) \tag{8.13}$$

To find the wave functions, we must carry out the expansion completely. We shall illustrate the method by going only as far as $n = 2$.

Expanding the left-hand side of Eq. (8.10) to this order we have

$$\left(\frac{m\omega}{\pi\hbar}\right)^{1/2} e^{-i\omega T/2}(1+\tfrac{1}{2}e^{-2i\omega T}+\cdots) \times \exp\left\{-\frac{m\omega}{2\hbar}\left[(x_b^2+x_a^2)(1+2e^{-2i\omega T}+\cdots)-4x_bx_a(e^{-i\omega T}+\cdots)\right]\right\} \tag{8.14}$$

or

$$\left(\frac{m\omega}{\pi\hbar}\right)^{1/2} e^{-i\omega T/2}(1+\tfrac{1}{2}e^{-2i\omega T}+\cdots)e^{-(m\omega/2\hbar)(x_b^2+x_a^2)} \times\left[1+\frac{2m\omega}{\hbar}x_bx_ae^{-i\omega T}+\frac{1}{2}\left(\frac{2m\omega}{\hbar}\right)^2 x_b^2x_a^2e^{-2i\omega T} -\frac{m\omega}{\hbar}(x_b^2+x_a^2)e^{-2i\omega T}+\cdots\right] \tag{8.15}$$

From this we can pick out the coefficient of the lowest term. It is

$$\left(\frac{m\omega}{\pi\hbar}\right)^{1/2} e^{-i\omega T/2}e^{-(m\omega/2\hbar)(x_b^2+x_a^2)} = e^{-(i/\hbar)E_0T}\phi_0(x_b)\phi_0^*(x_a) \tag{8.16}$$

This means that $E_0=\frac{1}{2}\hbar\omega$ and

$$\phi_0(x)=\left(\frac{m\omega}{\pi\hbar}\right)^{1/4} e^{-m\omega x^2/2\hbar} \tag{8.17}$$

We have chosen $\phi_0(x)$ to be real. We could make it complex by including a factor $e^{i\delta}$, where δ is a real constant; however, it would make no difference to any physical result.

The next-order term in the expansion is

$$\left(\frac{m\omega}{\pi\hbar}\right)^{1/2} e^{-i\omega T/2}e^{-(m\omega/2\hbar)(x_b^2+x_a^2)}\frac{2m\omega}{\hbar}x_bx_ae^{-i\omega T} = e^{-(i/\hbar)E_1T}\phi_1(x_b)\phi_1^*(x_a) \tag{8.18}$$

which implies that $E_1=\frac{3}{2}\hbar\omega$ and

$$\phi_1(x)=\left(\frac{2m\omega}{\hbar}\right)^{1/2} x\phi_0(x) \tag{8.19}$$

The next term corresponds to $E_2=\frac{5}{2}\hbar\omega$. The part of the term depending on x_b and x_a is

$$\left(\frac{m\omega}{\pi\hbar}\right)^{1/2} e^{-(m\omega/2\hbar)(x_b^2+x_a^2)}\left[\frac{1}{2}+\frac{2m^2\omega^2}{\hbar^2}x_b^2x_a^2-\frac{m\omega}{\hbar}(x_b^2+x_a^2)\right] \tag{8.20}$$

This must be the same as $\phi_2(x_b)\phi_2^*(x_a)$. Since the expression in the brackets can be rewritten as

$$\frac{1}{2}\left(\frac{2m\omega}{\hbar}x_b^2-1\right)\left(\frac{2m\omega}{\hbar}x_a^2-1\right) \tag{8.21}$$

we find

$$\phi_2(x) = \frac{1}{\sqrt{2}}\left(\frac{2m\omega}{\hbar}x^2 - 1\right)\phi_0(x) \tag{8.22}$$

These results are the same as those obtained from the solution of the energy wave equation, Eqs. (8.7) and (8.8).

All of the wave functions may be obtained in this manner. However, it is a difficult algebraic problem to get the general form for $\phi_n(x)$ directly from this expansion. A less direct way is illustrated in the problem.

Problem 8-1 The amplitude to go from any state $\psi(x)$ to another state $\chi(x)$ is the transition amplitude $\langle\chi|1|\psi\rangle$ as defined in Eq. (7.1).

Suppose $\psi(x)$ and $\chi(x)$ are expanded in terms of the orthogonal functions $\phi_n(x)$, the energy solutions to the wave equation associated with the kernel $K(b,a)$, as discussed in Sec. 4-2. Thus

$$\psi(x) = \sum_n \psi_n\phi_n(x) \qquad \chi(x) = \sum_n \chi_n\phi_n(x) \tag{8.23}$$

Using the coefficients ψ_n and χ_n and Eq. (4.59), show that the transition amplitude can be written as

$$\int_{-\infty}^{\infty}\int_{-\infty}^{\infty} \chi^*(x_b)K(x_b,T;x_a,0)\psi(x_a)\,dx_a\,dx_b = \sum_n \chi_n^*\psi_n e^{-(i/\hbar)E_nT} \tag{8.24}$$

Next, suppose we choose a special pair of functions $\psi(x)$ and $\chi(x)$ for which the expansion on the right-hand side of Eq. (8.24) is simple. Then after obtaining the functions ψ_n we could get some information about the wave functions $\phi_n(x)$ from the expansions of Eq. (8.23). Suppose we choose the functions $\psi(x)$ and $\chi(x)$ in the following way

$$\psi(x) = \left(\frac{m\omega}{\pi\hbar}\right)^{1/4} e^{-(m\omega/2\hbar)(x-a)^2} \tag{8.25}$$

$$\chi(x) = \left(\frac{m\omega}{\pi\hbar}\right)^{1/4} e^{-(m\omega/2\hbar)(x-b)^2} \tag{8.26}$$

These functions represent gaussian distributions centered about a and b respectively. We shall call $\psi_n = \psi_n(a)$ and $\chi_n = \psi_n(b)$. Determine the transition amplitude $\langle\chi|1|\psi\rangle$, where $\psi(x)$ and $\chi(x)$ are given by Eqs. (8.25) and (8.26), and the kernel is that for a harmonic oscillator, Eq. (8.1). Perform the integrals in Eq. (8.24) to get

$$\exp\left\{-\frac{i\omega T}{2} - \frac{m\omega}{4\hbar}(a^2+b^2-2abe^{-i\omega T})\right\} = \sum_n \psi_n^*(b)\psi_n(a)e^{-(i/\hbar)E_nT} \tag{8.27}$$

From this result show that $E_n = \hbar\omega(n + \frac{1}{2})$ and that

$$\psi_n(a) = \left(\frac{m\omega}{2\hbar}\right)^{n/2} \frac{a^n}{\sqrt{n!}} \exp\left\{-\frac{m\omega a^2}{4\hbar}\right\} \tag{8.28}$$

Use this result in Eq. (8.23) and write for $\phi_n(x)$ the form given by Eq. (8.7) considering the $H_n(y)$ still unknown. From this derive the generating function of Eq. (8.9) for these functions $H_n(y)$.

8-2 THE POLYATOMIC MOLECULE

In the preceding section we derived the wave functions and energy levels which describe the simple harmonic oscillator. In this section we begin our investigation of systems of interacting oscillators with the study of polyatomic molecules. We begin the analysis by assigning coordinates describing the position of each atom in the molecule. The position of any particular atom a will be given by the three cartesian coordinates x_a, y_a, and z_a, whose origin lies at the equilibrium position for the atom. If the mass of the atom is m_a, the kinetic energy of the whole molecule is given by

$$\sum_a \frac{1}{2} m_a(\dot{x}_a^2 + \dot{y}_a^2 + \dot{z}_a^2) \tag{8.29}$$

where the summation is carried out over all atoms in the molecule.

It will be more convenient for this general discussion to avoid the vector aspects of this description by making the following modification. We suppose there are N atoms in the molecule. We shall define $n = 3N$ coordinates in the following way:

$$\begin{array}{lll} q_1 = \sqrt{m_a}\, x_a & q_2 = \sqrt{m_a}\, y_a & q_3 = \sqrt{m_a}\, z_a \\ q_4 = \sqrt{m_b}\, x_b & q_5 = \sqrt{m_b}\, y_b & \cdots \end{array} \tag{8.30}$$

In terms of these new coordinates the kinetic energy is

$$\text{K.E.} = \frac{1}{2}\sum_{j=1}^{n} \dot{q}_j^2 \tag{8.31}$$

The potential energy is the function $V(q_1, q_2, \ldots, q_n)$ of all the displacements q_j. We can expand V in a Taylor series around the equilibrium position $q_j = 0$. Thus

$$V(q_1, q_2, \ldots, q_n) = V(0, 0, \ldots, 0) + \sum_{j=1}^{n} q_j V_j(0, 0, \ldots, 0) + \frac{1}{2} \sum_{j=1}^{n} \sum_{k=1}^{n} q_j q_k V_{jk}(0, 0, \ldots, 0) + \cdots \tag{8.32}$$

where

$$V_j = \frac{\partial V}{\partial q_j} \qquad V_{jk} = \frac{\partial^2 V}{\partial q_j \, \partial q_k} \tag{8.33}$$

The first term is the potential energy at equilibrium. It is a constant independent of the q_j. We shall assign it the value zero by shifting the zero-level of potential energy. The second term contains the factor $V_j(0, 0, \ldots, 0)$, which is the potential gradient or force associated with the coordinate q_j and evaluated at equilibrium position. This factor is therefore zero. To put this another way, since equilibrium corresponds to a minimum of potential energy, the first-order change for displacements about equilibrium must vanish.

The factors $V_{jk}(0, 0, \ldots, 0)$ appearing in the third term comprise a set of constants whose values depend on the structure of the molecule. Call these constants v_{jk}. Now suppose we neglect all higher-order terms. In this approximation the potential energy involves each coordinate quadratically. Even if the potential is not a pure quadratic function of the coordinates, our approximation will be valid for small displacements. It is by this approximation that we represent our molecule as a system of harmonic oscillators.

Combining Eqs. (8.31) and (8.32), we can write the lagrangian as

$$L = \frac{1}{2} \sum_{j=1}^{n} \dot{q}_j^2 - \frac{1}{2} \sum_{j=1}^{n} \sum_{k=1}^{n} v_{jk} q_j q_k \tag{8.34}$$

Next, we introduce this lagrangian into the path integral which defines the kernel° describing the motion of the atoms in the molecule,

$$K = \int \cdots \iint \exp \left\{ \frac{i}{\hbar} \left[\frac{1}{2} \sum_{j=1}^{n} \int \dot{q}_j^2(t) \, dt - \frac{1}{2} \sum_{j=1}^{n} \sum_{k=1}^{n} v_{jk} \int q_j(t) q_k(t) \, dt \right] \right\} \times \mathcal{D}q_1(t) \, \mathcal{D}q_2(t) \cdots \mathcal{D}q_n(t) \tag{8.35}$$

All of these path integrals are gaussian, and thus they can be solved by the methods discussed in Sec. 3-5. To carry out that solution, we shall have to find those paths $\overline{q}_j(t)$ which give a stationary value for the action integral. Variation with respect to each $q_j(t)$ gives these paths as solutions of

$$\ddot{\overline{q}}_j(t) = -\sum_{k=1}^{n} v_{jk}\overline{q}_k(t) \tag{8.36}$$

This last equation says that the force on any single atom in a particular direction is some linear combination of the displacements of all the atoms.

Such systems of interacting oscillators have been analyzed to a great extent from a classical point of view. Since in many problems of quantum mechanics we obtain the classical action as the first step in solving the kernel, all of this classical work is of great value to us. One important result of the classical analysis is the following. There are special ways to distort the molecule so that, as time goes on, the motion is of the simple periodic sinusoidal type. The pattern of distortions remains the same, and only the amount of the distortion varies sinusoidally with time. Different patterns of distortion, or, as we say, different modes, correspond in general to different frequencies. There may be some with zero frequency, and some groups of modes may all have the same frequency. The important fact is this: Any small displacement motion of the molecule can be built up as a linear combination of such modes. This kind of motion is called a *normal mode.*

If there are N atoms in the molecule, then the molecule has $n = 3N$ modes of motion. Thus, for example, the molecule CO_2 has nine modes, as shown by Fig. 8-1, where the motion of each atom is indicated by an arrow. Only modes 1 to 4 are periodic (i.e., have a non-zero frequency) and the direction of motion during the first half-cycle is indicated. For the second half-cycle, reverse all arrows.

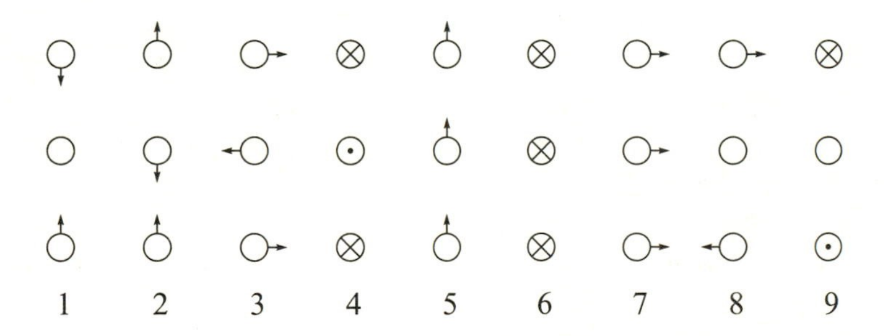

Fig. 8-1 Normal modes of the CO_2 molecule. The symbol ⊙ means motion out of the plane of the paper, and ⊗ means motion into the plane. Modes 1 to 4 are periodic; modes 5 to 7 are continuous translations; and modes 8 and 9 are continuous rotations.

We shall next derive the mathematical description of the modes. This derivation is, of course, part of classical physics rather than quantum mechanics. Consider a particular mode of frequency ω. All of the coordinates $q_j(t)$ move together and at the same frequency. There must be some special set of initial displacements a_j, different for each mode, such that if all initial velocities are zero, the subsequent motion of any coordinate can be written as

$$q_j(t) = a_j \cos \omega t \tag{8.37}$$

Substituting this equation into Eq. (8.36) gives

$$\omega^2 a_j = \sum_{k=1}^{n} v_{jk} a_k \tag{8.38}$$

This last formula is actually as set of n equations for the n unknowns a_j. Since it is homogeneous, it has a solution only if the determinant of coefficients vanishes. Thus we require

$$\begin{bmatrix} v_{11} - \omega^2 & v_{12} & \cdots & v_{1n} \\ v_{21} & v_{22} - \omega^2 & \cdots & v_{2n} \\ \vdots & \vdots & \ddots & \vdots \\ v_{n1} & v_{n2} & \cdots & v_{nn} - \omega^2 \end{bmatrix} = 0 \tag{8.39}$$

This equation has n solutions for ω^2. For a particular solution, say ω_α, we can get solutions for the set of equations (8.38). We shall call these $a_{j\alpha}$. The sizes of the solutions $a_{j\alpha}$ are determined relative to each other, but the overall magnitude of the whole set is arbitrary. We shall choose this magnitude so that

$$\sum_{j=1}^{n} a_{j\alpha}^2 = 1 \tag{8.40}$$

We can repeat this process for all of the n modes, $\alpha = 1, 2, \ldots, n$. We determine n values of ω_α, and for each value of α we obtain the solutions for the n constants $a_{j\alpha}$.

Any possible motion of the system is a linear combination of these modes. We can write an expression for a general type of motion as

$$q_j(t) = \sum_{\alpha=1}^{n} C_\alpha a_{j\alpha} \cos(\omega_\alpha t + \delta_\alpha) \tag{8.41}$$

Here the constant of amplitude C_α and the constant phase δ_α depend on the initial conditions. That such an expression does represent the motion of the system is easily verified by substituting Eq. (8.41) into Eq. (8.36).

It is convenient to use the complex notation for Eq. (8.41). That is,

$$q_j(t) = \Re e\left\{\sum_{\alpha=1}^{n} C_\alpha a_{j\alpha} e^{i\omega_\alpha t} e^{i\delta_\alpha}\right\} = \Re e\left\{\sum_{\alpha=1}^{n} c_\alpha a_{j\alpha} e^{i\omega_\alpha t}\right\} \tag{8.42}$$

The complex constants c_α depend on the initial conditions, and they can be determined as follows. If the initial positions and velocities are $q_j(0)$ and $\dot{q}_j(0)$, respectively, we have

$$q_j(0) = \Re e\left\{\sum_{\alpha=1}^{n} c_\alpha a_{j\alpha}\right\} = \sum_{\alpha=1}^{n} \Re e\{c_\alpha\} a_{j\alpha} \tag{8.43}$$

$$\dot{q}_j(0) = \Re e\left\{\sum_{\alpha=1}^{n} i c_\alpha a_{j\alpha} \omega_\alpha\right\} = -\sum_{\alpha=1}^{n} \Im m\{c_\alpha\} a_{j\alpha}\omega_\alpha$$

Since the constants $a_{j\alpha}$ are all real, this pair of equations determines both the real and imaginary parts of c_α.

We can solve Eqs. (8.43) in a simple way by using an important property to be expressed in Eq. (8.48), which we now derive. For any particular α the constants $a_{j\alpha}$ satisfy

$$\omega_\alpha^2 a_{j\alpha} = \sum_{k=1}^{n} v_{jk} a_{k\alpha} \tag{8.44}$$

If we multiply this equation by $a_{j\beta}$ and sum over all values of j, we find

$$\omega_\alpha^2 \sum_{j=1}^{n} a_{j\alpha} a_{j\beta} = \sum_{k=1}^{n}\sum_{j=1}^{n} v_{jk} a_{k\alpha} a_{j\beta} \tag{8.45}$$

Since the coefficients v_{jk} are symmetrical, the left-hand side of Eq. (8.45) will be the same if α and β are interchanged. This means

$$(\omega_\alpha^2 - \omega_\beta^2)\sum_{j=1}^{n} a_{j\alpha} a_{j\beta} = 0 \tag{8.46}$$

Thus if the frequencies ω_α and ω_β are different, it must be that

$$\sum_{j=1}^{n} a_{j\alpha} a_{j\beta} = 0 \tag{8.47}$$

If the two frequencies are the same, then the constants $a_{j\alpha}$ are not determinate. Instead, we have the freedom to make an arbitrary choice which can be made in such a way that Eq. (8.47) is satisfied for $\alpha \neq \beta$.

Thus finally, making us of the normalization established in Eq. (8.40), we can write

$$\sum_{j=1}^{n} a_{j\alpha}a_{j\beta} = \delta_{\alpha\beta} \tag{8.48}$$

where $\delta_{\alpha\beta}$ is the Kronecker delta.

We can now easily find the real part of c_α from Eqs. (8.43). Multiply the first of Eqs. (8.43) by $a_{j\beta}$ and sum over values of j. All terms on the right-hand side vanish except that for $\alpha = \beta$, which gives

$$\Re e\{c_\beta\} = \sum_{j=1}^{n} a_{j\beta}q_j(0) \tag{8.49}$$

In a similar manner we can find

$$\Im m\{c_\beta\} = -\frac{1}{\omega_\beta}\sum_{j=1}^{n} a_{j\beta}\dot{q}_j(0) \tag{8.50}$$

Thus a complete description of any arbitrary motion of the system can be determined from a knowledge of the normal modes of the system and the initial conditions of the motion.

8-3 NORMAL COORDINATES

We can analyze the motion of the system in another way. Let us choose a new set of coordinates $Q_\alpha(t)$, which are a particular linear combination of the old coordinates, namely

$$Q_\alpha(t) = \sum_{j=1}^{n} a_{j\alpha}q_j(t) \tag{8.51}$$

Alternatively, the old coordinates can be given in terms of the new by

$$q_j(t) = \sum_{\alpha=1}^{n} a_{j\alpha}Q_\alpha(t) \tag{8.52}$$

Using Eq. (8.48), we can write the kinetic energy as

$$\text{K.E.} = \frac{1}{2}\sum_{j=1}^{n}\dot{q}_j^2 = \frac{1}{2}\sum_{j=1}^{n}\sum_{\alpha=1}^{n}\sum_{\beta=1}^{n} a_{j\alpha}a_{j\beta}\dot{Q}_\alpha\dot{Q}_\beta = \frac{1}{2}\sum_{\alpha=1}^{n}\dot{Q}_\alpha^2 \tag{8.53}$$

The potential energy is

$$V = \frac{1}{2}\sum_{j=1}^{n}\sum_{k=1}^{n} v_{jk}q_jq_k = \frac{1}{2}\sum_{j=1}^{n}\sum_{k=1}^{n}\sum_{\alpha=1}^{n}\sum_{\beta=1}^{n} v_{jk}a_{j\alpha}a_{k\beta}Q_\alpha Q_\beta \tag{8.54}$$

From Eq. (8.38) we have

$$\sum_{k=1}^{n} v_{jk} a_{k\beta} = \omega_\beta^2 a_{j\beta} \tag{8.55}$$

which means that the potential energy can be written as (using Eq. 8.48)

$$V = \frac{1}{2} \sum_{\alpha=1}^{n} \sum_{\beta=1}^{n} \omega_\beta^2 Q_\alpha Q_\beta \sum_{j=1}^{n} a_{j\alpha} a_{j\beta} = \frac{1}{2} \sum_{\alpha=1}^{n} \omega_\alpha^2 Q_\alpha^2 \tag{8.56}$$

Thus the lagrangian of Eq. (8.34) can be written in terms of the new variables as

$$L = \frac{1}{2} \sum_{\alpha=1}^{n} (\dot{Q}_\alpha^2 - \omega_\alpha^2 Q_\alpha^2) \tag{8.57}$$

The lagrangian in this form represents a set of harmonic oscillators which no longer interact. That is, the variables are separated. Each oscillator has unit mass and its own particular frequency ω_α. The equation of motion for a particular oscillator is

$$\ddot{Q}_\alpha(t) = -\omega_\alpha^2 Q_\alpha(t) \tag{8.58}$$

This means that each mode oscillates freely at its own frequency independent of any other mode. By comparing Eqs. (8.49) and (8.50) with (8.51) we see that the real part of c_β and the imaginary part of $-c_\beta \omega_\beta$ are just the initial coordinate $Q_\beta(0)$ and the initial velocity $\dot{Q}_\beta(0)$, respectively, of the β mode. Thus the complicated molecule is equivalent to a simple set of independent harmonic oscillators.

This new set of coordinates Q_α, which permits us to describe the system as a set of independent oscillators, is called a set of *normal coordinates.* Using the lagrangian given by Eq. (8.57), we can write the path integral describing the motion of the system in terms of normal coordinates as

$$K = \int \cdots \int \exp\left\{ \frac{i}{\hbar} \frac{1}{2} \sum_{\alpha=1}^{n} \int [\dot{Q}_\alpha^2(t) - \omega_\alpha^2 Q_\alpha^2(t)]\, dt \right\} \mathcal{D}Q_1(t) \cdots \mathcal{D}Q_n(t) \tag{8.59}$$

This last result can be obtained directly from Eq. (8.35) by the explicit substitution $q_j(t) = \sum_\alpha a_{j\alpha} Q_\alpha(t)$. The exponent simplifies just as in the classical case, while $\mathcal{D}q_1 \cdots \mathcal{D}q_n = \mathcal{D}Q_1 \cdots \mathcal{D}Q_n$, at least within a constant factor. (Since the transformation of coordinates is linear, the jacobian is constant. Any such constant can be absorbed within the definition of the normalizing factors for the path integral $\mathcal{D}Q_1 \cdots \mathcal{D}Q_n$.)

This form of the path integral can be broken down into a product of path integrals. Thus

$$K = \prod_{\alpha=1}^{n} \int \exp\left\{\frac{i}{\hbar}\frac{1}{2}\int [\dot{Q}_\alpha^2(t) - \omega_\alpha^2 Q_\alpha^2(t)]\,dt\right\} \mathcal{D}Q_\alpha(t) \tag{8.60}$$

where each path integral now describes only one mode and each mode is a simple one-dimensional oscillator, for which we have already obtained a solution. In this manner any problem of interacting harmonic oscillators can be analyzed.

Since the path integral for the kernel can be separated into a product of path integrals, it follows that the wave function for the system in a given energy state can be written as the product of wave functions of each mode as discussed in Sec. 3-8.

As shown in Sec. 8-1, the energy wave functions for each separate mode are proportional to $e^{-(i/\hbar)E_n t}$, where E_n is the energy of the mode. A product of such wave functions is then proportional to $\exp\{-(i/\hbar)(\sum_n E_n)t\}$ From this it follows that the total energy of the system of oscillators is equal to the sum of all the separate energies. The energy in the α mode is $\hbar\omega_\alpha(m_\alpha + \frac{1}{2})$, where m_α is an integer. The energy of the whole system is then

$$E = \hbar\omega_1(m_1 + \tfrac{1}{2}) + \hbar\omega_2(m_2 + \tfrac{1}{2}) + \cdots + \hbar\omega_n(m_n + \tfrac{1}{2}) \tag{8.61}$$

where $m_1, m_2, \ldots, m_n$ are all integers (including zero). All independent choices are allowable because the excitation of oscillator 1 and oscillator 2 can be in different degrees.

If $\phi_m(Q)$ is the harmonic oscillator wave function for the mth energy eigenstate, then the wave function for the complete system is

$$\phi_{m_1}(Q_1)\phi_{m_2}(Q_2)\cdots\phi_{m_n}(Q_n) = \prod_{\alpha=1}^{n} \phi_{m_\alpha}(Q_\alpha) \tag{8.62}$$

Each $\phi_{m_\alpha}(Q_\alpha)$ is as given in Eq. (8.7) with ω replaced by ω_α. In this way classical physics, whereby we determine the normal modes, and quantum mechanics, whereby we determine the energy levels and wave functions for a single mode, are combined to give a complete solution for the energy levels and eigenfunctions of a polyatomic molecule.

We can express the wave functions in terms of the original coordinates $q_j(t)$ by using the transformation equations (8.51). For example, the lowest energy state of a system, which has the energy $\frac{1}{2}\sum_{\alpha=1}^{n} \hbar\omega_\alpha$, has the (unnormalized) wave function

$$\Phi_0 = \prod_{\alpha=1}^{n} \exp\left\{-\frac{\omega_\alpha Q_\alpha^2}{2\hbar}\right\} = \exp\left\{-\frac{1}{2\hbar}\sum_{\alpha=1}^{n}\omega_\alpha Q_\alpha^2\right\}$$

$$= \exp\left\{-\frac{1}{2\hbar}\sum_{\alpha=1}^{n}\sum_{j=1}^{n}\sum_{k=1}^{n}\omega_\alpha a_{j\alpha}a_{k\alpha}q_j q_k\right\} \tag{8.63}$$

That is, the wave function is an exponential function of the quadratic form $-\frac{1}{2}\sum_{j=1}^{n}\sum_{k=1}^{n} M_{jk}q_j q_k$ where the matrix element M_{jk} is

$$M_{jk} = \frac{1}{\hbar}\sum_{\alpha=1}^{n}\omega_\alpha a_{j\alpha}a_{k\alpha} \tag{8.64}$$

Problem 8-2 Show that the matrix

$$\tau_{jk} = \sum_{\alpha}\frac{a_{j\alpha}a_{k\alpha}}{\omega_\alpha}$$

is the reciprocal square root of the v_{jk} matrix. That is, show

$$\sum_{l=1}^{n}\sum_{m=1}^{n}\tau_{jl}\tau_{lm}v_{mk} = \delta_{jk} \tag{8.65}$$

It may happen that some of the frequencies ω_α are zero. For example, for the molecule CO_2 the modes 5 to 9, as pictured in Fig. 8-1, all have frequency zero. They correspond to a translation or a rotation of the whole molecule, motions for which there is no restoring force. Since there is no restoring force, the assumption that the coordinates Q_α are small is not generally true. A more exact analysis of the translation or rotation kinetic energy must be undertaken. Since such motions are not of interest in the present discussion, we shall assume that these modes, and their coordinates, either do not exist or are never excited, so that we have been dealing with modes for which $\omega_\alpha \neq 0$. If for particular values of α the solutions ω_α^2 come out negative (so that ω_α is imaginary), the system is in unstable equilibrium for motions in this mode, like a pencil balanced on its point. Instead of being simple harmonic, the motion is exponentially divergent, and again the coordinates Q_α do not stay small. This case again is of no interest in the present discussion, and we shall assume that there are no such modes.

8-4 THE ONE-DIMENSIONAL CRYSTAL

A Simple Model. We can think of a crystal as a large polyatomic molecule spread out in a three-dimensional array. We can begin learning about it by first studying a simpler one-dimensional line of equal atoms equally spaced, as in Fig. 8-2. Let the mass of each atom be m and let the displacement of the jth atom from its equilibrium position be $q_j/\sqrt{m}$. We suppose that the motions are restricted to lie along the line of the array, i.e., longitudinal motions only. Next, suppose each atom interacts only with its two neighbors, and that the potential energy of interaction between a pair of adjacent atoms separated by a distance R is $V(R)$. That is, we suppose the atoms are connected together by a set of springs. The equilibrium separation gives a minimum value to the potential. We shall assign this minimum the value 0. Suppose ΔR is the difference between the equilibrium displacement and some particular displacement. We can expand the potential in a power series in terms of ΔR, in a manner analogous to that of Eq. (8.32). We shall restrict our attention to those displacements which are so small that all terms higher than the second order in this expansion can be neglected. Between the jth and $(j+1)$st atoms the change in separation away from the equilibrium separation is $(q_{j+1}-q_j)/\sqrt{m} = \Delta R_{j,j+1}$. We shall call the second derivative of the potential with respect to the displacement $m\nu^2$ (the same for all atoms in the string). Then the potential energy associated with this pair is

$$V_{j,j+1} = \tfrac{1}{2}\nu^2(q_{j+1}-q_j)^2 \tag{8.66}$$

and the lagrangian can be written as

$$L = \frac{1}{2}\sum_{j=1}^{N}\dot{q}_j^2 - \frac{\nu^2}{2}\sum_{j=1}^{N-1}(q_{j+1}-q_j)^2 \tag{8.67}$$

If the first and last atoms are unattached, then the term for $j = N$ in the expression for potential energy must be omitted.

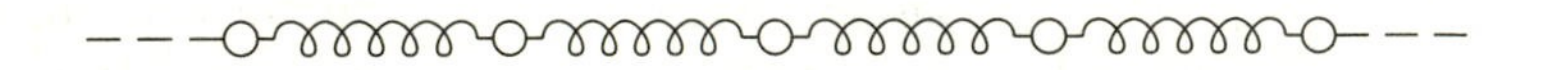

Fig. 8-2 A model of a one-dimensional "crystal," with mass particles evenly spaced along a line and springs connecting neighboring particles.

Based on this lagrangian the classical equations of motion for the atoms along the line are

$$\begin{aligned}\ddot{q}_j(t) &= \nu^2[(q_{j+1}(t) - q_j(t)) - (q_j(t) - q_{j-1}(t))] \\ &= \nu^2[q_{j+1}(t) - 2q_j(t) + q_{j-1}(t)]\end{aligned} \tag{8.68}$$

for all j except the end points $j = 1$ and $j = N$. Now this fact that the end particles have to be given separate consideration is just a minor annoyance for most problems. Usually we are interested in the gross properties for a large solid and are not concerned with surface or boundary effects. In such cases the main results desired are really independent of the actual boundary conditions (e.g., whether or not the end atoms are left free or are tied down, etc.). To avoid this problem, theoretical physicists use a trick of assuming a special set of simple boundary conditions, called periodic boundary conditions, so that these end points do not require special consideration in the analysis. Unfortunately, these special boundary conditions occur in actuality only rarely, if at all, but for phenomena which are independent of boundary effects the trick is useful.

The idea is to imagine that the string of atoms goes on beyond N, but a displacement of the $(N+j)$th atom is always exactly equal to that of the jth atom. Thus the boundary condition is

$$q_{N+1}(t) = q_1(t) \qquad \dot{q}_{N+1}(t) = \dot{q}_1(t) \tag{8.69}$$

This boundary condition would be right if our string were tied in a circle like a pearl necklace. However, in three dimensions there is no such picture to represent the boundary condition, and it must be considered completely artificial.

The value of this particular boundary condition is this. Most general ways of terminating the string (e.g., tying the last atom to a rigid wall, leaving the last atom free, etc.) result in a reflection of any wave traveling down the string. Only if the last atom is tied to another string of atoms of identical characteristics will no such reflection occur. Thus the boundary condition is analogous to tying a transmission line to a characteristic impedance in order to avoid reflections. The characteristic impedance is equivalent to an infinity of more line. In the present case we accomplish this end by tying the string to itself. We call these boundary conditions periodic, because anything the happens at the point k in the string is repeated again at the point $N+k$ and again at $2N+k$, etc. With this boundary condition, Eq. (8.68) for the motion of the atoms is valid for all of the atoms.

Solving the Classical Equations of Motion. We assume that the displacements $q(t)$ are periodic with frequency ω. Then we must solve

$$-\omega^2 q_j(t) = \nu^2[q_{j+1}(t) - 2q_j(t) + q_{j-1}(t)] \tag{8.70}$$

We could write down this set of equations in a determinant, and it turns out that the determinantal equation so obtained can be evaluated by theorems in mathematics. But this just means that the equations can be solved directly, and it is easier to solve them that way.

We shall restrict the symbol i to mean $\sqrt{-1}$, and not use it for an index. Normal mode solutions are of the form

$$q_j(t) = \Re e\{Ae^{-i(Kj-\omega t)}\} = \Re e\{a_j e^{i\omega t}\} \tag{8.71}$$

where K is a constant taking on a discrete set of values. This solution can be verified by substitution into Eq. (8.70). The frequency is given by

$$\omega^2 = -\nu^2[e^{-iK} - 2 + e^{iK}] = 4\nu^2 \sin^2 \frac{K}{2} \tag{8.72}$$

This gives the values of ω in terms of K, but not all values of K are allowed. The periodic boundary condition implies that $K = 2\pi\alpha/N$, where $\alpha = 0, 1, 2, \ldots, N-1$. (The case $\alpha = 0$ is just a translation, and we can omit it if desired. Furthermore, the case given by $\alpha' = N + \alpha$ is the same as the case given by α.) Thus for any particular choice of α we have the frequency

$$\omega_\alpha = 2\nu \left| \sin \frac{\pi\alpha}{N} \right| \tag{8.73}$$

and the amplitude for the jth coordinate at that frequency is

$$a_{j\alpha} = Ae^{-i2\pi\alpha j/N} \tag{8.74}$$

The constants $a_{j\alpha}$ determined in the last equation are complex. They could be made real by combining solutions for α and $-\alpha$ (or α and $N - \alpha$). However, it is more convenient to leave them in the complex form. It is also sometimes convenient to consider both positive and negative values of α, and so if N is odd, for example, to consider the range of α to be $-\frac{1}{2}(N-1)$ to $+\frac{1}{2}(N-1)$ rather than 0 to $N-1$.

The relative displacements of the atoms in the string depend on the size of α. The situation for two values of α, one with α small and the other with $\alpha = N/2$, is shown in Fig. 8.3.

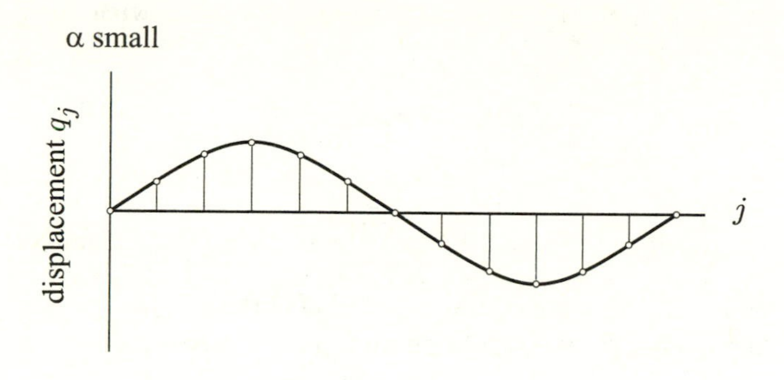

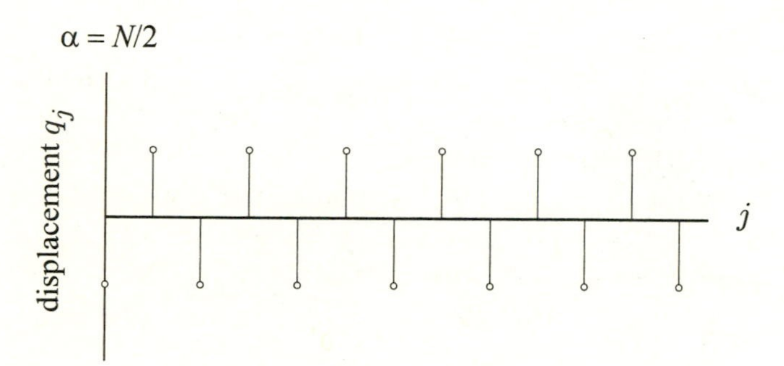

Fig. 8-3 The displacement of atoms along a string is plotted as the ordinate against the equilibrium positions j equally spaced along the abscissa. In the upper case the wavelength is long compared to the spacing between atoms (α small). In the lower case $\alpha = N/2$, and the displacements no longer give the appearance of a smooth sine wave.

Although the relative magnitudes of the various constants $a_{j\alpha}$ are determined by Eq. (8.74), the overall magnitude, determined by the constant A, is still arbitrary. We establish this with a normalizing equation analogous to Eq. (8.48). Thus choose A so that

$$\sum_{j=1}^{N} a_{j\alpha}^{*} a_{j\beta} = \delta_{\alpha\beta} \tag{8.75}$$

which implies that

$$A = \frac{1}{\sqrt{N}} \tag{8.76}$$

We are now in a position to represent the various modes with their normal coordinates as

$$Q_\alpha(t) = \sum_{j=1}^{N} a_{j\alpha} q_j(t) = \frac{1}{\sqrt{N}} \sum_{j=1}^{N} q_j(t) e^{-i2\pi\alpha j/N} \tag{8.77}$$

where an arbitrary motion is $q_j(t) = \Re e \left\{ \sum_{\alpha=0}^{N-1} c_\alpha a_{j\alpha} e^{i\omega_\alpha t} \right\}$, in analogy with Eq. (8.42). These normal coordinates are also complex, but we can ensure that the lagrangian derived from them is real by writing it as

$$L = \frac{1}{2} \sum_{\alpha=0}^{N-1} (\dot{Q}_\alpha^* \dot{Q}_\alpha - \omega_\alpha^2 Q_\alpha^* Q_\alpha) \tag{8.78}$$

Perhaps this use of complex coordinates Q_α needs a word of explanation. Since q_j, the physical coordinates, are real, Eq. (8.77) implies $Q_\alpha^* = Q_{-\alpha}$ so that, although two real numbers are required to specify each complex coordinate Q_α, only N independent real numbers are needed for all of them. If one prefers real coordinates, one can define instead two real quantities as coordinates by writing

$$Q_\alpha = \frac{1}{\sqrt{2}} (Q_\alpha^c - i Q_\alpha^s)$$

$$Q_\alpha^c = \frac{1}{\sqrt{2}} (Q_\alpha + Q_{-\alpha}) \tag{8.79}$$

$$Q_\alpha^s = \frac{i}{\sqrt{2}} (Q_\alpha - Q_{-\alpha}) \tag{8.80}$$

A term such as the kinetic energy is expressed in real variables as

$$\tfrac{1}{2}[(\dot{Q}_\alpha^c)^2 + (\dot{Q}_\alpha^s)^2] = \dot{Q}_\alpha \dot{Q}_{-\alpha} = \dot{Q}_\alpha \dot{Q}_\alpha^* \tag{8.81}$$

(The factor $\frac{1}{2}$ reappears in Eq. (8.78) because we sum there over all α, plus and minus, counting thereby each term twice, $Q_{-\alpha}^* Q_{-\alpha} = Q_\alpha Q_\alpha^*$.) Thus quadratic expressions derived previously for real quantities appear now as products of one complex number by its conjugate (for example, Eq. 8.75).

Problem 8-3 Show that Q_α^c, Q_α^s are normal coordinates corresponding to standing wave normal modes $\cos(2\pi\alpha j/N)$ and $\sin(2\pi\alpha j/N)$, in the sense that (for N odd)

$$q_j(t) = \sqrt{\frac{2}{N}} \left[\frac{1}{2} Q_0^c(t) + \sum_{\alpha=1}^{(N-1)/2} \left[Q_\alpha^c(t) \cos \frac{2\pi\alpha j}{N} - Q_\alpha^s(t) \sin \frac{2\pi\alpha j}{N} \right] \right] \tag{8.82}$$

Problem 8-4° Show that the ground-state wave function for the lagrangian of Eq. (8.78) can be written

$$\Phi_0 = A\exp\left\{-\frac{1}{2\hbar}\sum_{\alpha=1}^{N-1}\omega_\alpha Q_\alpha^* Q_\alpha\right\} \tag{8.83}$$

(where A is a constant) by starting with the wave function in terms of the real variables Q_α^c and Q_α^s.

Problem 8-5 A transition element which employs the same wave function as both the initial and final states is called an *expectation value.*[1] Thus the expectation value of F for the ground state Φ_0 of Eq. (8.83) is

$$\langle\Phi_0|F|\Phi_0\rangle = \int\cdots\iint \Phi_0^* F\Phi_0\,dQ_1\,dQ_2\cdots dQ_{N-1} \tag{8.84}$$

(The integral over complex variables is defined as equal to the corresponding integral over real normal coordiantes Q_α^c and Q_α^s.) Show that the following expectation values° are correct (for $\alpha \neq 0$):

$$\begin{aligned}
\langle\Phi_0|Q_\alpha|\Phi_0\rangle &= \langle\Phi_0|Q_\alpha^*|\Phi_0\rangle = 0\\
\langle\Phi_0|Q_\alpha^2|\Phi_0\rangle &= \langle\Phi_0|Q_\alpha^{*2}|\Phi_0\rangle = 0\\
\langle\Phi_0|Q_\alpha^* Q_\alpha|\Phi_0\rangle &= \frac{\hbar}{2\omega_\alpha}\langle\Phi_0|1|\Phi_0\rangle\\
\langle\Phi_0|Q_\alpha^* Q_\beta|\Phi_0\rangle &= 0 \quad \text{if } \alpha \neq \beta
\end{aligned} \tag{8.85}$$

Thus with the lagrangian written in terms of normal coordinates we have reduced the system to a set of independent simple harmonic oscillators. The quantum-mechanical part of the solution follows in a straightforward manner just as it did for the case of the polyatomic molecule. All that we need to know is the quantum-mechanical solution for an independent simple harmonic oscillator.

Problem 8-6 Show that the constants $a_{j\alpha}$ are the same even if the coupling is not just to the nearest neighbors but extents with strengths λ_k to atoms k spaces away. Assuming λ_k falls rapidly enough for large k, find the values of the frequency ω_α when such a coupling is present, i.e., when the potential energy, instead of being given by Eq. (8.66), is given by a similar equation, but one which contains the relative displacements of all pairs of atoms, each one multiplied by the appropriate λ_k, that is, $V = (\nu^2/2)\sum_j\sum_k \lambda_k(q_{j+k}-q_j)^2$.

[1] Compare this definition of *expectation value* with the definition of the *expected value* of an operator given in Sec. 5-3, particularly in Eq. (5.46).

8-5 THE APPROXIMATION OF CONTINUITY

The particular modes which we have determined here are those in which each atom oscillates with a phase difference behind the one next in line. There is a wave of oscillation passing down the line of atoms. If the phase difference between adjacent atoms is small, then the wavelength is long.

Of special interest is the behavior of the atoms in the long-wavelength modes. If the wavelength greatly exceeds the spacing between atoms, this spacing is unimportant. In this case the motion can be very well described by the fictitious "continuous medium" concept. A line of atoms can be replaced by a continuous rod with certain average properties, such as the mass per unit length $\rho = m/d$. More physically, a real rod is actually a discrete set of atoms. In this section we shall develop the approximation of continuity, wherein a line of atoms is replaced by a continuous string.

For a particular mode of motion the phase difference between adjacent atoms is $2\pi\alpha/N$, so that a wavelength contains N/α atoms, or if d is the equilibrium separation distance between neighboring atoms, the wavelength is $\lambda = Nd/\alpha$. The wave number is

$$k = \frac{2\pi}{\lambda} = \frac{2\pi\alpha}{Nd} \tag{8.86}$$

The wave aspect is made more clear in the mathematical representation of the motion by a slight change of notation. We shall refer to each mode by its k value instead of by its α value. Then a summation° α over the modes means a sum over discrete values of k. These values are the integers multiplied by $2\pi/L$, where $L = Nd$ is the length of the string. Suppose $x_j = jd$ is the equilibrium position of the jth atom. Then the equations describing the motion of the atom become

$$a_{jk} = \frac{1}{\sqrt{N}} e^{-ikx_j} \tag{8.87}$$

$$Q_k = \frac{1}{\sqrt{N}} \sum_{j=1}^{N} q_j e^{-ikx_j} \tag{8.88}$$

$$q_j = \frac{1}{\sqrt{N}} \sum_{k=1}^{N} Q_k e^{ikx_j} \tag{8.89}$$

and

$$\omega_k = 2\nu \left| \sin \frac{kd}{2} \right| \tag{8.90}$$

We now assume that the separation between atoms is very small compared to the length over which disturbances change. Using the symbols we have already defined, such situations as this are described by $kd \ll 1$. If we call the product $\nu d = c$, then for kd small we have $\omega \approx kc$. In this situation we can think of the coordinates q_j as being functions of position along the line of atoms. That is, we can specify the displacement of the jth atom, as shown in Fig. 8-3. For long waves the displacements $q(x_j)$ and $q(x_{j+1})$ are nearly equal, and we can consider the function $q(x)$ as a smooth continuous function defining displacement as a function of equilibrium position along the line. The normal coordinate $Q(k)$ is a Fourier transform of $q(x)$. That is, Eq. (8.88) can be replaced by

$$Q(k) = \frac{\sqrt{N}}{L} \int_0^L q(x) e^{-ikx}\, dx \tag{8.91}$$

This replacement is based on the approximate relation

$$\sum_{j=1}^{N} (\quad)_j \approx \frac{N}{L} \int_0^L (\quad)\, dx \tag{8.92}$$

which becomes more valid as the spacing between discrete points becomes very small.

A similar relation, namely,

$$\sum_{k=1}^{N} (\quad)_k \approx \frac{L}{2\pi} \int_0^{2\pi/d} (\quad)\, dk \tag{8.93}$$

leads to the inverse transform

$$q(x) = \frac{L}{2\pi\sqrt{N}} \int_0^{2\pi/d} Q(k) e^{ikx}\, dk \tag{8.94}$$

To make these quantities of more direct physical significance, let the actual displacement of the j atom be u_j. That is, $q_j = \sqrt{m}\, u_j$, where m is the mass of one atom and is equal to ρd. Let the Fourier transform of u be U. Thus

$$U(k) = \int_0^L u(x) e^{-ikx}\, dx \tag{8.95}$$

while the inverse transform is

$$u(x) = \int_{-\infty}^{\infty} U(k) e^{ikx} \frac{dk}{2\pi} \tag{8.96}$$

The new normal coordinate is then $U(k)$, and it is related to the previous normal coordinate $Q(k)$ by

$$U(k) = \frac{L}{\sqrt{mN}} Q(k) \tag{8.97}$$

The expression for the kinetic energy in terms of $u(x,t)$ can be worked out with the help of Eq. (8.92) to be

$$\text{K.E.} = \frac{\rho}{2} \int_0^L \left(\frac{\partial u}{\partial t} \right)^2 dx \tag{8.98}$$

To determine the potential energy in terms of all the new variables, we need to express the difference in the displacements of two adjacent atoms as a continuous function of position. Using our approximation of continuity, we can write

$$q_{j+1} - q_j = \sqrt{m}\,[u(x_{j+1}, t) - u(x_j, t)] \approx \sqrt{m}\, d \frac{\partial u}{\partial x} \tag{8.99}$$

That means that the potential energy is

$$V = \frac{\nu^2}{2} \frac{N}{L} \int_0^L md^2 \left(\frac{\partial u}{\partial x} \right)^2 dx = \frac{\rho c^2}{2} \int_0^L \left(\frac{\partial u}{\partial x} \right)^2 dx \tag{8.100}$$

In the last equation we have used the constant $c = \nu d$. This constant is actually a measure of the elasticity. We can define it physically in the following manner: Suppose we stretch the line of atoms, which has length L, by a fractional increase of amount ϵ, that is, to the new length $L(1+\epsilon)$. (We are considering a static stretch, not a vibration.) This means that we make the separation between each pair of atoms equal to $d(1+\epsilon)$ instead of d. Thus the difference in displacements of adjacent atoms becomes

$$q_{j+1} - q_j = \epsilon d \sqrt{m} \tag{8.101}$$

Using Eq. (8.66), this means the potential energy put into the string by the stretching is

$$V = \frac{\nu^2}{2} \epsilon^2 d^2 mN = \frac{\rho c^2}{2} \epsilon^2 L \tag{8.102}$$

Thus the force that is needed to stretch the string is, in the limit of small ϵ,

$$\frac{\partial V}{\partial(\epsilon L)} = \rho c^2 \epsilon \tag{8.103}$$

This last equation gives the stress in the string, while the strain (stretch per unit length) is of course ϵ. Thus we have

$$\frac{\text{stress}}{\text{strain}} = \rho c^2 = \text{elastic constant} \tag{8.104}$$

Combining Eqs. (8.98) and (8.100), we can construct the lagrangian as

$$L = \frac{\rho}{2}\int_0^L \left(\frac{\partial u}{\partial t}\right)^2 dx - \frac{\rho c^2}{2}\int_0^L \left(\frac{\partial u}{\partial x}\right)^2 dx \tag{8.105}$$

The fundamental modes which we are considering have the form e^{-ikx}, and the normal coordinates are $U(k,t)$. The reader can show that, for a long string, the lagrangian can be expressed in terms of these normal coordinates as

$$L = \frac{\rho}{2}\int_{-\infty}^{\infty} \left|\frac{\partial U(k,t)}{\partial t}\right|^2 \frac{dk}{2\pi} - \frac{\rho c^2}{2}\int_{-\infty}^{\infty} k^2 |U(k,t)|^2 \frac{dk}{2\pi} \tag{8.106}$$

We can consider the system described by this lagrangian as a set of harmonic oscillators, one oscillator for each value of k. In our present approximation of continuity, k is a continuous variable with an infinite number of values. We can reintroduce the picture of discrete atoms by remembering that the integral over k is really a sum over discrete values of k, where the various discrete values of k are spaced a distance $2\pi/L$ apart, with L the length of the string, and the number of such values is equal to the number of atoms in the string.

We can get equations of motion in terms of the continuous variables by finding the extremum of the action integral $\int_0^T L\,dt$. Using the form of L given by Eq. (8.105), the resulting equation of motion is

$$\rho\frac{\partial^2 u}{\partial t^2} = \rho c^2 \frac{\partial^2 u}{\partial x^2} \tag{8.107}$$

Following a line of argument demonstrated by Eq. (8.99), we can see that this equation of motion is analogous to the previous equation of motion which we derived, namely, Eq. (8.68). Equation (8.107) has the solution

$$u(x,t) = a(x)e^{i\omega t} \tag{8.108}$$

in analogy with Eq. (8.71), where

$$-\omega^2 a(x) = c^2 \frac{d^2 a(x)}{dx^2} \tag{8.109}$$

in analogy with Eq. (8.70), and

$$a(x) = e^{-ikx} \tag{8.110}$$

in analogy with Eq. (8.74).

Combining Eqs. (8.109) and (8.110), we see that $\omega = kc$. This is the analogue of Eq. (8.90), and, as a matter of fact, in the limit of small k, Eq. (8.90) reduces to this relation.

The motion described by Eq. (8.108), with the value of $a(x)$ given by Eq. (8.110), is that of a traveling wave moving with velocity c. That is to say, c is the speed of sound along the line of atoms. Actually, a real system shows dispersion; that is, ω is not exactly proportional to k. For wavelengths which are of the same order as the atomic spacing this lack of proportionality becomes important, as shown by Eq. (8.90).

8-6 QUANTUM MECHANICS OF A LINE OF ATOMS

The behavior of the atoms in a string can be described in terms of modes of motion. Each mode is a harmonic oscillator. The energy state of any particular mode is determined by the quantum number for that mode. Each mode is identified by its wave number k or its frequency ω. A mode of frequency ω can have the energy values $\frac{1}{2}\hbar\omega$, $\frac{3}{2}\hbar\omega$, $\frac{5}{2}\hbar\omega$, ..., or in other words 0, $\hbar\omega$, $2\hbar\omega$, ... above the ground state energy $\frac{1}{2}\hbar\omega$. For these cases we would say that there are 0, 1, 2, ... *phonons* of wave number k (or frequency ω) present.

It is possible to have several different modes excited simultaneously. For example, we could have (1) the mode of wave number k_1 excited to its first level above the ground state, (2) the mode of wave number k_2 excited to its first level also, and (3) the mode of wave number k_3 excited to its second level above its ground state. The state of the complete system would then have the total energy $\hbar(\omega_1 + \omega_2 + 2\omega_3)$ above the ground energy. We would say that there are four phonons present: one phonon of wave number k_1, one of wave number k_2, and two of wave number k_3.

The ground state of the entire system has the energy

$$E_{\text{gnd}} = \sum_k \frac{\hbar\omega_k}{2} \tag{8.111}$$

Using the approximation of continuity (see Eq. 8.93) and letting $\omega = kc$, this becomes

$$E_{\text{gnd}} = \frac{L}{2\pi}\int_0^{k_{\max}} \frac{\hbar kc}{2}\, dk \tag{8.112}$$

If the upper limit $k_{\max}$ on the integral over k goes to infinity, then the integral diverges. However, the form $\omega = kc$ used in this expression is valid only for long waves (i.e., small values of k).

We can make a better determination of the ground-state energy by using the correct expression for ω and establishing a reasonable upper limit for the integral over k. Thus, using Eq. (8.90) for ω_k, we can write the ground-state energy as

$$E_{\text{gnd}} = \sum_{k=-k_{\max}}^{k_{\max}} \hbar\nu \left| \sin \frac{kd}{2} \right| \tag{8.113}$$

where

$$k_{\max} = \frac{\pi}{d} \tag{8.114}$$

This can be rewritten as

$$E_{\text{gnd}} = \sum_{\alpha=-N/2}^{N/2} \hbar\nu \left| \sin \frac{\pi\alpha}{N} \right| = 2\hbar\nu\, \Im m \left\{ \sum_{\alpha=0}^{N/2} e^{i\pi\alpha/N} \right\} \tag{8.115}$$

For a very large N this sum can be approximated by an integral to give

$$E_{\text{gnd}} = 2\hbar\nu \frac{N}{\pi} = \frac{2\hbar c L}{\pi d^2} \tag{8.116}$$

This result shows that the energy is proportional to the length of the string, but apparently it has no limit as the spacing d approaches zero. That is, the ground-state energy is infinite for a continuous medium. Of course, for real matter the energy is finite.

It is very convenient to measure, instead of the total energy, the excess energy above the ground state. There are two reasons for this: (1) Really, the ground-state energy is not known, nor is it usually interesting to the physical problem in question. For example, the true ground-state energy includes all of the energy of the electrons attached to the atoms. (2) When dealing with the excitation of only long waves, the approximation of continuity is very useful, and it gives a good approximation to the excitation energies. However, this approximation gives an invalid result for the ground-state energy, since it neglects the separation d (i.e., treats d as 0). Thus we must avoid the necessity of evaluating the ground-state energy if we are to use the approximation of continuity.

8-7 THE THREE-DIMENSIONAL CRYSTAL

There is no difference in principle between a realistic three-dimensional crystal and the one-dimensional example which we have been considering. However, the detailed evaluation of the various modal frequencies is much harder. Results can be obtained in terms of the wave number $\mathbf{k}$, which is now a vector with components k_x, k_y, and k_z. The frequency, written in terms of these components, is generally very complicated. There is more than one solution for each value of $\mathbf{k}$ because of the possibility of various polarizations (directions of vibration). Furthermore, a real crystal often consists not of an array of atoms equally spaced, but rather of an array of unit cells, each unit cell consisting of a group of atoms in some characteristic geometrical arrangement. If there are several atoms (say p) in such a unit cell (and this example can be illustrated with a one-dimensional model), then there are $3p$ frequencies for each value of $\mathbf{k}$.

In the three-dimensional crystal we can still use the approximation of continuity to good advantage. In this approximation the true lattice structure of the crystal generally makes itself felt through the existence of different properties in different directions (e.g., anisotropic compressibility). The symmetry of the lattice is reflected by the symmetry of the elastic constants. Furthermore, the fundamental modes have vibration directions (polarization directions) which are not necessarily either parallel to or perpendicular to the direction of propagation of the wave.

For the present discussion, we shall assume that our substance displays the same elastic constants in all directions. (In general, it is not necessary for any crystal, even one as symmetric as a cubic crystal, to do this.) Then the waves are of two kinds, longitudinal and transverse. These two kinds of waves have different wave velocities, which we shall label c_L for the longitudinal and c_T for the transverse. For each $\mathbf{k}$ there are three modes. One of these has the frequency $\omega_L = c_L k$ (where k is the absolute magnitude of $\mathbf{k}$). Since, by hypothesis, there is no directional effect, the frequency is a function only of the absolute magnitude of the wave number and does not depend upon its specific components. There are two transverse modes (i.e., modes in which the direction of motion of the atoms is perpendicular to the direction of motion of the wave), both of which have the frequency $\omega_T = c_T k$.

Every separate mode, and that includes every separate direction of polarization, behaves like an independent oscillator.

Suppose we are dealing with a crystal of volume V. Let us compute the number of modes whose wave numbers lie in the k-space volume element $d^3\mathbf{k} = dk_x\, dk_y\, dk_z$ centered about the point $\mathbf{k}$. We assume the crystal is rectangular with edge lengths L_x, L_y, and L_z. We use the results obtained from the one-dimensional example to see that the discrete values of k_x are spaced apart a distance $2\pi/L_x$. So, in the range of wave number dk_x there are $dk_x\, L_x/2\pi$ discrete values of k_x. Applying this same reasoning to the other directions, we find that the number of discrete values of $\mathbf{k}$ included in the interval is

$$\frac{dk_x\, dk_y\, dk_z}{(2\pi)^3} L_x L_y L_z = \frac{d^3\mathbf{k}}{(2\pi)^3} V \tag{8.117}$$

This same result is obtained (in the limit of large crystals) for any shape.

For the general case the modal frequency $\omega_{\mathbf{k}}$ is, as we have mentioned, a very complicated function of $\mathbf{k}$ with several branches (values for the same $\mathbf{k}$), but its determination is a problem of classical physics; then the forms of oscillation in the fundamental modes are known, as are the normal coordinates describing these modes. The quantum-mechanical problem is then reduced to the solution of a simple set of oscillators, and all the properties can be worked out easily. The excitation of each mode is called the excitation of a phonon.

As a very simple special example, we shall consider the longitudinal modes of oscillation in an isotropic solid (i.e., sound or, in particular, longitudinal sound). We can start as we did in the one-dimensional example with the atoms in the crystal discretely spaced and later pass to the long-wavelength limit, or approximation of continuity.

A complete solution would show us all the effects of dispersion, the complicated branches, and the transverse waves. It is a very interesting study. However, one need not carry out all of the steps in order to obtain the proper quantum-mechanical form of the continuity approximation. One can make use directly of the results of classical physics. The entire procedure, starting with discretely spaced point masses, then passing to the long-wavelength limit, is just as useful and just as valid in quantum mechanics as it is in classical physics. The lagrangian has the same form so long as one restriction is imposed, i.e., that the potential can be adequately represented by a quadratic function of the displacements. The reason for the similarity between the results of the classical and quantum-mechanical approach is that the procedure consists only of various linear transformations, e.g., transforming to normal coordinates followed by certain approximations, such as the approximation of continuity. These transformations and approximations can be done in quantum mechanics exactly as they are done in classical physics.

The equations derived from classical physics are as follows. Suppose $\mathbf{u}(\mathbf{r}, t)$ represents the displacement of a particle whose equilibrium position is at $\mathbf{r}$. We assume that we are working in the long-wavelength region, so that the approximation of continuity applies. A plane-wave mode is easiest to describe in terms of the Fourier transform given by

$$\mathbf{U}(\mathbf{k}, t) = \iiint_V \mathbf{u}(\mathbf{r}, t) e^{-i\mathbf{k}\cdot\mathbf{r}}\, d^3\mathbf{r} \tag{8.118}$$

where $\mathbf{r}$ is a spatial vector having the components x, y, z. The normal coordinates of the various modes depend on the relationship between the direction of $\mathbf{U}$ and the direction of the wave vector $\mathbf{k}$. That is, the coordinate $U_x(\mathbf{k}, t)$ of the vector $\mathbf{U}$ does not necessarily represent a normal mode. For an isotropic material the three modes of a given $\mathbf{k}$ have the following normal coordinates:

$$U_0(\mathbf{k}, t) = \frac{\mathbf{k}}{k} \cdot \mathbf{U}(\mathbf{k}, t) \tag{8.119}$$

(that is, the component of $\mathbf{U}$ in the direction of $\mathbf{k}$) and

$$U_1(\mathbf{k}, t) = \mathbf{e}_1 \cdot \mathbf{U}(\mathbf{k}, t) \tag{8.120}$$

$$U_2(\mathbf{k}, t) = \mathbf{e}_2 \cdot \mathbf{U}(\mathbf{k}, t) \tag{8.121}$$

where $\mathbf{e}_1$ and $\mathbf{e}_2$ are two unit vectors perpendicular to $\mathbf{k}$ and perpendicular to each other. For the present study we shall restrict our attention to just that part of the kinetic and potential energy which arises from the longitudinal modes given by Eq. (8.119) and omit the transverse oscillations.

Using the results of classical physics, the lagrangian for the longitudinal modes can be written as

$$L = \frac{\rho}{2} \iiint_{-\infty}^{\infty} \left[\left| \frac{\partial U_0(\mathbf{k}, t)}{\partial t} \right|^2 - c^2 k^2 |U_0(\mathbf{k}, t)|^2 \right] \frac{d^3\mathbf{k}}{(2\pi)^3} \tag{8.122}$$

Here we have introduced the speed of sound $c = \omega/k$, which is a function of the direction of propagation. This is a direct generalization from the one-dimensional example. In terms of the original variables $\mathbf{u}(\mathbf{r}, t)$ the lagrangian is

$$L = \frac{\rho}{2} \iiint_V \left[\left(\frac{\partial \mathbf{u}}{\partial t} \right)^2 - c^2 (\boldsymbol{\nabla} \cdot \mathbf{u})^2 \right] d^3\mathbf{r} \tag{8.123}$$

The first term on the left-hand side of this equation is the kinetic energy, given by one-half the mass times the square of the velocity. The second term is the energy of compression given by $\boldsymbol{\nabla} \cdot \mathbf{u}$, which is the compressional strain. No energy of sheer strain is included here because we have disregarded transverse elastic waves.

Variation of the lagrangian with respect to $\mathbf{u}$ produces the classical equations of motion as

$$\frac{1}{c^2}\frac{\partial^2 \mathbf{u}}{\partial t^2} = \nabla(\nabla \cdot \mathbf{u}) \tag{8.124}$$

If we define a compressional strain function equal to the divergence of $\mathbf{u}$, that is,

$$\phi(\mathbf{r}, t) = \nabla \cdot \mathbf{u}(\mathbf{r}, t) \tag{8.125}$$

we have the result

$$\frac{1}{c^2}\frac{\partial^2 \phi}{\partial t^2} = \nabla^2 \phi \tag{8.126}$$

which is the classical wave equation.

The Fourier transform of Eq. (8.124), using the kernel $e^{-i\mathbf{k}\cdot\mathbf{r}}$ and taking the component of the result parallel to $\mathbf{k}$, gives

$$\frac{1}{c^2}\frac{\partial^2 U_0(\mathbf{k}, t)}{\partial t^2} = -k^2 U_0(\mathbf{k}, t) \tag{8.127}$$

This is the equation of a single harmonic oscillator, and it shows us that $U_0(\mathbf{k}, t)$ is indeed a normal coordinate.

The quantum-mechanical results from the lagrangian given by Eq. (8.123) can be obtained easily. The energy levels of the mode in question are given by $n\hbar(kc)$ above the ground level. Let us ask for the amplitude to go from a given initial set of coordinates $\mathbf{u}(\mathbf{r}, 0)$ to a given final set of coordinates $\mathbf{u}(\mathbf{r}, T)$. It is

$$K[\mathbf{u}(\mathbf{r}, T), T; \mathbf{u}(\mathbf{r}, 0), 0] = \int \exp\left\{\frac{i}{\hbar}\frac{\rho}{2}\int_0^T \iiint_V \left[\left(\frac{\partial \mathbf{u}}{\partial t}\right)^2 - c^2(\nabla \cdot \mathbf{u})^2\right] d^3\mathbf{r}\, dt\right\} \mathcal{D}^3\mathbf{u}(\mathbf{r}, t) \tag{8.128}$$

The path integral of Eq. (8.128) is carried out over the paths $\mathbf{u}(\mathbf{r}, t)$ defined in terms of all three components of the vector $\mathbf{r}$, as well as the time t. It is subject, of course, to the condition that the function $\mathbf{u}(\mathbf{r}, t)$ take on a given form at both the initial and final points. This is an interesting extension of our original path integral idea. Up to now we have dealt with integrands which were functionals of one (or perhaps a few) function $x(t)$ of one variable t, and we have carried out the integration over all such paths, or functions. Now we must integrate a functional of the function $\mathbf{u}(\mathbf{r}, t)$ of four variables x, y, z, and t and carry out the path integration over all values of this function. We can accomplish this by the regular techniques which we have described before, for our integrand is still a gaussian functional.

The first step in the solution of the path integral is to find the path which leads to a stationary value for the integral appearing in the exponent, the one which satisfies Eq. (8.124), or, more conveniently, the wave equation given by Eq. (8.126). We must impose the required boundary conditions at the times $t = 0$ and $t = T$. Satisfying the boundary conditions is not a difficult problem; however, it is a little different from the usual problem in classical physics in which the coordinate and its derivative are given at $t = 0$, that is, $\mathbf{u}(\mathbf{r}, 0)$ and $(\partial \mathbf{u}/\partial t)_{t=0}$.

We could proceed along this line and solve the problem. However, we have learned from previous examples that it is much easier to transform the problem into normal coordinates before carrying out the path integral. Such a transformation gives us (using Eq. A.11)

$$K = \int_{U_1(0)}^{U_1(T)} \exp\left\{\frac{i}{\hbar}\frac{\rho}{2}\frac{1}{V}\sum_{\mathbf{k}}\int_0^T [|\dot{U}_1|^2 - k^2c^2|U_1|^2]\,dt\right\} \mathcal{D}U_1(\mathbf{k}, t) \tag{8.129}$$

where the boundary values are given by

$$\begin{aligned} U_0(T) = U_0(\mathbf{k}, T) &= \frac{\mathbf{k}}{k}\cdot \iiint_V \mathbf{u}(\mathbf{r}, T) e^{-i\mathbf{k}\cdot\mathbf{r}}\, d^3\mathbf{r} \\ U_0(0) = U_0(\mathbf{k}, 0) &= \frac{\mathbf{k}}{k}\cdot \iiint_V \mathbf{u}(\mathbf{r}, 0) e^{-i\mathbf{k}\cdot\mathbf{r}}\, d^3\mathbf{r} \end{aligned} \tag{8.130}$$

This is once more the simpler type of path integral, where the path is described in terms of only the one variable t. Since the path integral can be written as a product of path integrals, each one defining the motion of a normal mode, we find that we have already solved the problem. The result is (see Eq. 8.1)

$$\begin{aligned} K = \prod_{\mathbf{k}} \left(\frac{\rho k c}{2\pi i \hbar V \sin kcT}\right)^{1/2} \exp\Bigg\{ &\frac{i\rho k c}{2\hbar V \sin kcT} \\ &\times \Big([U_0^2(\mathbf{k}, T) + U_0^2(\mathbf{k}, 0)] \cos kcT - 2U_0(\mathbf{k}, T)U_0(\mathbf{k}, 0)\Big)\Bigg\} \end{aligned} \tag{8.131}$$

In the products over the components of $\mathbf{k}$, the x component, for example, takes on the values $2\pi n_x/L_x$, where n_x is an integer running from 0 to $N_x = L_x/d$. Here d is the spacing between atoms, and the sample under study has edge lengths L_x, L_y, and L_z. Of course, the approximation of continuity implies a zero spacing between the atoms, which means that the product is unbounded. However, we shall disregard such problems and concentrate only on the form of those terms showing dependence on initial and final coordinates. Thus, disregarding the radical which multiplies the exponential term of Eq. (8.131), we can write this equation approximately as

$$K \sim \exp\left\{\frac{i}{\hbar}\frac{\rho}{2}\int\int\int_{-\infty}^{\infty}\frac{kc}{\sin kcT} \times \left(\left[U_0^2(\mathbf{k},T)+U_0^2(\mathbf{k},0)\right]\cos kcT - 2U_0(\mathbf{k},T)U_0(\mathbf{k},0)\right)\frac{d^3\mathbf{k}}{(2\pi)^3}\right\} \tag{8.132}$$

The dependence of the amplitude on the boundary values $U_0(\mathbf{k},0)$ and $U_0(\mathbf{k},T)$ is contained in this last result. For any choice of these functions [and they, in turn, depend on $\mathbf{u}(\mathbf{r},0)$ and $\mathbf{u}(\mathbf{r},T)$, as shown by Eqs. (8.130)] the integration in Eq. (8.132) can be carried out, formally, and a final answer obtained. In this manner all questions about the quantum-mechanical behavior of the system can be answered, at least in principle.

8-8 QUANTUM FIELD THEORY

Suppose we are dealing with waves or modes which are described by continuous functions, like $\mathbf{u}(\mathbf{r},t)$, for which there is no atomic substructure or for which the wavelengths are long enough that we can neglect such a substructure. In this case we say that $\mathbf{u}(\mathbf{r},t)$ is a field, i.e., a function of each point in space. In the example we have just considered the field is the displacement field of sound. In this terminology the equations of motion are called the field equations. In the present chapter we have been dealing only with linear field equations. The lagrangians can be called the lagrangians for the field. The normal coordinates $\mathbf{U}(\mathbf{k},t)$ are the coordinates for the normal modes of the field. The description of these modes as quantum oscillators is called *quantizing the field*. The resultant theory is called quantum field theory, to distinguish it from the classical analysis of the equations.

As we have seen, almost all of the effort in quantum field theory is devoted to solving the classical equations of motion to find the normal modes, an activity completely within the realm of classical physics. The "quantization" consists then of no more than the additional remark that each of the normal modes is a quantum oscillator, with energy levels $(n+\frac{1}{2})\hbar\omega$. Presented in this way, quantum field theory seems to be just a special consequence of the Schrödinger equation, and not an extra theory at all.

That is, or should be, the case for any situation in which the field variables (like sound displacement or pressure) are defined ultimately in terms of some combination of the basic mechanical variables. These basic variables describe the position of the particles, atoms, electrons, and also nuclei, which comprise the material carrying the field. For example, in the case of sound we assume that Schrödinger's equation describes the motion of the constituent parts, or atoms, in a crystal. Then we easily deduce that the long-wavelength sound waves obey the classical linear field equations, and we find that the modes are quantized.

In a few cases the classical equations of some field pertaining to a system are known, even though the quantum-mechanical derivation starting from Schrödinger's equation has not yet been made. For example, the equations describing the oscillations of a drop of nuclear matter have been guessed by classical analogy.[1] In such a situation it is an excellent guess that the modes of the field will turn out to be quantized oscillators if and when the complete quantum-mechanical derivation is worked out. Actually, not many such examples are left. Nearly all cases have by now been worked out.

Another type of field equation, fundamentally different from that described above, exists in quantum mechanics. An example of this type is Maxwell's set of electromagnetic equations, a set of linear field equations. These equations lead to a wave equation which is analogous to the one we developed for sound, although there are different polarization conditions. Just as an organ pipe has standing waves, or modes, so an electromagnetic field in a cavity can be described classically in terms of fundamental modes of oscillation. It is a natural inference that these oscillations are also quantized in the sense that each mode can have the energy levels $n\hbar\omega$ above the ground level, etc. This is the fundamental assumption of the quantum theory of electromagnetism. It is not a strict deduction from the Schrödinger equation for matter, because the electromagnetic field is not understood as a long-wavelength approximation of an atomic medium. Today, we do not think of any particular medium, but take the equations of Maxwell for granted. We simply assume they are to be quantized in the simple direct manner described above. We shall discuss this example in more detail in Chap. 9.

The assumption of quantization for the electromagnetic fields turns out to be consistent with all experiments carried out so far, although there are some theoretical difficulties. These difficulties are associated with the extension of the scheme to modes of very short wavelengths. There are various effects which lead to diverging integrals if the integra-

[1] M.S. Plesset, On the Classical Model of Nuclear Fission, *Am. J. Phys.*, vol. 9, pp. 1–10, 1941.

tions are carried towards zero wavelength. The corresponding difficulties do not really arise in a vibrating crystal because if we wish to carry the analysis into the very short wavelength region, where the wavelengths are comparable to the atomic spacing, we must drop the approximation of continuity. Then in the case of a crystal we find that there are only a finite number of modes in any finite volume, while in electrodynamics the number of modes in any volume is infinite.

When the various modes of a field are excited, we say there are "things" present which have different names for different cases. For sound or crystal vibrations we call them *phonons*, for the electromagnetic field *photons*, for meson field theory *mesons*, etc. Even electrons can be represented as being excitations of a field, but it is a field of a very different kind from what we have been discussing. It is called a Fermi field; the particles obey the exclusion principle, and the lagrangian is quantized not by representing it as a set of harmonic oscillators, but in a different way. Fields quantized as modes of harmonic oscillators are called Bose particles; they obey Bose or symmetric statistics. This just means that if one has two particles, one of wave vector $\mathbf{k}_1$ and one of $\mathbf{k}_2$, there is only one state. There is no new state where the first has $\mathbf{k}_2$ and the second has $\mathbf{k}_1$. This is because for our field there is only one state with $\mathbf{k}_1$ and $\mathbf{k}_2$ each excited to the first level. It has energy $\hbar\omega_1 + \hbar\omega_2$, and it is meaningless to ask: After an exchange, which excitation is which? In the next chapter we discuss this in more detail for the case of photons of the electromagnetic field.

Problem 8-7 It is believed that neutral particles of spin zero (like neutral pions) can, when free, be represented by a field $\phi(\mathbf{r}, t)$ with a lagrangian

$$L = \frac{1}{2}\int\left[\left(\frac{\partial\phi}{\partial t}\right)^2 - c^2(\boldsymbol{\nabla}\phi)^2 + \left(\frac{\mu c^2}{\hbar}\right)^2\phi^2\right]d^3\mathbf{r} \tag{8.133}$$

where μ is some constant. Show that this field has quantized states corresponding to waves $e^{i\mathbf{k}\cdot\mathbf{r}}$, where the energy of excitation is

$$\hbar\omega = \sqrt{(\hbar kc)^2 + (\mu c^2)^2} \tag{8.134}$$

If $\hbar\mathbf{k} = \mathbf{p}$ is considered as the momentum of each excitation the energy is

$$E = \sqrt{(|\mathbf{p}|c)^2 + (\mu c^2)^2} \tag{8.135}$$

This is the relativistic formula for the energy of a particle of momentum $\mathbf{p}$ and mass μ. (*Note:* For p^2 small it is approximately

$$E = \mu c^2 + \frac{p^2}{2\mu} + \cdots$$

the rest energy μc^2 plus the kinetic energy $p^2/2\mu$.)

We interpret the state of the field when the mode $\mathbf{k}_1$ is excited to the second quantum level, $\mathbf{k}_2$ to the first, etc., as the state of a system containing two particles with momentum $\hbar\mathbf{k}_1$, one with $\hbar\mathbf{k}_2$, etc. The ground state is considered the state in which no particles are present, and it is called the vacuum state. Excitation or deexcitation of the field oscillators corresponds to creation or annihilation of particles, and this is the way that such processes are represented in relativistic quantum field theory.

8-9 THE FORCED HARMONIC OSCILLATOR

In this chapter we have dealt with the simple harmonic oscillator or with systems that could be reduced to a set of such oscillators. But the oscillators have been free, not interacting with anything else. We must develop our analysis further if we wish to deal with such linear systems in interaction with other systems or driven by external forces. Examples of such systems include polyatomic molecules in varying external fields, colliding polyatomic molecules, crystals through which an electron is passing and exciting the oscillator modes, and other interactions of the modes with external fields. We shall not discuss the problem of interaction in general; instead, we use as a prototype the example of the interaction of atomic systems and charges with the electromagnetic field. We do this in the next chapter. Other cases may be analyzed by direct analogy.

These problems involve two aspects: (1) the resolution of the field into its component independent oscillators and (2) the interaction of each oscillator with external potentials or other systems. The resolution into oscillators has been exhaustively studied so far in this chapter.

To prepare the complete machinery for such problems, it remains only to analyze the behavior of a single oscillator disturbed by an external potential. We shall put these pieces together in the next chapter.

In this section we go back to the study of a single harmonic oscillator, but coupled linearly to some external potential or disturbance. The lagrangian for such a system is given by

$$L(\dot{x}, x, t) = \frac{M}{2}\dot{x}^2 - \frac{M\omega^2}{2}x^2 + f(t)x \tag{8.136}$$

where $f(t)$ is the external force. We assume for convenience that it is turned on only during a certain time interval T from $t = 0$ to $t = T$, so that the oscillator is free initially at $t = 0$ and free finally at $t = T$. In

Prob. 3-11 we completely solved this problem, obtaining the amplitude $K(b,a)$ that the oscillator goes from point x_a at $t=0$ to point x_b at $t=T$. But for the present applications it is convenient to find as well the amplitude G_{mn} that the oscillator initially in energy state n is found at time T in energy state m. This representation is often more convenient than the coordinate representation.

In Sec. 8-1 we determined the wave functions $\phi_n(x)$ for the free harmonic oscillator, and in Prob. 3-11 we evaluated the kernel describing the motion of a forced harmonic oscillator. This means that we can determine the amplitude G_{mn} by direct substitution into

$$G_{mn} = e^{(i/\hbar)E_m T} \int_{-\infty}^{\infty}\int_{-\infty}^{\infty} \phi_m(x_b) K(x_b,T;x_a,0)\phi_n(x_a)\,dx_a\,dx_b \tag{8.137}$$

For the case $m=n=0$ this integral is a gaussian somewhat lengthy to evaluate but presenting no special problems. The result° is

$$G_{00} = \exp\left\{-\frac{1}{2\hbar M\omega}\int_0^T\int_0^t f(t)f(s)e^{-i\omega(t-s)}\,ds\,dt\right\} \tag{8.138}$$

If m and n are not equal to 0, then the integral is somewhat more complicated. However, we can use the same sort of trick that we used in Prob. 8-1. We shall ask for the amplitude that a forced harmonic oscillator goes from the state ψ to the state χ, where these two states are defined in Prob. 8-1. This amplitude is (using Eq. 8.28)

$$\begin{aligned} F(b,a) &= \sum_{m=0}^{\infty}\sum_{n=0}^{\infty} G_{mn}\psi_m^*(b)\psi_n(a)e^{-(i/\hbar)E_m T} \\ &= \exp\left\{-\frac{M\omega}{4\hbar}(b^2+a^2)\right\} \\ &\quad\times \sum_{m=0}^{\infty}\sum_{n=0}^{\infty} G_{mn}\frac{b^m a^n}{\sqrt{m!n!}}\left(\frac{M\omega}{2\hbar}\right)^{(m+n)/2} e^{-i(m+1/2)\omega T} \end{aligned} \tag{8.139}$$

If we can work out $F(b,a)$, we can get G_{mn} through multiplying F by $\exp\{(M\omega/4\hbar)(b^2+a^2)\}$ and developing the resulting expression in a power series in a and b. That is, we want first to solve

$$\begin{aligned} F(b,a) &= \left(\frac{M\omega}{\pi\hbar}\right)^{1/2} \\ &\quad\times \int_{-\infty}^{\infty}\int_{-\infty}^{\infty} e^{-(M\omega/2\hbar)(x_b-b)^2}K(x_b,T;x_a,0)e^{-(M\omega/2\hbar)(x_a-a)^2}\,dx_a\,dx_b \end{aligned} \tag{8.140}$$

where $K(x_b,T;x_a,0)$ is the kernel for a forced harmonic oscillator, Eq. (3.66). The variables appear only quadratically in the exponent of this integrand, so that the integration can be performed easily. Some of the resulting algebra is a little lengthy; however, eventually one finds

$$F(b,a) = \exp\left\{-\frac{i\omega T}{2} - \frac{M\omega}{4\hbar}(a^2 + b^2 - 2abe^{-i\omega T}) \right. \tag{8.141}$$
$$\left. + i\sqrt{\frac{M\omega}{2\hbar}}(a\beta + b\beta^* e^{-i\omega T}) - \frac{1}{2\hbar M\omega}\int_0^T \int_0^t f(t)f(s)e^{-i\omega(t-s)}\, ds\, dt\right\}$$

where

$$\beta = \frac{1}{\sqrt{2\hbar M\omega}}\int_0^T f(t)e^{-i\omega t}\, dt \tag{8.142}$$

$$\beta^* = \frac{1}{\sqrt{2\hbar M\omega}}\int_0^T f(t)e^{+i\omega t}\, dt \tag{8.143}$$

The value of G_{00} can be obtained easily from Eq. (8.141) by setting $a = b = 0$. The result is the same as Eq. (8.138). Next we multiply by the exponential function, as described below Eq. (8.139), and find, by putting

$$x = \sqrt{\frac{M\omega}{2\hbar}}\, a \qquad y = \sqrt{\frac{M\omega}{2\hbar}}\, be^{-i\omega T}$$

that

$$\sum_{m=0}^{\infty}\sum_{n=0}^{\infty} G_{mn}\frac{y^m x^n}{\sqrt{m!n!}} = G_{00}\exp\{xy + i\beta x + i\beta^* y\} \tag{8.144}$$

By expanding the right-hand side in powers of x and y and comparing terms, we obtain the final result

$$G_{mn} = \frac{G_{00}}{\sqrt{m!n!}}\sum_{r=0}^{l}\frac{m!}{(m-r)!r!}\frac{n!}{(n-r)!r!}r!(i\beta)^{n-r}(i\beta^*)^{m-r} \tag{8.145}$$

where l is the smaller of m or n.

This completely solves the problem of a forced harmonic oscillator. We shall discuss it further and make use of it in the next chapter.

9

Quantum Electrodynamics

IN this chapter we shall discuss the interaction between charged particles and an electromagnetic field. We have seen one example of such an interaction in Sec. 7-6, where the electromagnetic field variables entered into the potential term of the lagrangian. The electromagnetic term introduced in that section is the vector potential $\mathbf{A}$. Section 7-6 deals only with the motion in a definite given field. It does not tell us anything about how the field $\mathbf{A}$ arises or how it is affected by the moving particles. That is, the formulation of the problem does not contain any analysis of the dynamics of the field. Such an approach, using given potentials, is only an approximation. It is valid when these potentials arise from such large pieces of apparatus that the motion of the particle does not affect the potential.

In this chapter we shall be concerned not only with the way in which potentials affect the motion of the particle, but also with the way in which the particle affects the potentials. We shall start with the classical approach and use Maxwell's equations to describe the dynamics of the electromagnetic field. These equations express the field in terms of the charge and current density of the matter present.

We have found in preceding chapters that the quantum-mechanical laws which correspond to some classical system can be easily determined if only we can express the classical laws in the form of a least-action principle. Thus we have found that if the extremum of some action S, varied with respect to some variable x, corresponds to the classical equation of motion, then the quantum-mechanical laws are expressed as follows: The quantum-mechanical amplitude for any given situation, corresponding to the action S, is the path integral of $e^{iS/\hbar}$ integrated over all possible paths of the variable x which fit the conditions of the situation.

It is vital to our present approach that classical electrodynamics, as expressed by Maxwell's equations, can be written as a principle of least action. An action S exists which can be expressed in terms of the vector and scalar potentials $\mathbf{A}$ and ϕ. The determination of an extremum for this action, by variation of the field variables $\mathbf{A}(\mathbf{r},t)$ and $\phi(\mathbf{r},t)$, leads to a formulation of electrodynamics equivalent to Maxwell's equations. Hence, quantum electrodynamics results from the rule that the amplitude for an event is

$$K(b,a) = \int_a^b e^{iS[\mathbf{A},\phi]/\hbar}\, \mathcal{D}\mathbf{A}(\mathbf{r},t)\, \mathcal{D}\phi(\mathbf{r},t) \tag{9.1}$$

where the path integral is over all values of $\mathbf{A}$ and ϕ at each point of space and time, subject to the boundary conditions at the initial and final points of the event (cf. Eq. 8.128).

9-1 CLASSICAL ELECTRODYNAMICS

Maxwell's Equations. We shall begin our study of electrodynamics from the customary classical fundamentals, i.e., from Maxwell's equations. We shall assume the magnetic permeability and dielectric constant are those for free space. Then, with $\mathbf{E}$ as the electric field vector, $\mathbf{B}$ as the magnetic field vector, c as the speed of light, ρ as the charge density, and $\mathbf{j}$ as the current density, Maxwell's equations are

$$\nabla \cdot \mathbf{E} = 4\pi\rho \tag{9.2}$$

$$\nabla \cdot \mathbf{B} = 0 \tag{9.3}$$

$$\nabla \times \mathbf{E} = -\frac{1}{c}\frac{\partial \mathbf{B}}{\partial t} \tag{9.4}$$

$$\nabla \times \mathbf{B} = \frac{1}{c}\left(\frac{\partial \mathbf{E}}{\partial t} + 4\pi\mathbf{j}\right) \tag{9.5}$$

These equations make sense only if charge is conserved, that is,

$$\nabla \cdot \mathbf{j} = -\frac{\partial \rho}{\partial t} \tag{9.6}$$

Equation (9.3) implies that $\mathbf{B}$ is the curl of some vector $\mathbf{A}$:

$$\mathbf{B} = \nabla \times \mathbf{A} \tag{9.7}$$

This relation does not fully determine $\mathbf{A}$; we still may specify its divergence. We choose

$$\nabla \cdot \mathbf{A} = 0 \tag{9.8}$$

This choice is not recommended if it is desired to keep the full relativistic four-dimensional symmetry of the equations in evidence. (It is not that the results using Eq. (9.8) are not relativistically invariant; for the results are independent of the choice of $\nabla \cdot \mathbf{A}$. It is, rather, that the invariance does not appear obvious at first glance.) In our case we shall deal with matter in the nonrelativistic approximation anyway (for we do not have a simple path integral for the Dirac equation). We wish to illustrate the properties of the quantized electromagnetic field, and the results are least cumbersome with the choice of Eq. (9.8).

Substitution into Eq. (9.4) shows that $\mathbf{E} + (1/c)\,\partial\mathbf{A}/\partial t$ has zero curl, so it must be the gradient of some scalar potential

$$\mathbf{E} = -\nabla\phi - \frac{1}{c}\frac{\partial \mathbf{A}}{\partial t} \tag{9.9}$$

From Eqs. (9.2), (9.8), and (9.9) we see that

$$\boldsymbol{\nabla}\cdot\mathbf{E} = -\nabla^2\phi = 4\pi\rho \tag{9.10}$$

If there is no charge and no current density, the equations are easily solved. In Eq. (9.10) $\rho = 0$, so $\phi = 0$ and $\mathbf{E} = -(1/c)\,\partial\mathbf{A}/\partial t$. In Eq. (9.5) with $\mathbf{j} = 0$ this gives [*note:* $\boldsymbol{\nabla}\times(\boldsymbol{\nabla}\times\mathbf{A}) = \boldsymbol{\nabla}(\boldsymbol{\nabla}\cdot\mathbf{A}) - \nabla^2\mathbf{A}$]

$$\nabla^2\mathbf{A} - \frac{1}{c^2}\frac{\partial^2\mathbf{A}}{\partial t^2} = 0 \tag{9.11}$$

Thus, each component of $\mathbf{A}$ satisfies a wave equation.

If we assume $\mathbf{A}$ is a running plain wave, that is,

$$\mathbf{A}(\mathbf{r}, t) = \mathbf{a}_{\mathbf{k}}(t)e^{i\mathbf{k}\cdot\mathbf{r}} \tag{9.12}$$

the equation for the amplitude $\mathbf{a}_{\mathbf{k}}$ is $\ddot{\mathbf{a}}_{\mathbf{k}} = -k^2c^2\mathbf{a}_{\mathbf{k}}$, which implies that $\mathbf{a}_{\mathbf{k}}$ is a simple harmonic oscillator with frequency $\omega = kc$ for each component direction of $\mathbf{a}$. Actually there are only two independent transverse waves; the component of $\mathbf{a}_{\mathbf{k}}$ in the direction of $\mathbf{k}$ must be zero. This is the implication of Eq. (9.8), which can be rewritten as

$$\mathbf{k}\cdot\mathbf{a}_{\mathbf{k}} = 0 \tag{9.13}$$

Thus the field in free space is equivalent to a set of free harmonic oscillators with two transverse waves for each value of $\mathbf{k}$.

Problem 9-1 Show that $\mathbf{E}$, $\mathbf{B}$, and $\mathbf{k}$ are mutually perpendicular for this plane-wave solution.

Solution with Charges and Currents Present. We shall expand the solutions for $\mathbf{A}$, ϕ, and the current and charge density in plane waves, writing

$$\begin{aligned}
\mathbf{A}(\mathbf{r}, t) &= \sqrt{4\pi}\,c\int \mathbf{a}_{\mathbf{k}}(t)e^{i\mathbf{k}\cdot\mathbf{r}}\frac{d^3\mathbf{k}}{(2\pi)^3}\\
\phi(\mathbf{r}, t) &= \int \phi_{\mathbf{k}}(t)e^{i\mathbf{k}\cdot\mathbf{r}}\frac{d^3\mathbf{k}}{(2\pi)^3}\\
\mathbf{j}(\mathbf{r}, t) &= \int \mathbf{j}_{\mathbf{k}}(t)e^{i\mathbf{k}\cdot\mathbf{r}}\frac{d^3\mathbf{k}}{(2\pi)^3}\\
\rho(\mathbf{r}, t) &= \int \rho_{\mathbf{k}}(t)e^{i\mathbf{k}\cdot\mathbf{r}}\frac{d^3\mathbf{k}}{(2\pi)^3}
\end{aligned} \tag{9.14}$$

Problem 9-2 Explain why the charge density corresponding to a single charge e located at the point $\mathbf{x}(t) = (x(t), y(t), z(t))$ at time t is

$$\rho(\mathbf{r}, t) = \mathrm{e}\delta(r_x - x(t))\delta(r_y - y(t))\delta(r_z - z(t)) = \mathrm{e}\delta^3(\mathbf{r} - \mathbf{x}(t))$$

Show that

$$\rho_{\mathbf{k}}(t) = \mathrm{e}e^{-i\mathbf{k}\cdot\mathbf{x}(t)} \tag{9.15}$$

Explain why the current density is $\mathbf{j}(\mathbf{r}, t) = \mathrm{e}\dot{\mathbf{x}}(t)\delta^3(\mathbf{r} - \mathbf{x}(t))$. If we have a number of charges e_i located at $\mathbf{x}_i(t)$, the values of $\rho_{\mathbf{k}}$ and $\mathbf{j}_{\mathbf{k}}$ are

$$\rho_{\mathbf{k}} = \sum_i \mathrm{e}_i e^{-i\mathbf{k}\cdot\mathbf{x}_i(t)} \qquad \mathbf{j}_{\mathbf{k}} = \sum_i \mathrm{e}_i \dot{\mathbf{x}}_i(t) e^{-i\mathbf{k}\cdot\mathbf{x}_i(t)} \tag{9.16}$$

If the expansions of $\mathbf{E}$ and $\mathbf{B}$ are

$$\mathbf{E}(\mathbf{r}, t) = \int \mathbf{E}_{\mathbf{k}}(t) e^{i\mathbf{k}\cdot\mathbf{r}} \frac{d^3\mathbf{k}}{(2\pi)^3} \qquad \mathbf{B}(\mathbf{r}, t) = \int \mathbf{B}_{\mathbf{k}}(t) e^{i\mathbf{k}\cdot\mathbf{r}} \frac{d^3\mathbf{k}}{(2\pi)^3}$$

then, using Eqs. (9.9) and (9.7), the expansion coefficients satisfy

$$\mathbf{E}_{\mathbf{k}} = -i\mathbf{k}\phi_{\mathbf{k}} - \sqrt{4\pi}\,\dot{\mathbf{a}}_{\mathbf{k}} \quad \text{and} \quad \mathbf{B}_{\mathbf{k}} = \sqrt{4\pi}\, c\, i(\mathbf{k} \times \mathbf{a}_{\mathbf{k}})$$

From Eqs. (9.8) and (9.10), the coefficient of expansion of $\boldsymbol{\nabla} \cdot \mathbf{E}$ is $i\mathbf{k}\cdot\mathbf{E}_{\mathbf{k}} = k^2\phi_{\mathbf{k}}$, so we have

$$k^2\phi_{\mathbf{k}} = 4\pi\rho_{\mathbf{k}} \tag{9.17}$$

or $\phi_{\mathbf{k}} = 4\pi\rho_{\mathbf{k}}/k^2$. The function $\phi_{\mathbf{k}}$ is completely determined in terms of the charge density $\rho_{\mathbf{k}}$; there are no dynamic differential equations to solve, involving, for example, $\ddot{\phi}_{\mathbf{k}}$.

Problem 9-3 Prove that the relation $\phi_{\mathbf{k}} = 4\pi\rho_{\mathbf{k}}/k^2$ simply means that $\phi_{\mathbf{k}}$ at any instant is the Coulomb potential from the charges at that instant, so that, for example, if ρ comes from a number of charges e_i at distances R_i from a point, the potential at the point is $\phi = \sum_i \mathrm{e}_i/R_i$. This is just the content of Eq. (9.10).

Equation (9.5) still remains to be solved, that is,

$$i\mathbf{k} \times \mathbf{B}_{\mathbf{k}} = \frac{1}{c}\dot{\mathbf{E}}_{\mathbf{k}} + \frac{1}{c}4\pi\mathbf{j}_{\mathbf{k}} \tag{9.18}$$

But (using $\mathbf{k}\cdot\mathbf{a}_{\mathbf{k}} = 0$)

$$i\mathbf{k} \times \mathbf{B}_{\mathbf{k}} = -\sqrt{4\pi}\, c\mathbf{k} \times (\mathbf{k} \times \mathbf{a}_{\mathbf{k}}) = \sqrt{4\pi}\, ck^2\mathbf{a}_{\mathbf{k}}$$

and $\dot{\mathbf{E}}_{\mathbf{k}} = -i\mathbf{k}\dot{\phi}_{\mathbf{k}} - \sqrt{4\pi}\,\ddot{\mathbf{a}}_{\mathbf{k}}$, and using Eq. (9.17) to express $\dot{\phi}_{\mathbf{k}}$ as $4\pi\dot{\rho}_{\mathbf{k}}/k^2$, we get

$$\ddot{\mathbf{a}}_{\mathbf{k}} + k^2c^2\mathbf{a}_{\mathbf{k}} = \sqrt{4\pi}\left(\mathbf{j}_{\mathbf{k}} - \frac{i\mathbf{k}\dot{\rho}_{\mathbf{k}}}{k^2}\right) = \sqrt{4\pi}\,\mathbf{j}'_{\mathbf{k}} \tag{9.19}$$

where we can call $\mathbf{j}'_{\mathbf{k}} = \mathbf{j}_{\mathbf{k}} - i\mathbf{k}\dot{\rho}_{\mathbf{k}}/k^2$ the transverse part of $\mathbf{j}_{\mathbf{k}}$. The law of conservation of current, expressed by Eq. (9.6), says that $\dot{\rho}_{\mathbf{k}} = -i\mathbf{k}\cdot\mathbf{j}_{\mathbf{k}}$, so

$$\mathbf{j}'_{\mathbf{k}} = \mathbf{j}_{\mathbf{k}} - \frac{\mathbf{k}(\mathbf{k}\cdot\mathbf{j}_{\mathbf{k}})}{k^2} \tag{9.20}$$

which means that $\mathbf{j}'_{\mathbf{k}}$ is $\mathbf{j}_{\mathbf{k}}$ less its component in the direction of $\mathbf{k}$. Clearly, $\mathbf{k}\cdot\mathbf{j}'_{\mathbf{k}} = 0$.

We have certainly reduced Maxwell's equations to a very simple form — aside from the instantaneous Coulomb interaction between particles, we have no more than the equations for two transverse waves for each value of $\mathbf{k}$, the amplitude of each being a harmonic oscillator driven by the component of current in the corresponding direction. That is, if we choose two directions perpendicular to $\mathbf{k}$, say 1 and 2, and call the components of $\mathbf{a}_{\mathbf{k}}$ in these directions $a_{1,\mathbf{k}}$ and $a_{2,\mathbf{k}}$, Maxwell's equations are

$$\ddot{a}_{1,\mathbf{k}} + k^2c^2a_{1,\mathbf{k}} = \sqrt{4\pi}\, j_{1,\mathbf{k}} \tag{9.21}$$

$$\ddot{a}_{2,\mathbf{k}} + k^2c^2a_{2,\mathbf{k}} = \sqrt{4\pi}\, j_{2,\mathbf{k}} \tag{9.22}$$

where $j_{1,\mathbf{k}}$ and $j_{2,\mathbf{k}}$ are the components of $\mathbf{j}_{\mathbf{k}}$ in these directions. (Why do we not need to say "of $\mathbf{j}'_{\mathbf{k}}$"?)

The Least-action Principle. The hypothesis of quantum electrodynamics[1] is that the oscillators defined in Eqs. (9.21) and (9.22) are quantum oscillators. To carry out the quantization, we must find the principle of least action which gives these defining equations of motion as well as the equations of motion of the particles in the field. The action is

$$S = S_1 + S_2 + S_3 \tag{9.23}$$

where

$$S_1 = \sum_i \frac{m_i}{2} \int |\dot{\mathbf{x}}_i|^2\, dt \tag{9.24}$$

is the action of all the particles, disregarding the field (if there are nonelectric forces between the particles, they are to be included in S_1),

$$\begin{aligned} S_2 &= -\iint \left[\rho(\mathbf{r},t)\phi(\mathbf{r},t) - \frac{1}{c}\mathbf{j}(\mathbf{r},t)\cdot\mathbf{A}(\mathbf{r},t)\right] d^3\mathbf{r}\, dt \\ &= -\sum_i e_i \int \left[\phi(\mathbf{x}_i(t),t) - \frac{1}{c}\dot{\mathbf{x}}_i(t)\cdot\mathbf{A}(\mathbf{x}(t),t)\right] dt \end{aligned} \tag{9.25}$$

[1] It should be pointed out here that some physicists use the term "quantum electrodynamics" to include electron-positron pair theory. In the present chapter we do not cover such problems. So for us, quantum electrodynamics means the quantum theory of the electromagnetic field.

is the action of interaction of field and particles, and

$$S_3 = \frac{1}{8\pi}\iint [E^2 - B^2]\, d^3\mathbf{r}\, dt$$
$$= \frac{1}{8\pi}\iint \left[\left|-\boldsymbol{\nabla}\phi - \frac{1}{c}\frac{\partial \mathbf{A}}{\partial t}\right|^2 - |\boldsymbol{\nabla}\times\mathbf{A}|^2\right] d^3\mathbf{r}\, dt \tag{9.26}$$

is the action of the field. The variables are $\mathbf{A}(\mathbf{r},t)$, $\phi(\mathbf{r},t)$, and $\mathbf{x}_i(t)$.

Problem 9-4 In Sec. 2-1 we discussed the mechanisms for obtaining the mechanical equations of motion from the form of the action S by obtaining the extremum S_{cl} under the condition $\delta S = 0$ for variations of the coordinates, $\delta\mathbf{x}$. Show how Maxwell's equations can be derived from the action S defined in Eq. (9.23) by requiring $\delta S = 0$ for first-order variations of $\mathbf{A}$ and ϕ.

Since the dynamic equations are simplest in terms of the variables $\mathbf{a_k}$, it is worthwhile to express the action in these variables. Substitution of the expansion given in Eqs. (9.14) into S_3 gives

$$S_3 = \frac{1}{2}\iint \left[\left|\dot{\mathbf{a}}_\mathbf{k} + i\mathbf{k}\frac{\phi_\mathbf{k}}{\sqrt{4\pi}}\right|^2 - c^2|\mathbf{k}\times\mathbf{a}_\mathbf{k}|^2\right] \frac{d^3\mathbf{k}\, dt}{(2\pi)^3}$$
$$= \frac{1}{2}\iint \left[|\phi_\mathbf{k}|^2\frac{k^2}{4\pi} + \dot{\mathbf{a}}_\mathbf{k}^*\cdot\dot{\mathbf{a}}_\mathbf{k} - k^2c^2\mathbf{a}_\mathbf{k}^*\cdot\mathbf{a}_\mathbf{k}\right] \frac{d^3\mathbf{k}\, dt}{(2\pi)^3} \tag{9.27}$$

and S_2 becomes

$$S_2 = -\iint \left[\rho_{-\mathbf{k}}\phi_\mathbf{k} - \sqrt{4\pi}\,\mathbf{j}_{-\mathbf{k}}\cdot\mathbf{a}_\mathbf{k}\right] \frac{d^3\mathbf{k}\, dt}{(2\pi)^3} \tag{9.28}$$

Upon substitution of $\phi_\mathbf{k} = 4\pi\rho_\mathbf{k}/k^2$, the terms in $\phi_\mathbf{k}$ in S_2 and S_3 add to give

$$S_c = -\frac{4\pi}{2}\iint \frac{\rho_\mathbf{k}\rho_{-\mathbf{k}}}{k^2}\frac{d^3\mathbf{k}\, dt}{(2\pi)^3} = -\frac{1}{2}\int\sum_i\sum_j \frac{e_i e_j}{|\mathbf{x}_i - \mathbf{x}_j|}\, dt \tag{9.29}$$

using Eqs. (9.16), since $\int (4\pi/k^2)e^{i\mathbf{k}\cdot\mathbf{r}}\, d^3\mathbf{k}/(2\pi)^3 = 1/r$. This is just the Coulomb interaction between the charges, which is usually considered in analyzing atoms when electromagnetic radiation effects are neglected. That is, we shall include this interaction in the action S_{mat} of the matter,

$$S_{\text{mat}} = S_1 + S_c = \int\sum_i\left(\frac{m_i}{2}\dot{\mathbf{x}}_i^2 - \frac{1}{2}\sum_j\frac{e_i e_j}{|\mathbf{x}_i - \mathbf{x}_j|}\right) dt \tag{9.30}$$

and write $S = S_{\text{mat}} + S_{\text{int}} + S_{\text{rad}}$. We have thus divided the action of the electromagnetic field S_2 into two parts; one contributes to an instantaneous Coulomb interaction, and the remainder we shall call the radiation field S_{rad}. (The radiation field takes care of all corrections to the instantaneous field, such as, for example, that the total effects are retarded and act no faster than the speed of light.) The action of the radiation field is S_3 less the terms involving $\phi_{\mathbf{k}}$. That is,

$$S_{\text{rad}} = \frac{1}{2}\iint (\dot{a}^*_{1,\mathbf{k}}\dot{a}_{1,\mathbf{k}} - k^2c^2a^*_{1,\mathbf{k}}a_{1,\mathbf{k}} + \dot{a}^*_{2,\mathbf{k}}\dot{a}_{2,\mathbf{k}} - k^2c^2a^*_{2,\mathbf{k}}a_{2,\mathbf{k}})\frac{d^3\mathbf{k}\,dt}{(2\pi)^3} \tag{9.31}$$

which is just the action of the radiation oscillators. The action of interactions of these oscillators with the particles is

$$S_{\text{int}} = \sqrt{4\pi}\iint (j_{1,-\mathbf{k}}a_{1,\mathbf{k}} + j_{2,-\mathbf{k}}a_{2,\mathbf{k}})\frac{d^3\mathbf{k}\,dt}{(2\pi)^3} \tag{9.32}$$

Clearly, the variation of the total action of S with respect to the $a_{1,\mathbf{k}}$ and $a_{2,\mathbf{k}}$ gives the equations of motion (9.21) and (9.22).

Written more explicitly, the action S_{int} is

$$S_{\text{int}} = \sqrt{4\pi}\sum_j \mathrm{e}_j\iint (\dot{x}_{1j}a_{1,\mathbf{k}} + \dot{x}_{2j}a_{2,\mathbf{k}})e^{i\mathbf{k}\cdot\mathbf{x}_j(t)}\frac{d^3\mathbf{k}\,dt}{(2\pi)^3} \tag{9.33}$$

where x_{1j} and x_{2j} are the components of $\mathbf{x}_j$ in the direction transverse to $\mathbf{k}$. Thus all the laws of nonrelativistic mechanics and of electrodynamics are contained in the proposition that S, the sum of Eqs. (9.30), (9.31), and (9.33), is stationary for variations in the paths of the variables $\mathbf{x}_j(t)$, $a_{1,\mathbf{k}}(t)$, and $a_{2,\mathbf{k}}(t)$. Quantum electrodynamics results from integrating $e^{iS/\hbar}$ over these paths, and it is described in the next section.

Problem 9-5 The momentum in the field is given by

$$\frac{1}{4\pi c}\int \mathbf{E}\times\mathbf{B}\,d^3\mathbf{r}$$

In the absence of matter (so $\phi_{\mathbf{k}} = 0$), show this is $i\int \mathbf{k}(\mathbf{a}^*_{\mathbf{k}}\cdot\dot{\mathbf{a}}_{\mathbf{k}})\,d^3\mathbf{k}/(2\pi)^3$.

9-2 THE QUANTUM MECHANICS OF THE RADIATION FIELD

We begin by discussing the quantum mechanics of the radiation field in empty space. There is no matter present, so the total action is that of the radiation field alone

$$S = S_{\text{rad}} \tag{9.34}$$

as given by Eq. (9.31). It is evidently the action of a set of harmonic oscillators. We have seen some examples of expressions like Eq. (9.31) in Chap. 8. We make the assumption discussed in Sec. 8-8 that quantum electrodynamics results from considering these as quantum-mechanical oscillators.

The modes of our system are running waves, two for each value of $\mathbf{k}$ (polarization 1 and 2) with frequency $\omega = kc$. For one of the modes, say $a_{1,\mathbf{k}}$, the available energy levels are

$$E_{1,\mathbf{k}} = (n_{1,\mathbf{k}} + \tfrac{1}{2})\hbar kc \tag{9.35}$$

where $n_{1,\mathbf{k}}$ is any positive integer or zero.

If $n_{1,\mathbf{k}} = 1$, we say there is one photon present of polarization 1 and momentum $\hbar\mathbf{k}$; in general, we say $n_{1,\mathbf{k}}$ such photons are present. The energy of a single photon of this kind is $\hbar kc$.

Later on when we consider the interactions of matter with the radiation field, we shall find that the matter absorbs or emits one photon at a time of energy $\hbar\omega$. Of course, this is the same as Planck's original hypothesis.

It is quite striking and surprising that the states n of the oscillators can also be described by imagining that there are n "particles" or "photons" present. It is clear, of course, that the energy values agree. But there is one further subtle point that must be noted before the oscillator states can be completely successfully described as particles. Suppose, for example, that just two of the n_j differ from zero, say, $n_a = 1$, $n_b = 1$. This single state we may wish to represent by saying that we have one photon in level a and another in level b. But at first sight this way of speaking might seem to imply that there were two states available, both of the same energy. For we could also expect to be able to put the first photon in level b and the second in level a. The way out of this can be seen when we consider the example of alpha particles. Suppose we have two alpha particles with coordinates x and y, and say the x particle is in a level represented by $f(x)$ and the y particle is in a level $g(y)$. Thus the wave function for the system would be

$$\psi(x,y) = f(x)g(y) \tag{9.36}$$

a function of the two variables x, y. But another state might have y in the level f and x in g, leading to another state of wave function

$$\psi(x,y) = g(x)f(y) \tag{9.37}$$

which differs from the first. But if the particles are truly identical, like alpha particles, the two states are indistinguishable. As we described in

Sec. 1-3, it turns out to be a rule of quantum mechanics (not derivable from the Schrödinger equation) that for alpha particles the amplitudes for two cases which differ only by exchange of the alpha particles must always be added. The only allowed wave function is in this case

$$\psi(x, y) = f(x)g(y) + g(x)f(y) \tag{9.38}$$

(suitably normalized: if f and g are orthonormal, the factor is $1/\sqrt{2}$; if $f = g$ and they are normalized, it is $1/2$). In general $\psi(x, y) = \psi(y, x)$ for alpha particles and for other particles obeying Bose statistics. There is, for such particles, only one state: one particle in level f, the other in level g.

It turns out that all the results are consistent if, when we consider oscillator excitation states as representing numbers of photons, we also say that photons are Bose particles. Then the single state $n_a = 1$, $n_b = 1$ represents the situation that there are two photons, one in a, one in b. Exchange does not produce a new state.

For electrons of parallel spin or other Fermi particles we must subtract the amplitudes when the identity of the particles is reversed.

$$\psi(x, y) = f(x)g(y) - g(x)f(y) \tag{9.39}$$

The wave function $\psi(x, y) = -\psi(y, x)$ is antisymmetric in general for Fermi particles. This is, of course, also just one state. But for Fermi particles, two identical particles cannot occupy the same level. If we put $f = g$ into Eq. (9.39), we get zero. Two photons, like two alpha particles, *can* occupy the same level; for photons it corresponds to the $n = 2$ oscillator levels.

There is one particular situation with matter present which, in the ideal case, can be handled nearly as simply as the matter-free case. That is the case of a cavity resonator (or a wave guide) where the walls may be idealized as perfect conductors. Then classically, as is well known, there are a number of possible oscillator modes with more or less complicated distributions of electric fields. The classical action is then reducible to a set of free oscillators, but the variables now represent the amplitudes of the various modes, rather than the amplitudes of plane running waves. These oscillators are then analyzed as quantum oscillators, and we speak of the number of photons in each mode.

9-3 THE GROUND STATE

Vacuum Energy. The state of the electromagnetic field of lowest possible energy, which we shall call the ground state or the vacuum state, is that in which there are no photons in any mode. This means that the energy in each mode is $\hbar\omega/2$, where ω is the frequency of the mode. Now if we were to sum this ground-state energy over all of the infinite number of possible modes of ever-increasing frequency which exist even for a finite box, the answer would be infinity. This is the first symptom of the difficulties which beset quantum electrodynamics.

In the present case, for the vacuum state, the trouble is easily fixed. Suppose we choose to measure energy from a different zero point. Since there is no physical effect resulting from a constant energy, the result of any experiment we perform will be insensitive to the arbitrary choice of the zero point in energy. Therefore, we assign to the vacuum state the energy zero. Then the total energy in any state of the electromagnetic field is given by

$$E = \sum_j n_j \hbar\omega_j \tag{9.40}$$

where the sum is taken over all the modes j of the field.

Unfortunately, it is really not true that the zero point of energy can be assigned completely arbitrarily. Energy is equivalent to mass, and mass has a gravitational effect. Even light has a gravitational effect, for light is deflected by the sun. So, if the law that action equals reaction has qualitative validity, then the sun must be attracted by the light. This means that a photon of energy $\hbar\omega$ has a gravity-producing effect, and the question is: Does the ground-state energy term $\hbar\omega/2$ also have an effect? The question stated physically is: Does a vacuum act like a uniform density of mass in producing a gravitational field?

Since most of the space is a vacuum, any effect of the vacuum-state energy of the electromagnetic field would be large. We can estimate its magnitude. First, it should be pointed out that some other infinities occurring in quantum-electrodynamic problems are avoided by a particular assumption called the *cutoff rule.* This rule states that those modes having very high frequencies (short wavelengths) are to be excluded from consideration. The rule is justified on the grounds that we have no evidence that the laws of electrodynamics are obeyed for wavelengths shorter than any which have yet been observed. In fact, there is a good reason to believe that the laws cannot be extended to the short-wavelength region. Mathematical representations which work quite well at longer wavelengths lead to divergences if extended into the short-wavelength region. The wavelengths in question are of the order of the Compton wavelength of the proton; $1/2\pi$ times this wavelength is $\hbar/m_p c \approx 2 \times 10^{-14}$ cm.

For our present estimate suppose we carry out sums over wave numbers only up to the limiting value $k_{\max} = m_p c/\hbar$. Approximating the sum over levels by an integral, we have, for the vacuum-state energy per unit volume,

$$\frac{E_0}{\text{unit vol}} = 2\int_0^{k_{\max}} \frac{\hbar kc}{2}\frac{4\pi k^2\,dk}{(2\pi)^3} = \frac{\hbar c k_{\max}^4}{8\pi^2} \tag{9.41}$$

(Note the first factor of 2, for there are two modes for each $\mathbf{k}$.) The equivalent mass of this energy is obtained by dividing the result by c^2. This gives

$$\frac{m_0}{\text{unit vol}} = 2 \times 10^{15}\ \text{g/cm}^3 \tag{9.42}$$

Such a mass density would, at first sight at least, be expected to produce very large gravitational effects which are not observed. It is possible that we are calculating in a naive manner, and, if all of the consequences of the general theory of relativity (such as the gravitational effects produced by the large stresses implied here) were included, the effects might cancel out; but nobody has worked all this out. It is possible that some cutoff procedure that not only yields a finite energy density for the vacuum state but also provides relativistic invariance may be found. The implications of such a result are at present completely unknown.

For the present we are safe in assigning the value zero to the vacuum-state energy density. Up to the present time no experiments that would contradict this assumption have been performed. As we progress further into the field of quantum electrodynamics we shall find other divergent integrals which are more difficult to circumvent.

Vacuum Wave Function. The wave function for the set of oscillators is just the product of the wave functions for each mode. For the ground state the wave function of the oscillator 1, $\mathbf{k}$ is (see Eq. 8.83) proportional to $\exp\{-(kc/2\hbar)\bar{a}_{1,\mathbf{k}}^*\bar{a}_{1,\mathbf{k}}\}$, where

$$\bar{a}_{1,\mathbf{k}} \equiv a_{1,\mathbf{k}}/\sqrt{\text{Vol}}$$

and "Vol" represents the volume of the normalizing box (see Sec. 4-3). Thus the wave function for the entire system in the ground state, or vacuum state, is, within a normalization constant,

$$\Phi_0 = \exp\left\{-\sum_{\mathbf{k}} \frac{kc}{2\hbar}(\bar{a}_{1,\mathbf{k}}^*\bar{a}_{1,\mathbf{k}} + \bar{a}_{2,\mathbf{k}}^*\bar{a}_{2,\mathbf{k}})\right\} \tag{9.43}$$

Problem 9-6 Show, using sine and cosine modes and real variables, that this expression using complex variables is indeed correct (cf. Prob. 8-4).

Problem 9-7 Show, for the vacuum state, the expectation value of $\bar{a}^*_{1,\mathbf{k}}\bar{a}_{1,\mathbf{q}}$ is $(\hbar/2kc)\delta_{\mathbf{k},\mathbf{q}}$ and that of $\bar{a}_{1,\mathbf{k}}\bar{a}_{1,\mathbf{q}}$ is $(\hbar/2kc)\delta_{-\mathbf{k},\mathbf{q}}$. Develop a formula for the expectation of $(\bar{a}^*_{1,\mathbf{k}}\bar{a}_{1,\mathbf{k}})^r$ for integral r and explain thereby how the expectation of such quantities as $(\bar{a}^*_{1,\mathbf{k}}\bar{a}_{1,\mathbf{k}})^r(\bar{a}^*_{1,\mathbf{q}}\bar{a}_{1,\mathbf{q}})^s$ can be got for $\mathbf{q} \neq \mathbf{k}$. Show that the expectation of $(\bar{a}_{1,\mathbf{k}})^2$ or $(\bar{a}^*_{1,\mathbf{k}})^2$ vanishes. Show that the expectation of the product of any odd number of $\bar{a}$'s is zero and that you can compute the expectation value of any product of $\bar{a}$'s or $\bar{a}^*$'s for the vacuum state.

Problem 9-8 For the state for which there is just one photon present in level 1, $\mathbf{k}$, all of the factors in the wave function are ϕ_0 except one, which is ϕ_1. But for an oscillator $\phi_1(x) = \sqrt{2}\,x\phi_0(x)$. The wave function representing an excited running wave is a linear superposition of the state with the cosine mode excited and i times the state with the sine wave excited, so show that the unnormalized wave function for just one photon present in level 1, $\mathbf{k}$ is $\bar{a}^*_{1,\mathbf{k}}\Phi_0$. The normalization° is $\int \Phi^*_0\bar{a}_{1,\mathbf{k}}\bar{a}^*_{1,\mathbf{k}}\Phi_0\,d\bar{a}$, or the expectation of $\bar{a}_{1,\mathbf{k}}\bar{a}^*_{1,\mathbf{k}}$ for the vacuum, which we have seen in the preceding problem is $\hbar/2kc$. Hence the normalized one-photon state is $\sqrt{2kc/\hbar}\,\bar{a}^*_{1,\mathbf{k}}\Phi_0$.

9-4 INTERACTION OF FIELD AND MATTER

To deal with the interaction of the radiation field with matter is not difficult in a formal way. Evidently from the action expression of Eqs. (9.30), (9.31), and (9.33) we see we must deal with the matter system interacting with the radiation oscillators and must calculate amplitudes from

$$\text{Amplitude} = \iiint \exp\left\{\frac{i}{\hbar}(S_{\text{mat}} + S_{\text{int}} + S_{\text{rad}})\right\} \prod_{i,\mathbf{k}} \mathcal{D}\mathbf{x}_i\,\mathcal{D}a_{1,\mathbf{k}}\,\mathcal{D}a_{2,\mathbf{k}} \tag{9.44}$$

The coordinates of the radiation oscillators can be integrated out immediately; for they appear only in quadratic expressions. We shall do this integration in the next subsection (starting at Eq. 9.60).

Emission from an Atom. Part of the complication in this problem is simply the confusion produced by so many coordinates and states. So we shall begin by dealing with a simple problem just to get more used to what is involved. We shall solve the problem of the probability of emission of light by a single atom, using perturbation theory (assuming the interaction S_{int} of light and matter is small and expanding it only to the first order).

If S_{int} is neglected, the radiation and matter are independent systems. Let the states of the atom alone have energy E_N for various values of N with wave functions $\psi_N(\mathbf{x})$, where $\mathbf{x}$ represents the $\mathbf{x}_i$ of all the particles of the atom. The state of the radiation can be defined by giving the values of all the integers $n_{1,\mathbf{k}}$ and $n_{2,\mathbf{k}}$. The energy levels of the combined system are

$$E = E_N + \sum_{\mathbf{k}} (n_{1,\mathbf{k}} + n_{2,\mathbf{k}})\hbar kc \tag{9.45}$$

The wave function for this state is a product

$$\Psi = \psi_N(\mathbf{x})\Phi(n_{1,\mathbf{k}}, n_{2,\mathbf{k}}) \tag{9.46}$$

where $\Phi(n_{1,\mathbf{k}}, n_{2,\mathbf{k}})$ is the wave function for the radiation field (a product of harmonic oscillator wave functions).

To deal with atomic radiation of a photon, we consider as the initial state that the atom is in some level M and no photons are present (all $n_{1,\mathbf{k}}$ and $n_{2,\mathbf{k}}$ equal 0). This wave function is

$$\Psi_a = \psi_M(\mathbf{x})\Phi_0 \tag{9.47}$$

with Φ_0 from Eq. (9.43). In the final state the atom is in another level N, but now a photon is present, say of momentum $\hbar\mathbf{q}$ and polarization 1. According to Prob. 9-8 the wave function of the radiation alone is proportional to $\bar{a}^*_{1,\mathbf{q}}\Phi_0$; the complete final wave function is

$$\Psi_b = \psi_N(\mathbf{x})\sqrt{\frac{2qc}{\hbar}}\bar{a}^*_{1,\mathbf{q}}\Phi_0 \tag{9.48}$$

Now to find the transition probability per second (to first order) we see, according to Eq. (6.79), we shall need the matrix element V_{ba} of the perturbation potential between these states. The perturbation action is S_{int} as defined in Eq. (9.32), and the corresponding potential is

$$V = -\sqrt{4\pi}\sum_{\mathbf{k}} \bar{j}_{1,-\mathbf{k}}\bar{a}_{1,\mathbf{k}} \tag{9.49}$$

where $\bar{j}_{1,\mathbf{k}} \equiv j_{1,\mathbf{k}}/\sqrt{\text{Vol}}$ depends on the atomic variables, as in Prob. 9-2. This matrix element is

$$V_{ba} = -\iint \psi_N^* \sqrt{\frac{2qc}{\hbar}} \Phi_0^* \bar{a}_{1,\mathbf{q}} \sqrt{4\pi} \sum_{\mathbf{k}} \bar{j}_{1,-\mathbf{k}} \bar{a}_{1,\mathbf{k}} \psi_M \Phi_0 \, d\mathbf{x} \prod_{\mathbf{k}'} d\bar{a}_{1,\mathbf{k}'} \tag{9.50}$$

$$= -\sqrt{\frac{8\pi qc}{\hbar}} \sum_{\mathbf{k}} \int \psi_N^* \bar{j}_{1,-\mathbf{k}} \psi_M \, d\mathbf{x} \int \Phi_0^* \bar{a}_{1,\mathbf{q}} \bar{a}_{1,\mathbf{k}} \Phi_0 \prod_{\mathbf{k}'} d\bar{a}_{1,\mathbf{k}'} \tag{9.51}$$

because only the currents j depend on $\mathbf{x}$. The expectation values of the a's for the vacuum state were worked out in Prob. 9-7, that is,

$$\int \Phi_0^* \bar{a}_{1,\mathbf{q}} \bar{a}_{1,\mathbf{k}} \Phi_0 \prod_{\mathbf{k}'} d\bar{a}_{1,\mathbf{k}'} = 0$$

unless $\mathbf{k} = -\mathbf{q}$, in which case it is $\hbar/2qc$. Let us write the matrix element $\int \psi_N^* \mathbf{j} \psi_M \, d\mathbf{x}$ as $(\mathbf{j})_{NM}$. Our matrix element is therefore

$$V_{ba} = -\sqrt{2\pi\hbar/qc}\,(\bar{j}_{1,\mathbf{q}})_{NM}.$$

The probability of transition per second is then (see Eq. 6.94)

$$\left(\frac{2\pi}{\hbar}\right)\left(\frac{2\pi\hbar}{qc}\right) |\bar{j}_{1,\mathbf{q}}|^2_{NM} \delta(E_N - E_M + \hbar qc) \tag{9.52}$$

Ordinarily we are not interested in the problem of exciting one particular photon but would rather see the probability of emission of any photon (of polarization 1) into some small solid angle $d\Omega$. We must sum $\mathbf{q}$ over all values which correspond to this direction. The number of values of $\mathbf{q}$ per unit volume is $d^3\mathbf{q}/(2\pi)^3$, or if $\mathbf{q}$ is in the specified direction, we require the integral of $q^2\,dq\,d\Omega/(2\pi)^3$, so that we find that the probability of a transition per second is

$$\frac{dP}{dt} = \int \frac{(2\pi)^2}{qc} |j_{1,\mathbf{q}}|^2_{NM} \delta(E_N - E_M + \hbar qc) q^2 \frac{dq\,d\Omega}{(2\pi)^3} \tag{9.53}$$

The integral on q gives

$$\frac{dP}{dt} = \frac{\omega}{2\pi\hbar c^3} |j_{1,\mathbf{q}}|^2_{NM} \, d\Omega \tag{9.54}$$

for the rate of emission of light of polarization 1 in direction $\mathbf{q}$ into the solid angle $d\Omega$. The frequency emitted satisfies

$$\omega = qc = \frac{E_M - E_N}{\hbar} \tag{9.55}$$

Problem 9-9 For a complicated system moving nonrelativistically

$$(j_{1,\mathbf{k}})_{NM} = \sum_i (\mathrm{e}_i \mathbf{e}_1 \cdot \dot{\mathbf{x}}_i e^{-i\mathbf{k}\cdot\mathbf{x}_i})_{NM} \tag{9.56}$$

where $\mathbf{e}_1$ is a unit vector in the direction of the polarization of the light and e_i and $\mathbf{x}_i$ are the charge and position of the ith particle. Assume the wavelength of the light is very large compared with the size of the atom, i.e., that the absolute square of the wave function describing the position of the ith electron falls to 0 over a distance small compared with $1/k$. Show that we can then approximate $e^{-i\mathbf{k}\cdot\mathbf{x}_i}$ by unity and write the matrix element as

$$(j_{1,\mathbf{k}})_{NM} = i\omega\mathbf{e}_1\cdot\boldsymbol{\mu}_{NM} \tag{9.57}$$

where

$$\boldsymbol{\mu}_{NM} = \sum_i (e_i\mathbf{x}_i)_{NM} \tag{9.58}$$

The function $\boldsymbol{\mu}_{NM}$ is called the *matrix element of the electric dipole moment* of the atom, and the approximation used to derive Eq. (9.57) is called the *dipole approximation.* Show that the total probability to emit light in any direction per unit time is

$$\frac{dP}{dt} = \frac{4\omega^3}{3\hbar c^3}|\boldsymbol{\mu}_{NM}|^2 \tag{9.59}$$

(Integrate Eq. (9.54) over all directions, remembering that $\mathbf{e}_1$ is perpendicular to $\mathbf{k}$ and that there are two possible directions of polarization.)

Elimination of Electromagnetic Field Variables. Since the radiation field is represented by a quadratic action functional, we can integrate out all its coordinates. We shall do so here. We must integrate all the variables $a_{1,\mathbf{k}}$, $a_{2,\mathbf{k}}$ in Eq. (9.44). We must specify the initial and final states of the radiation field. First we shall take the simplest case that initially and finally we have a vacuum, the oscillators all going from 0 to 0 photon number. Our amplitude can be written

$$\text{Amplitude} = \int e^{(i/\hbar)S_{\text{mat}}} X(\mathbf{x}_i) \prod_i \mathcal{D}\mathbf{x}_i \tag{9.60}$$

where

$$X(\mathbf{x}_i) = \iint e^{(i/\hbar)(S_{\text{int}}+S_{\text{rad}})} \prod_{\mathbf{k}} \mathcal{D}a_{1,\mathbf{k}}\,\mathcal{D}a_{2,\mathbf{k}} \tag{9.61}$$

is a function of the $\mathbf{x}_i$'s which appear on the right-hand side of the equation in the current variables j. Since the action is a sum of contributions $\sum_{\mathbf{k}}(S_{1,\mathbf{k}} + S_{2,\mathbf{k}})$ from each mode, where

$$S_{1,\mathbf{k}} = \int \Bigg[\sqrt{\pi}(\bar{j}^*_{1,\mathbf{k}}\bar{a}_{1,\mathbf{k}} + \bar{j}_{1,\mathbf{k}}\bar{a}^*_{1,\mathbf{k}}) + \frac{1}{2}\dot{\bar{a}}^*_{1,\mathbf{k}}\dot{\bar{a}}_{1,\mathbf{k}} - \frac{k^2c^2}{2}\bar{a}^*_{1,\mathbf{k}}\bar{a}_{1,\mathbf{k}} - \frac{\hbar kc}{2}\Bigg]\,dt \tag{9.62}$$

clearly X is a product of corresponding factors. The integral for one typical mode,

$$X_{1,\mathbf{k}} = \int \exp\left\{\frac{i}{\hbar}\int\left[\sqrt{\pi}(\bar{j}^*_{1,\mathbf{k}}\bar{a}_{1,\mathbf{k}} + \bar{j}_{1,\mathbf{k}}\bar{a}^*_{1,\mathbf{k}}) + \frac{1}{2}\dot{\bar{a}}^*_{1,\mathbf{k}}\dot{\bar{a}}_{1,\mathbf{k}} - \frac{k^2c^2}{2}\bar{a}^*_{1,\mathbf{k}}\bar{a}_{1,\mathbf{k}} - \frac{\hbar kc}{2}\right]dt\right\}\mathcal{D}\bar{a}_{1,\mathbf{k}}$$

$$= \exp\left\{-\frac{\pi}{\hbar kc}\int_{t_a}^{t_b}\int_{t_a}^{t_b}\bar{j}_{1,\mathbf{k}}(t)\bar{j}^*_{1,\mathbf{k}}(s)e^{-ikc|t-s|}\,ds\,dt\right\} \tag{9.63}$$

is a type of path integral which we have already done many times, except for the complication of complex variables, which can first be reduced to real variables. In fact, this is exactly the problem discussed in Sec. 8-9. The interaction function $f(t)$ of Eq. (8.136) is here related to $\sqrt{\pi}\,\bar{j}_{1,\mathbf{k}}(t)$, and $\omega = kc$. The final expression of Eq. (9.63) is equivalent to Eq. (8.138). The product of such factors for each $\mathbf{k}$ and polarization gives $X = e^{iI/\hbar}$, where

$$I = i\sum_{\mathbf{k}}\frac{\pi}{kc}\int_{t_a}^{t_b}\int_{t_a}^{t_b}[\bar{j}_{1,\mathbf{k}}(t)\bar{j}^*_{1,\mathbf{k}}(s) + \bar{j}_{2,\mathbf{k}}(t)\bar{j}^*_{2,\mathbf{k}}(s)]e^{-ikc|t-s|}\,ds\,dt \tag{9.64}$$

Thus the problem of a vacuum-to-vacuum transition is completely solved in terms of a path integral over the matter variables alone:

$$\text{Amplitude} = \int e^{(i/\hbar)(S_{\text{mat}}+I)}\prod_i \mathcal{D}\mathbf{x}_i \tag{9.65}$$

We shall discuss a number of consequences of this result. (The case that the initial or final state is not a vacuum is described in Sec. 9-7.)

It appears that the net result is simply this: The matter acts not with S_{mat} but with a modified action $S'_{\text{mat}} = S_{\text{mat}} + I$. The modification results from a reaction with the electromagnetic field. This is not true in a strictly classical sense, for the action I is a complex number. It can be shown that the classical physics which results from using the principle of least action, with the real part of S'_{mat} only, is exactly equivalent to the combination of Maxwell's equations and Newton's laws. But it does not correspond to the case that Maxwell's equations are solved by using just retarded waves. (In fact, a restriction to retarded waves cannot be represented by any principle of least action in which only matter coordinates appear. Instead it corresponds to using half the advanced and half the retarded solution.[1]) Our full quantum-mechanical complex expression for I is correct, and we shall now look at its consequences.

[1] J.A. Wheeler and R.P. Feynman, Interaction with the Absorber as the Mechanism of Radiation, *Rev. Mod. Phys.*, vol. 17, pp. 157–181, 1945.

First-order Perturbation Expansion. The integral over the **x**'s is too complicated to do exactly, but in the expression for the currents in I the charge e of the particles occurs. Thus I is proportional to e^2, which in dimensionless form, for the electron's charge, is the fine-structure constant

$$\frac{e^2}{\hbar c} = \frac{1}{137.039}$$

a small number whose exact value and meaning are unknown except experimentally. Thus we may expect that the effect of I is small. We already know that the Schrödinger theory gives atomic levels, for example, quite accurately. There can be only small errors arising from the neglect of I. Let us look at the effect of I in first order in e^2, corresponding to second order in e on the original action of Eq. (9.32). Let us take the transition amplitude λ_{MM} as defined in Sec. 6-5, where the matter system begins and ends in state M. If I is neglected, the zero order is

$$\lambda_{MM}{}^{0} = e^{-(i/\hbar)E_M T} \tag{9.66}$$

The first-order term is (where $\mathbf{x}$ represents all of the $\mathbf{x}_i$ variables)

$$\begin{aligned}
\lambda_{MM}{}^{1} &= \frac{i}{\hbar}\int_{t_a}^{t_b} \psi_M^*(\mathbf{x}_b) e^{(i/\hbar)S_{\text{mat}}} I \psi_M(\mathbf{x}_a)\,\mathcal{D}\mathbf{x}(t) \\
&= \frac{i}{\hbar}\sum_{\mathbf{k}}\int_{t_a}^{t_b} \psi_M^*(\mathbf{x}_b) e^{(i/\hbar)S_{\text{mat}}} \\
&\quad \times i\frac{\pi}{kc}\int_{t_a}^{t_b}\int_{t_a}^{t_b} [\bar{j}_{1,\mathbf{k}}(t)\bar{j}^*_{1,\mathbf{k}}(s) + \bar{j}_{2,\mathbf{k}}(t)\bar{j}^*_{2,\mathbf{k}}(s)] e^{-ikc|t-s|}\,ds\,dt \\
&\quad \times \psi_M(\mathbf{x}_a)\,\mathcal{D}\mathbf{x}(t)
\end{aligned} \tag{9.67}$$

Now terminate the integral on s at t, and double the result.° The evaluation of a similar expression was worked out in Sec. 5-1. For the present example, with large values of T, we get

$$\lambda_{MM}{}^{1} = -\frac{i}{\hbar}(\Delta E) T e^{-(i/\hbar)E_M T}$$

where

$$\begin{aligned}
\Delta E &= -\sum_N\sum_{\mathbf{k}} i\frac{2\pi}{kc}[(\bar{j}_{1,\mathbf{k}})_{MN}(\bar{j}^*_{1,\mathbf{k}})_{NM} + (\bar{j}_{2,\mathbf{k}})_{MN}(\bar{j}^*_{2,\mathbf{k}})_{NM}] \\
&\quad \times \int_0^\infty e^{(i/\hbar)(E_M - E_N - \hbar kc)\tau}\,d\tau \\
&= \sum_N \int \frac{2\pi\hbar}{kc}\,\frac{|(j_{1,\mathbf{k}})_{NM}|^2 + |(j_{2,\mathbf{k}})_{NM}|^2}{E_M - E_N - \hbar kc + i\epsilon}\,\frac{d^3\mathbf{k}}{(2\pi)^3}
\end{aligned} \tag{9.68}$$

This has a real and imaginary part and can be written as

$$\Delta E = \delta E - \frac{i\hbar\gamma}{2}$$

The real part δE represents a small shift in the energy levels of the atom, called the Lamb shift. Such a shift was discovered experimentally by Lamb and Retherford. This is

$$\delta E = \sum_N \int \frac{2\pi\hbar}{kc}[|(j_{1,\mathbf{k}})_{NM}|^2 + |(j_{2,\mathbf{k}})_{NM}|^2]$$
$$\times \text{P.P.}\left(\frac{1}{E_M - E_N - \hbar kc}\right)\frac{d^3\mathbf{k}}{(2\pi)^3} \tag{9.69}$$

and the imaginary part is

$$\frac{\hbar\gamma}{2} = \sum_N \int \frac{2\pi\hbar}{kc}[|(j_{1,\mathbf{k}})_{NM}|^2 + |(j_{2,\mathbf{k}})_{NM}|^2]$$
$$\times \pi\delta(E_M - E_N - \hbar kc)\frac{d^3\mathbf{k}}{(2\pi)^3} \tag{9.70}$$

The amplitude that the atom remains in the upper state with no photons emitted goes as $\exp\{-(i/\hbar)(E_M + \delta E - i\hbar\gamma/2)T\}$ and the probability as $e^{-\gamma T}$. That is, the probability to remain in state M decreases exponentially with the decay rate γ. Physically it should decrease because the atom in state M can emit a photon and fall to a lower state N. Comparison with Eq. (9.53) shows that γ in Eq. (9.70) is indeed the total rate of transition from state M to all lower states N.

9-5 A SINGLE ELECTRON IN A RADIATIVE FIELD

The Energy Correction. In order to study the electromagnetic energy correction δE, we shall consider the simplest case: that in which the matter system has only one moving charge (e.g., a hydrogen atom with an infinitely heavy nucleus or a free electron in empty space) whose coordinates we shall call $\mathbf{x}$. Thus $\mathbf{j_k} = \mathrm{e}\dot{\mathbf{x}}e^{-i\mathbf{k}\cdot\mathbf{x}}$. We have here a case where $\mathbf{j_k}$ contains $\dot{\mathbf{x}}$, and in considering second-order terms we must take appropriate care, as discussed in Sec. 7-3. There is an additional term to δE from the squared velocity term $\dot{\mathbf{x}}^2$. Expressing $\dot{\mathbf{x}}$ in terms of the momentum operator $\mathbf{p}$, as in Sec. 7-5, we obtain

$$\delta E = \frac{\mathrm{e}^2}{m^2}\sum_N \int \frac{2\pi\hbar}{kc}\frac{|\mathbf{p}_1 e^{-i\mathbf{k}\cdot\mathbf{x}}|^2_{NM} + |\mathbf{p}_2 e^{-i\mathbf{k}\cdot\mathbf{x}}|^2_{NM}}{E_M - E_N - \hbar kc}\frac{d^3\mathbf{k}}{(2\pi)^3}$$
$$+ \frac{\mathrm{e}^2}{m}\int \frac{2\pi\hbar}{kc}\frac{d^3\mathbf{k}}{(2\pi)^3} \tag{9.71}$$

Problem 9-10 Why do we not need to be careful to write $\frac{1}{2}[\mathbf{p}_1 e^{-i\mathbf{k}\cdot\mathbf{x}} + e^{-i\mathbf{k}\cdot\mathbf{x}}\mathbf{p}_1]$ in the matrix elements?

Let us take the simplest case of a free electron at rest. Any δE_R we get for energy in the field will represent a correction to the rest energy, or as can be shown from the relativity theory, to the mass, $\delta m = \delta E_R/c^2$. This is the so-called *electromagnetic mass correction.* For a free particle at rest, the states are plane waves. If the momenta in M and N are $\mathbf{p}_M$ and $\mathbf{p}_N$, the matrix element $(\mathbf{p}_1 e^{-i\mathbf{k}\cdot\mathbf{x}})_{NM}$ is zero unless $\mathbf{p}_N = \mathbf{p}_M - \hbar\mathbf{k}$, in which case it is $\mathbf{p}_{1N}$. Thus for an electron at rest initially, the matrix element is 0 and δE_R is just the last integral of Eq. (9.71), which is infinite!

Difficulties at Short Wavelengths. This in not the whole of it. When, at Eq. (9.29), we eliminated the term $4\pi\rho_\mathbf{k}\rho_{-\mathbf{k}}/k^2$ in S_c, we pointed out that this represented the interaction between point charges

$$\frac{1}{2}\sum_i\sum_j \frac{\mathrm{e}_i\mathrm{e}_j}{|\mathbf{x}_i - \mathbf{x}_j|}$$

but neglected to point out that the infinite terms $i = j$ must also be included in the sum. Indeed, for a single particle $\rho_\mathbf{k} = \mathrm{e}e^{-i\mathbf{k}\cdot\mathbf{x}}$, so $4\pi|\rho_\mathbf{k}|^2/k^2 = 4\pi\mathrm{e}^2/k^2$ and the term is

$$\delta E_c = 4\pi\mathrm{e}^2 \int \frac{1}{k^2}\frac{d^3\mathbf{k}}{(2\pi)^3}$$

The infinities here and in δE_R above do not cancel, and we are left with a real difficulty; our integrals over momentum $\mathbf{k}$ diverge quadratically. Quantum electrodynamics gives nonsensical results.

It is true that we are using a nonrelativistic treatment of the charged particle. The relativistic treatment of the matter (quantum electrodynamics is not altered) does not rid us of the divergent results, although the order of infinity may be changed. For a particle of spin 0, like a π meson, the order is unchanged; it is still a quadratic divergence. Here there is presumably an experimental value of the mass correction available. As far as is known, through other interactions, the sole difference between charged and neutral π mesons is the charge, i.e., the different way they couple to the electromagnetic field. So presumably the mass difference of the charged π meson with a mass m_π of 273.2 electron masses and the neutral π meson of 264.2 electron masses, that is 9.0 electron masses, or $0.034m_\pi$, or 4.6 MeV, represents energy in the electromagnetic field.

If we arbitrarily stop our integrals at some higher momentum $k_{\max}$ (which is not a relativistically invariant procedure), we get an energy $\hbar e^2(k_{\max})^2/2\pi m_\pi c$ from the last term of Eq. (9.71), which is the largest term if $\hbar k_{\max}/c$ is very much larger than the π meson mass m_π. If this equals $0.034 m_\pi c^2$, then $(e^2/2\pi\hbar c)(\hbar k_{\max}/m_\pi c)^2 = 0.034$, or

$$k_{\max} \approx \frac{5.4 m_\pi c}{\hbar} \approx \frac{0.8 m_p c}{\hbar}$$

where m_p is the mass of a proton. (The relativistic theory gives $\Delta E = 0.034 m_\pi c^2$ with a cutoff at about the same energy.) It is for this reason that we conclude that our present-day formulation of quantum electrodynamics (or of the "particles" with which photons interact) is faulty. The fault lies in the way we deal with energies beyond proton mass or with corresponding frequencies, or wave numbers. The difficult arises with modes whose wavelength is less than about $4\pi \times 10^{-14}$ cm.

For the electron of spin $\frac{1}{2}$ the Dirac theory shows that the electron should have a certain magnetic moment. It turns out that with such a magnetic moment the negative magnetic energy almost perfectly cancels the positive electric energy. The difference still diverges, although only logarithmically. If a cutoff is applied to integrals over wavelengths, at the wavelength limit suggested above, the correction to the electron mass is only about 3 per cent, but there is no way to test this, for we do not recognize a neutral counterpart to the electron.

For the proton the anomalous magnetic moment is so high that the magnetic energy exceeds the electric energy and the correction can be negative. The neutron is a magnet, too, so its correction is also negative. Since the proton moment is higher, the fact that the neutron is heavier than the proton might be explained. If the integrals are cut off at an energy of the order of a proton mass, the difference comes out correctly, but this is too crude a way to calculate such an accurately known energy as the 782.61 ± 0.40 keV† equivalent to this mass difference. These mass differences (of proton and neutron, of charged and neutral π meson, of positive, neutral, and negative sigma mesons, and of charged and neutral K mesons, etc.) present a serious challenge to modern physics and possibly point to the failure of quantum electrodynamics to give us a complete theory for calculating electromagnetic effects. We do not know whether it is truly quantum electrodynamics or our assumptions about the distribution of charge inside the particles which are at fault. Only when we have a more complete theory of these particles and their

---

†Page 354 of F. Everling, et al., Atomic Masses of Nuclides for $A \leq 70$, *Nucl. Phys.*, vol. 15, pp. 342–355, 1960.

interactions will we be able to determine the limitations, if any, of our present theory of quantum electrodynamics.

9-6 THE LAMB SHIFT

According to the Schrödinger equation, the second level of the hydrogen atom is degenerate. The $2s$ and $2p$ levels occur at the same energy. Likewise, for the Dirac equation there is a degenerate pair $2s_{1/2}$ and $2p_{1/2}$. But Lamb and Retherford found in 1946 that there is indeed a small separation (about 1 part in 3×10^6) with the $2s_{1/2}$ lying higher by a frequency of 1,057.1 megacycles.

Although theorists reasoned that such an energy difference might arise from effects of the term I, the infinities of the divergent integrals confused all attempts to calculate the difference until the work of Bethe and Weisskopf in 1947. They reasoned as follows:

First, since

$$\frac{1}{E_M - E_N - \hbar kc} = \frac{1}{\hbar kc}\frac{E_M - E_N}{E_M - E_N - \hbar kc} - \frac{1}{\hbar kc} \tag{9.72}$$

the energy (9.71) was expressed as the sum of three terms,

$$\delta E = \delta E' + \delta E'' + \delta E''' \tag{9.73}$$

where

$$\delta E' = \frac{2\pi e^2}{m^2c^2}\int \frac{1}{k^2}\sum_N \frac{(E_M - E_N)(|\mathbf{p}_1 e^{-i\mathbf{k}\cdot\mathbf{x}}|^2_{NM} + |\mathbf{p}_2 e^{-i\mathbf{k}\cdot\mathbf{x}}|^2_{NM})}{E_M - E_N - \hbar kc}\frac{d^3\mathbf{k}}{(2\pi)^3} \tag{9.74}$$

$$\delta E'' = -\frac{2\pi e^2}{m^2c^2}\int \frac{1}{k^2}\sum_N (|\mathbf{p}_1 e^{-i\mathbf{k}\cdot\mathbf{x}}|^2_{NM} + |\mathbf{p}_2 e^{-i\mathbf{k}\cdot\mathbf{x}}|^2_{NM})\frac{d^3\mathbf{k}}{(2\pi)^3} \tag{9.75}$$

$$\delta E''' = \frac{2\pi e^2\hbar}{mc}\int \frac{1}{k}\frac{d^3\mathbf{k}}{(2\pi)^3} \tag{9.76}$$

The term $\delta E'''$ and the infinity from the Coulomb term δE_c are independent of the state of the electron. They would (it was hoped) be made finite in some future theory. It would contribute some δm to the rest mass of the electron. If m_0 is the mechanical mass, the true experimental mass would be $m = m_0 + \delta m$, where $\delta mc^2 = \delta E''' + \delta E_c$. In the total energy (including the rest energy of the particles and the binding energy) of the hydrogen atom, such a rest-energy correction to the energy is, of course, expected, but we have already included it when we measure all binding energies relative to the free-particle ionized state. The δm term is thus identified, because it is the only term for an electron at rest, and it is independent of the motion or state of the electron.

The term $\delta E''$ could be simplified, for the sum over N could be taken to give $(p_1^2 + p_2^2)_{MM}$ (by the law of matrix multiplication). When $\mathbf{k}$ is integrated over all directions, this becomes $\frac{2}{3}(\mathbf{p}\cdot\mathbf{p})_{MM}$, and

$$\delta E'' = -\frac{(\mathbf{p}\cdot\mathbf{p})_{MM}}{2m}\frac{8\pi \mathrm{e}^2}{3mc^2}\int \frac{1}{k^2}\frac{d^3\mathbf{k}}{(2\pi)^3} \tag{9.77}$$

Again it was hoped that some day this term would be finite. It exists even for a free electron. It is interpreted as follows: The mechanical kinetic energy $p^2/2m_0$ would be altered (if the mass is altered) to the expression

$$\frac{p^2}{2m} \approx \frac{p^2}{2m_0}\left(1 - \frac{\delta m}{m_0}\right) \tag{9.78}$$

and the term $\delta E''$ must represent $-(p^2/2m_0)\,\delta m/m_0$. But we have already taken this term into account; for we calculate the Schrödinger energy levels with $p^2/2m$, where m is the experimental mass. The term is identified because it is the only extra term for a moving free electron and it is proportional to the kinetic energy.[1] Finally, even though these terms may be interpreted wrongly, when we calculate the difference of the values of δE for the $2s$ and $2p$ states, the terms will drop out, because $\delta E'''$ and δE_c are the same for all states and $\delta E''$ is also the same, since $(p^2/2m)_{MM}$ turns out to be the same for the two states $2s$ and $2p$.

In the remaining term $\delta E'$ the argument was made that the dipole approximation would suffice. Then the matrix elements are independent of $\mathbf{k}$, and since

$$\int \frac{1}{k^2}\frac{1}{E_M - E_N - \hbar kc}\frac{d^3\mathbf{k}}{(2\pi)^3} = \frac{1}{2\pi^2\hbar c}\ln\frac{\hbar k_{\max}c}{E_M - E_N} \tag{9.79}$$

we get

$$\delta E' = \frac{\mathrm{e}^2}{\pi m^2\hbar c^3}\sum_N\left[(E_M - E_N)\tfrac{2}{3}|\mathbf{p}_{NM}|^2\ln\frac{\hbar k_{\max}c}{E_M - E_N}\right] \tag{9.80}$$

Since the states and the matrix elements are known for hydrogen, the sum can be worked out. The only question is the value of $\hbar k_{\max}c$. Bethe

[1]The δm implied by Eq. (9.77) is $(8\pi\mathrm{e}^2/3c^2)\int(1/k^2)d^3\mathbf{k}/(2\pi)^3$ and is *not* equal to the δm obtained from $\delta E/c^2$ for a static electron. This is because we are limited to a nonrelativistic approximation. When a fully relativistic analysis is carried out, the two ways of calculating δm agree.

argued that the nonrelativistic approximation is at fault here and that, if the full relativistic calculation were made, $\hbar k_{\max} c$ would turn out to be of the order mc^2. Putting $\hbar k_{\max} c = mc^2$ gave about 1,000 megacycles, so Bethe knew he was on the right track.

The remaining problem was to make a relativistic calculation with the Dirac wave function and states. Only in this way could a precise determination of the effective $k_{\max}$ be made. This turned out to be quite confusing, for it was hard to identify the various infinite terms. It would not do to simply cut them off at some maximum momentum and take the difference; for this is not necessarily a relativistically invariant procedure because it deals with momentum and energy in different ways. (One consequence of this has already been pointed out in the footnote.) One method for resolving the confusion was developed by Schwinger, who showed how the relativistic symmetry could be kept clear throughout the calculation and the infinite terms identified. Another method worked out by Feynman was to give a relativistically invariant procedure to cut off the infinite integrals. Here we shall illustrate the latter method.

The total effect of the electromagnetic field, which this time includes the Coulomb interaction, is represented by an extra term $I + S_c$ in the action. The relativistic invariance of an expression for I like Eq. (9.64) will not be self-evident, since that formula is expressed in terms of $\mathbf{k}$ and t instead of either $\mathbf{r}$ and t or $\mathbf{k}$ and ω. Let us represent I in terms of wave number $\mathbf{k}$ and frequency ω variables. First note that, in light of Eq. (A.10),

$$\int_{-\infty}^{\infty} e^{-ikc|\tau|} e^{-i\omega\tau}\, d\tau = -\frac{2ikc}{k^2c^2 - \omega^2 - i\epsilon} \tag{9.81}$$

or

$$e^{-ikc|t-s|} = -\int_{-\infty}^{\infty} \frac{2ikc}{k^2c^2 - \omega^2 - i\epsilon} e^{i\omega(t-s)} \frac{d\omega}{2\pi} \tag{9.82}$$

Suppose we define

$$\begin{aligned} \mathbf{j}(\mathbf{k}, \omega) &= \int \mathbf{j}_{\mathbf{k}}(t) e^{+i\omega t}\, dt \\ &= \iint \mathbf{j}(\mathbf{r}, t) e^{-i(\mathbf{k}\cdot\mathbf{r} - \omega t)}\, d^3\mathbf{r}\, dt \end{aligned} \tag{9.83}$$

Then I becomes (for long time intervals T)

$$I = 2\pi \int \frac{|j_1(\mathbf{k}, \omega)|^2 + |j_2(\mathbf{k}, \omega)|^2}{k^2c^2 - \omega^2 - i\epsilon} \frac{d^3\mathbf{k}\, d\omega}{(2\pi)^4} \tag{9.84}$$

The relativistic symmetry of this expression in $\mathbf{k}$ and ω is already clear; for $\mathbf{k}^2c^2 - \omega^2$ is invariant to the Lorentz transformation. The currents, however, do not appear in a relativistically symmetrical manner.

We would have expected an invariant combination like $\mathbf{j}\cdot\mathbf{j} - c^2\rho^2$, since $\mathbf{j}$ and $c\rho$ form a four-vector. But if we define

$$\rho(\mathbf{k},\omega) = \int \rho_{\mathbf{k}}(t)e^{+i\omega t}\,dt$$
$$= \iint \rho(\mathbf{r},t)e^{-i(\mathbf{k}\cdot\mathbf{r}-\omega t)}\,d^3\mathbf{r}\,dt \tag{9.85}$$

the Coulomb portion of the action, Eq. (9.29), is

$$S_c = -2\pi \int \frac{|\rho(\mathbf{k},\omega)|^2}{k^2}\frac{d^3\mathbf{k}\,d\omega}{(2\pi)^4} = -2\pi \int \frac{|\rho|^2c^2 - (\omega|\rho|/k)^2}{k^2c^2 - \omega^2}\frac{d^3\mathbf{k}\,d\omega}{(2\pi)^4} \tag{9.86}$$

the last resulting simply by multiplying numerator and denominator by $c^2 - \omega^2/k^2$. But the law of conservation of current

$$-\frac{\partial\rho}{\partial t} = \boldsymbol{\nabla}\cdot\mathbf{j} \tag{9.87}$$

becomes

$$\omega\rho(\mathbf{k},\omega) = \mathbf{k}\cdot\mathbf{j}(\mathbf{k},\omega) \tag{9.88}$$

Alternatively, if we call j_3 the component of $\mathbf{j}$ in the direction of $\mathbf{k}$, $\omega\rho/k = j_3$ and we have altogether

$$I + S_c = 2\pi \int \frac{|j_1(\mathbf{k},\omega)|^2 + |j_2(\mathbf{k},\omega)|^2 + |j_3(\mathbf{k},\omega)|^2 - c^2|\rho(\mathbf{k},\omega)|^2}{k^2c^2 - \omega^2 - i\epsilon}\frac{d^3\mathbf{k}\,d\omega}{(2\pi)^4} \tag{9.89}$$

The sum of the three j terms is just $\mathbf{j}^*(\mathbf{k},\omega)\cdot\mathbf{j}(\mathbf{k},\omega)$, and the four-dimensional invariance is evident.

A suggestion is now made that in view of our present ignorance, convergence of the integrals can be made artificially by supplying an additional factor such as

$$\left(\frac{\Lambda^2}{k^2c^2 - \omega^2 + \Lambda^2 - i\epsilon}\right)^2$$

in the integrand of Eq. (9.89), where Λ is some very high frequency. For small values of k and ω this factor is unity, whereas for high values it cuts off the integral. Furthermore, such a factor clearly does not destroy the relativistic invariance of the expression. All physical quantities are to be calculated by assuming $I + S_c$ contains this cutoff factor. If they are insensitive to Λ for large Λ (like the Lamb shift), the theoretical value is to be trusted. If, on the other hand, the result depends sensitively on Λ (as does the charged and neutral π meson mass difference) no quantitative meaning can be given to the result, for the cutoff function is arbitrary and is not completely satisfactory. This is the present state of quantum electrodynamics.

Problem 9-11 Show that the cutoff function is not completely satisfactory by arguing that γ calculated in the manner of Sec. 9-4 would be altered by the cutoff, yet the probability of emission of a real photon would not be so altered because for such a photon $\omega = kc$ and the modifying cutoff factor is exactly 1. Thus the balance of probability would not result (i.e., the probability that the atom emits plus the probability that it does not emit would no longer add to unity). The difficulty suggested by this problem has never been solved. No modification of quantum electrodynamics at high frequencies is known which simultaneously makes all results finite, maintains relativistic invariance, and keeps the sum of probabilities over all alternatives equal to unity.

Problem 9-12 Transform $I + S_c$ into space coordinates by using

$$\int \frac{e^{i(\mathbf{k}\cdot\mathbf{r}-\omega t)}}{k^2c^2-\omega^2-i\epsilon}\frac{d^3\mathbf{k}\,d\omega}{(2\pi)^4} = \frac{1}{(2\pi)^2 c}\frac{i}{(r^2-c^2t^2+i\epsilon)} = \frac{1}{4\pi c}\delta_+(r^2-c^2t^2) \tag{9.90}$$

(*Note:* The function $i/[\pi(x+i\epsilon)]$ is often written as $\delta_+(x)$, and we have introduced that convention here.) Then find

$$I + S_c = \frac{1}{2c}\iint [\mathbf{j}(\mathbf{r}_1,t_1)\cdot\mathbf{j}(\mathbf{r}_2,t_2) - c^2\rho(\mathbf{r}_1,t_1)\rho(\mathbf{r}_2,t_2)] \times \delta_+(|\mathbf{r}_1-\mathbf{r}_2|^2 - c^2(t_1-t_2)^2)\,d^3\mathbf{r}_1\,dt_1\,d^3\mathbf{r}_2\,dt_2 \tag{9.91}$$

9-7 THE EMISSION OF LIGHT

In Sec. 9-4 we found an expression for the amplitude that the matter system would do something when interacting with an electromagnetic field, as shown in Eq. (9.60) and the following development. This derivation was restricted to the special case that the field is both initially and finally in the vacuum state with no photons present. The result was that the action S_{mat} in the path integrals must be replaced by an effective action $S'_{\text{mat}} = S_{\text{mat}} + I$. In a more general case, photons are present, both initially and finally. As an illustration suppose that initially no photons are present but in the final state there is just one photon of momentum $\hbar\mathbf{q}$ and polarization 1. The only change which this makes in our previous calculation is the change in the integral defining X, that is, Eq. (9.61). We shall now use

$$X' = \iint e^{(i/\hbar)(S_{\text{int}}+S_{\text{rad}})} \prod_{\mathbf{k}} \mathcal{D}a_{1,\mathbf{k}}\,\mathcal{D}a_{2,\mathbf{k}} \tag{9.92}$$

where the path integral is carried out between a vacuum initial state and a final state consisting of a vacuum plus one photon. Then every oscillator, except 1, $\mathbf{q}$, goes from the initial state $n=0$ to the final state $n=0$, so the factor $X_{1,\mathbf{k}}$ for all these oscillators is unchanged. Only the contribution from the single oscillator 1, $\mathbf{q}$ is altered; for it now becomes

$$X'_{1,\mathbf{q}} = \int \exp\left\{\frac{i}{\hbar}\int\left[\sqrt{\pi}(\bar{\jmath}^*_{1,\mathbf{q}}\bar{a}_{1,\mathbf{q}} + \bar{\jmath}_{1,\mathbf{q}}\bar{a}^*_{1,\mathbf{q}}) + \frac{1}{2}\dot{\bar{a}}^*_{1,\mathbf{q}}\dot{\bar{a}}_{1,\mathbf{q}} - \frac{q^2c^2}{2}\bar{a}^*_{1,\mathbf{q}}\bar{a}_{1,\mathbf{q}} - \frac{\hbar qc}{2}\right]dt\right\}\mathcal{D}\bar{a}_{1,\mathbf{q}} \tag{9.93}$$

This expression is the same as Eq. (9.63) except that the oscillator path is taken between the state $n=0$ and the state $n=1$ instead of $n=0$ to $n=0$ as in the previous expression. We worked out the behavior of a forced harmonic oscillator in Sec. 8-9, and we can use the results of that section to write

$$X'_{1,\mathbf{q}} = \left(i\sqrt{\frac{2\pi}{\hbar qc}}\int_{t_a}^{t_b}\bar{\jmath}_{1,\mathbf{q}}e^{iqct}\,dt\right)X_{1,\mathbf{q}} \tag{9.94}$$

where $X_{1,\mathbf{q}}$ is the $n=0$ to $n=0$ factor previously calculated. Therefore, evidently the complete factor X' is simply the original factor X multiplied by

$$i\sqrt{\frac{2\pi}{\hbar qc}}\int_{t_a}^{t_b}\bar{\jmath}_{1,\mathbf{q}}e^{iqct}\,dt$$

and we find for the amplitude

$$\text{Amplitude} = i\sqrt{\frac{2\pi}{\hbar qc}}\int e^{(i/\hbar)(S_{\text{mat}}+I)}\int_{t_a}^{t_b}\bar{\jmath}_{1,\mathbf{q}}e^{iqct}\,dt\,\mathcal{D}x \tag{9.95}$$

The perturbation theory expression which we previously evaluated (at Eq. 9.50) is equivalent to the transition element

$$i\sqrt{\frac{2\pi}{\hbar qc}}\int e^{(i/\hbar)S_{\text{mat}}}\int_{t_a}^{t_b}\bar{\jmath}_{1,\mathbf{q}}e^{iqct}\,dt\,\mathcal{D}x \tag{9.96}$$

so we see that the net result is the same as that given by the perturbation theory except that the transition amplitude must be calculated with the effective action $S'_{\text{mat}} = S_{\text{mat}} + I$ instead of with just S_{mat}. The effect of I is to change the energy levels a bit, as we discussed, but also to make the energy values complex. The result is that the emitted light gives a spectral line with a little width, which is called the natural line width. We shall not go any further into the details of this calculation but leave the subject and the generalizations to a number of photons both entering and leaving the system to those who wish to study quantum electrodynamics specifically in more detail.

9-8 SUMMARY

Review of the Approach. In this chapter we have a considerable amount of analysis of the quantum electromagnetic field. It is worthwhile looking back again to see the central ideas and results. The separation of the Coulomb interaction and the use of running waves are technical ways of accomplishing our ends, but the essential result is the formula of Eq. (9.89) (or its equivalent, Eq. 9.91). Let us review this result from the more general point of view exemplified by the ideas of Eq. (9.1).

Suppose we have a system which can be described by an action

$$S = S_1[\mathbf{x}] + S_2[\mathbf{x}, \mathbf{A}, \phi] + S_3[\mathbf{A}, \phi] \tag{9.97}$$

where $S_1[\mathbf{x}]$ is the action of the matter alone, $S_2[\mathbf{x}, \mathbf{A}, \phi]$ is the interaction of matter and field, and $S_3[\mathbf{A}, \phi]$ is the action of the field alone, and where $\mathbf{x}$ stands for all the coordinates of the matter, while $\mathbf{A}$, ϕ describe the field. Then the amplitude for any event results from evaluating a path integral like

$$K = \int \exp\left\{\frac{i}{\hbar}\left(S_1[\mathbf{x}] + S_2[\mathbf{x}, \mathbf{A}, \phi] + S_3[\mathbf{A}, \phi]\right)\right\} \mathcal{D}\mathbf{x}\, \mathcal{D}\mathbf{A}\, \mathcal{D}\phi \tag{9.98}$$

subject to the boundary conditions of the problem in question. In this summary we shall assume that conditions for the field are that initially and finally no photons are resent (i.e., ground state to ground state for the field), and we abbreviate this set of conditions as *gnd-gnd*. Later on we shall consider the consequences of integrating $\mathbf{x}$ first and $\mathbf{A}$, ϕ last. What we have done so far corresponds instead to integrating $\mathbf{A}$, ϕ first and reserving the $\mathbf{x}$ integral as a subsequent step.

Usually $S_2[\mathbf{x}, \mathbf{A}, \phi]$ is linear in the field variables $\mathbf{A}$, ϕ and can be written as

$$S_2 = -\iint \left[\rho(\mathbf{r}, t)\phi(\mathbf{r}, t) - \mathbf{j}(\mathbf{r}, t)\cdot\mathbf{A}(\mathbf{r}, t)/c\right] d^3\mathbf{r}\, dt \tag{9.99}$$

where ρ and $\mathbf{j}$ are the electric charge and current density, which depend on the $\mathbf{x}$ only. The integral over $\mathbf{A}$, ϕ is then easily performed because it is a gaussian integral. It is the burden of Eq. (9.91) to tell us the value of this integral, namely,

$$\int_{gnd}^{gnd} \exp\left\{\frac{i}{\hbar}\left[S_3[\mathbf{A},\phi] - \iint (\rho\phi - \mathbf{j}\cdot\mathbf{A}/c)\, d^3\mathbf{r}\, dt\right]\right\} \mathcal{D}\mathbf{A}\,\mathcal{D}\phi = e^{(i/\hbar)J} \tag{9.100}$$

where J, which we called $I + S_c$ in Eq. (9.91), is

$$J = \frac{1}{2c}\iint [\mathbf{j}(\mathbf{r}_1,t_1)\cdot\mathbf{j}(\mathbf{r}_2,t_2) - c^2\rho(\mathbf{r}_1,t_1)\rho(\mathbf{r}_2,t_2)] \times \delta_+(|\mathbf{r}_1-\mathbf{r}_2|^2 - c^2(t_1-t_2)^2)\, d^3\mathbf{r}_1\, dt_1\, d^3\mathbf{r}_2\, dt_2 \tag{9.101}$$

for *any* functions ρ, $\mathbf{j}$ of $\mathbf{r}$, t. The expression for Eq. (9.101) as an integral over momentum space appears in Eq. (9.89).

In the applications of Eq. (9.98) these ρ, $\mathbf{j}$ are some function of $\mathbf{x}$ and $\dot{\mathbf{x}}$, so we obtain the result that

$$K(gnd, gnd) = \int \exp\left\{\frac{i}{\hbar}(S_1[\mathbf{x}] + J[\mathbf{x}])\right\} \mathcal{D}\mathbf{x} \tag{9.102}$$

where $J[\mathbf{x}]$, a functional of $\mathbf{x}(t)$, is given by Eq. (9.101) with the correct ρ, $\mathbf{j}$ substituted. This summarizes the results for *gnd-gnd* transitions. We express the modifying effect of the field on the action of the particles by the addition of $J[\mathbf{x}]$ to $S_1[\mathbf{x}]$. The central formula for electrodynamics then is the general result of Eqs. (9.100) and (9.101).

General Formulation of Quantum Electrodynamics. It is also of interest to pursue these matters in a different direction, by integrating over the matter coordinates first, and leaving the field variables for later. We shall limit ourselves to a brief general description of what results from this procedure. If in Eq. (9.98) we contemplate integrating $\mathbf{x}$ first, the factor $e^{(i/\hbar)S_3}$ is a constant and can be left out. We can therefore write Eq. (9.98) this way: If we define

$$T[\mathbf{A},\phi] = \int \exp\left\{\frac{i}{\hbar}(S_1[\mathbf{x}] + S_2[\mathbf{x},\mathbf{A},\phi])\right\} \mathcal{D}\mathbf{x} \tag{9.103}$$

then

$$K = \int e^{(i/\hbar)S_3[\mathbf{A},\phi]}\, T[\mathbf{A},\phi]\, \mathcal{D}\mathbf{A}\,\mathcal{D}\phi \tag{9.104}$$

This K gives us the amplitude that the particle goes through a certain motion *and* the field undergoes a certain transition. Like all other amplitudes, it is the sum over all possible alternatives. Each separate alternative is constructed as the amplitude $T[\mathbf{A}, \phi]$ for the motion in a particular field $\mathbf{A}$, ϕ times the amplitude $e^{(i/\hbar)S_3}$ that the field is $\mathbf{A}$, ϕ. In carrying out the sum, we sum over all possible fields $\mathbf{A}$, ϕ.

This law, given by Eq. (9.104), is the general fundamental rule for all of quantum electrodynamics. It is a correct formulation even when the functional $T[\mathbf{A}, \phi]$, the amplitude for the motion of the particles in an external potential $\mathbf{A}$, ϕ, cannot be represented as a path integral. For example, for a relativistic particle with spin (described by the Dirac equation), the quantity $T[\mathbf{A}, \phi]$ cannot be described by a simple path integral based on any reasonable action. However, it is possible to calculate $T[\mathbf{A}, \phi]$ by other means, for example, from the Dirac equation. After the form of this functional has been derived, the amplitude K can be worked out, in principle, from Eq. (9.104).

In stating the law of quantum electrodynamics in the form of Eq. (9.104), we have isolated the behavior of the electromagnetic field from the behavior of the particle (or system of particles) on which it acts. That such an isolation can be carried out is an important result. For example, the functional $T[\mathbf{A}, \phi]$ may represent the behavior of a nucleus whose properties are not completely known. However, if we know only the behavior of the nucleus in an external field, then we can solve quantum-electrodynamic problems involving nuclei.

Of course, to use Eq. (9.104) strictly, T must be known as a functional of $\mathbf{A}$, ϕ for all $\mathbf{A}$, ϕ, but this much information is rarely available. Even if it is available, the path integral over $\mathbf{A}$, ϕ may not be easy. But in practice the formula is very useful. Sometimes T can be approximated by an exponential, linear in $\mathbf{A}$, ϕ, of exactly the form of Eq. (9.99). Then the result is obtained directly from the general formula of Eqs. (9.100) and (9.101). More often, T can be represented by a sum, or integral, over such exponential forms with various ρ, $\mathbf{j}$ and the result of Eq. (9.104) is a corresponding sum or integral over expressions containing $e^{(i/\hbar)J}$ with J in Eq. (9.101) involving the corresponding ρ, $\mathbf{j}$.

In most practical situations T can be expressed as a power series in $\mathbf{A}$, ϕ. The first few terms can be found from the theory of the matter considering $\mathbf{A}$, ϕ as a small perturbation. Subsequent substitution into Eq. (9.104) and integration over $\mathbf{A}$, ϕ gives a corresponding perturbation expansion (in powers of $\mathrm{e}^2/\hbar c$) for K. The necessary path integrals such as

$$\int e^{(i/\hbar)S_3[\mathbf{A},\phi]} A_i(\mathbf{r}_1, t_1) A_j(\mathbf{r}_2, t_2)\ \mathcal{D}\mathbf{A}\,\mathcal{D}\phi$$
$$= -i\hbar c\,\delta_{i,j}\,\delta_+(|\mathbf{r}_1 - \mathbf{r}_2|^2 - c^2(t_1 - t_2)^2)$$

can be discovered by expanding the general formula of Eqs. (9.100) and (9.101) on both sides in powers of ρ, $\mathbf{j}$ and comparing corresponding terms. We shall not go further into these matters here, but refer the reader to the literature (e.g., sec. 8 of R.P. Feynman, Mathematical Formulation of the Quantum Theory of Electromagnetic Interaction, *Phys. Rev.*, vol. 80, pp. 440–457, 1950).

10

Statistical Mechanics

IN preceding chapters we have discussed transitions in which a system goes from one known state to another. In most physically realistic situations the initial state is not completely known. The system may be in one or another state with different probabilities associated with each. In this case the final state is equally uncertain, being that set of states resulting from the various possible initial states with the corresponding probabilities. Or we may not be interested in the probability to go to just one specified final state, but rather the chance to end up in any one of a set of such states.

An especially interesting case of statistical uncertainty of states is that corresponding to thermal equilibrium at some temperature T. A quantum-mechanical system in thermal equilibrium can exist in one or another energy state. The results of quantum-statistical mechanics show that the probability that a system is in a state of energy E is proportional to $e^{-E/kT}$, where kT measures the temperature in natural energy units. (The conversion factor k, known as Boltzmann's constant, is $1.380\,65 \times 10^{-16}$ erg/K, or 1 eV per 11 605 K.)

In this book we shall neither derive nor discuss this exponential distribution law. We emphasize that the energy E is the energy of the entire system. If an energy level is degenerate, then each state at that particular level has equal probability. This means that the total probability for the system to have the particular energy *value* is enhanced by a factor corresponding to the number of states in the degenerate level.

The exponential law given above is not yet a true probability distribution, since it has not been normalized. The normalizing factor can be written $1/Z$, so that the probability that a system should be in the state of energy E_n (assumed nondegenerate for the time being) is

$$p_n = \frac{1}{Z} e^{-\beta E_n} \tag{10.1}$$

where $\beta = 1/kT$. This means

$$Z = \sum_n e^{-\beta E_n} \tag{10.2}$$

An equivalent normalization consists of defining an energy F such that

$$p_n = e^{-\beta(E_n - F)} \tag{10.3}$$

F is called the *Helmholtz free energy*. Its value is, of course, dependent on the temperature T, although the various energy values E_n do not depend on T. It is evident that

$$Z = e^{-\beta F} \tag{10.4}$$

10-1 THE PARTITION FUNCTION

The physical properties of a system in thermal equilibrium can be derived from the exponential distribution function. Suppose A is the measure of some property and that its mean value in the nth energy state is

$$A_n = \int \phi_n^* A \phi_n \, d\Gamma \tag{10.5}$$

where the integral is taken over the configuration space of the system. Then the statistical average for A for the whole system is

$$\bar{A} = \sum_n p_n A_n = \frac{1}{Z} \sum_n A_n e^{-\beta E_n} \tag{10.6}$$

For example, the average or expected value of the energy itself is

$$\begin{aligned} U &= \sum_n p_n E_n = \frac{1}{Z} \sum_n E_n e^{-\beta E_n} \\ &= \sum_n E_n e^{-\beta(E_n - F)} \end{aligned} \tag{10.7}$$

If the normalizing factor Z is known as a function of the temperature, the sum of Eq. (10.7) can be easily evaluated. From Eq. (10.2) we have

$$\sum_n E_n e^{-\beta E_n} = -\frac{\partial Z}{\partial \beta} = kT^2 \frac{\partial Z}{\partial T} \tag{10.8}$$

This means that

$$\begin{aligned} U &= \frac{kT^2}{Z} \frac{\partial Z}{\partial T} = kT^2 \frac{\partial \ln Z}{\partial T} = F - T \frac{\partial F}{\partial T} \\ &= \frac{\partial(\beta F)}{\partial \beta} \end{aligned} \tag{10.9}$$

We have written the derivatives with respect to the temperature as partial derivatives because other variables, such as the volume of the system or any external fields, which determine the energy levels are all held fixed.

It is interesting to see what happens to the expected value of the energy if some other variable such as the volume is changed. Suppose the system is in a particular state ϕ_n and we make a small change in the value of a certain parameter, say α. Using a first-order perturbation principle, we find that the first-order change in energy is equal to the expected value of the first-order change in the hamiltonian. That is,

$$\begin{aligned} E_n + \Delta E_n &= \int \phi_n^*(H + \Delta H)\phi_n \, d\Gamma \\ \Delta E_n &= \int \phi_n^* \Delta H \phi_n \, d\Gamma \end{aligned} \tag{10.10}$$

Using the language of classical physics, we would say that the ratio $-\Delta H/\Delta\alpha$ is the "force" associated with the parameter α. In case this parameter is the volume, the force is the pressure. That is, we define the concept of force by

force $\times$ change in parameter = −change in energy

or

$$f_\alpha = -\frac{\partial H}{\partial \alpha} \tag{10.11}$$

As an example, then, if P = pressure and V = volume,

$$-P\,\Delta V = \Delta E \tag{10.12}$$

We write the expected value of the force as

$$\begin{aligned} \bar{f}_\alpha &= -\overline{\left(\frac{\partial H}{\partial \alpha}\right)} = -\sum_n p_n \left(\frac{\partial H}{\partial \alpha}\right)_n = -\sum_n p_n \frac{\partial E_n}{\partial \alpha} \\ &= -\frac{1}{Z}\sum_n \frac{\partial E_n}{\partial \alpha} e^{-E_n/kT} = \frac{kT}{Z}\frac{\partial}{\partial \alpha}\left(\sum_n e^{-E_n/kT}\right) = \frac{kT}{Z}\frac{\partial Z}{\partial \alpha} \end{aligned} \tag{10.13}$$

so that

$$\bar{f}_\alpha = \frac{1}{\beta}\frac{\partial \ln Z}{\partial \alpha} \tag{10.14}$$

where β and other parameters are held constant. Using Eq. (10.4), we can write this as

$$\bar{f}_\alpha = -\frac{\partial F}{\partial \alpha} \tag{10.15}$$

If the parameter α is the volume V so that $\bar{f}_\alpha$ is the pressure P, we have

$$P = -\frac{\partial F}{\partial V} \tag{10.16}$$

When the volume is changed by an infinitesimal amount for a system at a constant temperature, two things happen simultaneously. First, each energy level shifts slightly. Second, for the system to stay in equilibrium at a constant temperature T (maintained by a bath, for example), the probability associated with each energy level changes slightly (because the energy of that level changes). If the only effect were a change in the energy of each level, then the change in the total energy of the system would correspond to this change averaged over all the levels. From our foregoing discussion this is the negative of pressure times the change in volume. However, to keep the temperature fixed, some readjustment of the probability of each level must occur. Thus the total energy must make an additional change which we will call dQ. This additional energy comes from the external system (the bath) which maintains the temperature, and it is called the heat exchanged. Thus

$$dU = -P\,dV + dQ \tag{10.17}$$

We can find dQ easily from the expression for U given in Eq. (10.7). When V is altered by the change dV, then each energy level E_n undergoes the change dE_n and the Helmholtz free energy changes by an amount dF. Thus the total energy changes by the amount

$$\begin{aligned} dU = {} & \sum_n dE_n\, e^{-\beta(E_n - F)} \\ & + \beta\, dF \sum_n E_n e^{-\beta(E_n - F)} - \beta \sum_n E_n\, dE_n\, e^{-\beta(E_n - F)} \end{aligned} \tag{10.18}$$

The first term in this expression is the expected value of dE_n, which is $-P\,dV$, as we have already explained. The remaining two terms constitute dQ. These two terms can also be expressed with the derivatives of the sum in Eq. (10.2), and ultimately in terms of F. In fact, we find

$$dQ = -T\frac{\partial^2 F}{\partial T\,\partial V}\,dV \tag{10.19}$$

That this is true can be seen also from Eq. (10.17), which gives

$$\begin{aligned} \frac{dQ}{dV} &= \frac{\partial U}{\partial V} + P = \frac{\partial}{\partial V}\left(F - T\frac{\partial F}{\partial T}\right) - \frac{\partial F}{\partial V} \\ &= -T\frac{\partial^2 F}{\partial T\,\partial V} \end{aligned} \tag{10.20}$$

Equation (10.19) gives the heat exchanged dQ in changing the volume by the amount dV while keeping the temperature constant. If we change any other parameter, we shall arrive at an analogous result. For example, if we change the temperature T while holding the volume V constant, the heat exchanged is equal to the change in total energy. That is,

$$dQ = \frac{\partial U}{\partial T}\,dT = \frac{\partial}{\partial T}\left(F - T\frac{\partial F}{\partial T}\right)dT = -T\frac{\partial^2 F}{\partial T^2}\,dT \tag{10.21}$$

In general, then, we have the result

$$dQ = -T\left(\frac{\partial^2 F}{\partial T\,\partial V}\,dV + \frac{\partial^2 F}{\partial T\,\partial \alpha}\,d\alpha + \frac{\partial^2 F}{\partial T^2}\,dT\right) \tag{10.22}$$

The right-hand side of Eq. (10.22) is of the form T times the total change in a quantity $\mathrm{S} = -(\partial F/\partial T)$, which is called the *entropy*. That is,

$$dQ = T\,d\mathrm{S} \tag{10.23}$$

$$\mathrm{S} = -\frac{\partial F}{\partial T} \tag{10.24}$$

$$U = F + T\mathrm{S} \tag{10.25}$$

It is evident that all the standard thermodynamic quantities — internal energy, entropy, pressure, etc. — can be evaluated if a single function, the partition function Z, is known in terms of the temperature, volume, external field, etc. The thermodynamic quantities are obtained simply by differentiating Z or, equivalently, the free energy F.

The determination of some physical quantities, even for a system in thermal equilibrium, requires more information than only the partition function. For example, suppose the system is in a configuration space with a coordinate x and we ask: What is the probability of finding the system at location x? We know that if the system is in the single state defined by the wave function $\phi_n(x)$, the probability of observing x is the absolute square of the wave function, $\phi_n^*(x)\phi_n(x)$. Thus, averaging over all possible states, the probability of observing x is

$$P(x) = \frac{1}{Z}\sum_n \phi_n^*(x)\phi_n(x)e^{-\beta E_n} \tag{10.26}$$

In the general case, if we are interested in any quantity A, then the expected value is given by

$$\bar{A} = \frac{1}{Z}\sum_n A_n e^{-\beta E_n} = \frac{1}{Z}\sum_n \int \phi_n^*(x)A\phi_n(x)\,dx\,e^{-\beta E_n} \tag{10.27}$$

It is evident that the expected values of all such quantities could be obtained if we knew the function

$$\rho(x', x) = \sum_n \phi_n(x')\phi_n^*(x)e^{-\beta E_n} \tag{10.28}$$

This suffices since the function A appearing in the integral of Eq. (10.27) is an operator which operates only on the $\phi_n(x)$ of that expression, and not on $\phi_n^*(x)$. Using the quantity $\rho(x', x)$, we can imagine A to act on x'

only, after which we set x' equal to x in the form $A\rho(x', x)$, and finally integrate over all values of x. This process is called finding the *trace* of $A\rho$.

From the definition of $\rho(x', x)$ it is clear that

$$P(x) = \frac{1}{Z}\rho(x, x) \tag{10.29}$$

and since the probability $P(x)$ is normalized, so that the integral over all of x gives 1, we have

$$Z = \int \rho(x, x)\, dx \equiv \text{trace}\{\rho\} \tag{10.30}$$

The quantity $\rho(x', x)$ is called the *density matrix.* [More precisely, it is called the "statistical density matrix for temperature T"; the term "density matrix" also has a wider use for general systems in or out of thermal equilibrium and is usually used for the normalized version of our function $\rho(x', x)$, that is, for the function we would write as $\rho(x', x)/Z$.] The general problem of statistical mechanics is to evaluate Eq. (10.28) to find the density matrix. If we are interested only in conventional thermodynamic variables, we need only the trace or diagonal sum of the density matrix, which gives us the partition function Z.

10-2 THE PATH INTEGRAL EVALUATION

The formulation of the density matrix given in Eq. (10.28) bears a close resemblance to the general expression for the kernel, which was derived in Chap. 4 and given in Eq. (4.59) as

$$K(x_b, t_b; x_a, t_a) = \sum_n \phi_n(x_b)\phi_n^*(x_a)e^{-(i/\hbar)E_n(t_b - t_a)} \tag{10.31}$$

The validity of this expression is restricted to situations in which the hamiltonian is constant in time and $t_b > t_a$. However, this situation is implied in statistical mechanics; for only if the hamiltonian is constant in time can equilibrium be achieved. The difference between the form of Eq. (10.31) and that of Eq. (10.28) is in the argument of the exponential. If the time difference $t_b - t_a$ of Eq. (10.31) is replaced by $-i\beta\hbar$, we see that the expression for the density matrix is formally identical to the expression for the kernel corresponding to an imaginary negative time interval.

We can develop the similarity between these two expressions from another point of view. Suppose we write the density matrix in a way which makes it look a little bit more like a kernel, thus, $k(x_b, u_b; x_a, u_a)$ for $\rho(x_b, x_a)$, where

$$k(x_b, u_b; x_a, u_a) = \sum_n \phi_n(x_b)\phi_n^*(x_a)e^{-[(u_b-u_a)/\hbar]E_n} \tag{10.32}$$

Then if $x_b = x'$, $x_a = x$, $u_b = \hbar\beta$, and $u_a = 0$, Eq. (10.32) becomes identical with Eq. (10.28).

If we differentiate k partially with respect to u_b, we get

$$\frac{\partial k(b,a)}{\partial u_b} = -\frac{1}{\hbar}\sum_n E_n\phi_n(x_b)\phi_n^*(x_a)e^{-[(u_b-u_a)/\hbar]E_n} \tag{10.33}$$

But now recall that $E_n\phi_n(x_b) = H\phi_n(x_b)$; so if we understand H_b to imply operations only upon the variables x_b, we can write

$$\frac{\partial k(b,a)}{\partial u_b} = -\frac{1}{\hbar}H_b k(b,a) \tag{10.34}$$

or, to put the same thing another way,

$$\frac{\partial \rho(b,a)}{\partial \beta} = -H_b\rho(b,a) \tag{10.35}$$

We notice that this differential equation for ρ is similar to the Schrödinger equation for the kernel K which was developed in Chap. 4 and given in Eq. (4.25). We can rewrite it here as

$$\frac{\partial K(b,a)}{\partial t_b} = -\frac{i}{\hbar}H_b K(b,a) \qquad \text{for } t_b > t_a \tag{10.36}$$

We found in Chap. 4 that the kernel $K(b,a)$ is Green's function for Eq. (10.36). In the same sense the density matrix $\rho(b,a)$ is Green's function for Eq. (10.35).

With simple hamiltonians involving only momenta and coordinates we have been able to write the kernel as a path integral. For example, in a one-particle, one-dimensional situation where the hamiltonian is given by

$$H = -\frac{\hbar^2}{2m}\frac{d^2}{dx^2} + V(x) \tag{10.37}$$

the solution for the kernel over a very short time interval

$$t_b - t_a = \epsilon$$

is

$$K(b,a) = \left(\frac{m}{2\pi i\hbar\epsilon}\right)^{1/2} \exp\left\{\frac{i}{\hbar}\left[\frac{m}{2}\frac{(x_b-x_a)^2}{\epsilon} - \epsilon V\left(\frac{x_b+x_a}{2}\right)\right]\right\} \tag{10.38}$$

which can be directly verified by substitution into Eq. (10.37) and taking the limit $\epsilon \to 0$. By building up a product of many kernels of the form (10.38), and summing over paths, and taking the limit as the time interval ϵ goes to 0 and the number of terms in the product becomes infinite, we have produced a path integral describing the kernel over a finite period of time.

We can produce a solution to Eq.(10.34) in the same manner. The solution for an infinitesimal interval of $u_b - u_a = \eta$ is given by substituting $\epsilon = -i\eta$ into Eq. (10.38). Thus

$$k(x_b, \eta; x_a, 0) = \left(\frac{m}{2\pi\hbar\eta}\right)^{1/2} \exp\left\{-\frac{1}{\hbar}\left[\frac{m}{2}\frac{(x_b - x_a)^2}{\eta} + \eta V\left(\frac{x_b + x_a}{2}\right)\right]\right\} \tag{10.39}$$

That this is a valid solution of Eq. (10.34) in the limit $\eta \to 0$ can be demonstrated by direct substitution.

The rule for the combination of functions defined for successive values of u is the same as the rule for the combination of kernels for successive intervals of time. That is,

$$k(b, a) = \int k(b, c)k(c, a)\, dx_c \tag{10.40}$$

That this result still holds follows from the fact that Eq. (10.33) is a first-order derivative in u. This rule can be used to obtain the path integral to define $k(b, a)$ as

$$k(x_b, u_b; x_a, u_a) = \frac{1}{a}\int \cdots \int \exp\left\{-\frac{1}{\hbar}\sum_{i=0}^{N-1}\left[\frac{m}{2}\frac{(x_{i+1} - x_i)^2}{\eta} + \eta V(x_i)\right]\right\}\prod_{i=1}^{N-1}\frac{dx_i}{a} \tag{10.41}$$

The normalizing constant a now becomes

$$a = \left(\frac{2\pi\hbar\eta}{m}\right)^{1/2} \tag{10.42}$$

and the integral is carried out over all paths going from x_a to x_b (that is, for $i = 0$, $x_i = x_a$ and for $i = N$, $x_i = x_b$) in the interval $u_b - u_a = N\eta$.

The result of this derivation is that if we consider a "path" $x(u)$ as a function which gives a coordinate in terms of the parameter u, and if we call $\dot{x}$ the derivative dx/du, then

$$\rho(x_b, x_a) = \int \exp\left\{-\frac{1}{\hbar}\int_0^{\beta\hbar}\left[\frac{m}{2}\dot{x}^2(u) + V(x(u))\right] du\right\} \mathcal{D}x(u) \tag{10.43}$$

This is a very amusing result, because it gives the complete statistical behavior of a quantum-mechanical system as a path integral without the appearance of the ubiquitous i so characteristic of quantum mechanics. (Incidentally, this is not so for a system moving in a magnetic field.) This path integral of Eq. (10.43) is much easier to work with and visualize than the complex integrals which we have studied previously. Here, it is easy to see why some paths contribute very little to the integral — these are the paths for which the exponent is very large and thus the integrand is negligibly small. Furthermore, it is not necessary to think about whether or not nearby paths cancel each other's contributions, since in the present case all contributions add together with some being large and others small.

The parameter u is not the true time in any sense. It is just a parameter in an expression for the density matrix ρ. However, if we wish to think through analogy, we can consider u as the time for a certain path, an in so doing we can state the result given by Eq. (10.43) in a vivid pictorial way. What we are doing is providing a physical analogue for the mathematical expression. We shall call u the "time," leaving the quotation marks to remind us that it is not real time (although u does have the dimensions of time). Likewise $\dot{x}$ will be called the "velocity," $m\dot{x}^2/2$ the "kinetic energy," etc. Then Eq. (10.43) says that the density matrix for a temperature $1/k\beta$ is given in the following way:

Consider all the possible paths, or "motions," by which the system can travel between the initial and final configurations in the "time" $\beta\hbar$. The density matrix ρ is a sum of contributions from each motion, the contribution from a particular motion being the "time" integral of the "energy" divided by $\hbar$ for the path in question.

The partition function is derived by considering only those paths in which the final configuration is the same as the initial configuration, and we sum over all possible initial configurations.

Problem 10-1 Show that the density matrix for a one-dimensional harmonic oscillator is

$$\rho(x', x) = \left(\frac{m\omega}{2\pi\hbar\sinh\beta\hbar\omega}\right)^{1/2} \times \exp\left\{-\frac{m\omega}{2\hbar\sinh\beta\hbar\omega}[(x'^2 + x^2)\cosh\beta\hbar\omega - 2x'x]\right\} \tag{10.44}$$

This answer can be compared with the results of Prob. 3-8. Show also that the free energy is $kT\ln[2\sinh(\hbar\omega/2kT)]$. Check this latter value by a direct evaluation of the sum of Eq. (10.2).

The Classical Approximation. If the temperature is not too low (how low is too low will be discussed below Eq. (10.49)), $\beta\hbar$ is very small. Thus, in calculating the partition function for which $x_b = x_a$, each path starts from x_a and in a very short "time" is back at x_a again. In fact, the paths cannot ever wander very far from x_a, because traveling far away and returning again in the short "time" available requires a high "velocity" and a large "kinetic energy." For such a path the exponential function appearing in Eq. (10.43) becomes negligibly small, and it will contribute a negligible amount to the sum over all paths. Under these circumstances the paths $x(u)$ which must be considered in evaluating $V(x(u))$ never move very far from the initial point x_a. Thus to a first approximation we can write $V(x(u)) \approx V(x_a)$ for all paths. In this approximation the potential energy is independent of the path, and the exponential function dependent on the potential can be taken outside the integral. Thus for temperatures which are not too low we have

$$\rho(x_a, x_a) = e^{-\beta V(x_a)} \int_{x_a}^{x_a} \exp\left\{-\frac{1}{\hbar}\frac{m}{2}\int_0^{\beta\hbar} \dot{x}^2(u)\,du\right\} \mathcal{D}x(u) \tag{10.45}$$

In this last expression the path integral is that for a free particle. It can be solved in the same way that we solved the path integral defining the kernel for the motion of a free particle in Chap. 3. The result is

$$\int_{x_a}^{x_b} \exp\left\{-\frac{m}{2\hbar}\int_0^{\beta\hbar} \dot{x}^2(u)\,du\right\} \mathcal{D}x(u)$$
$$= \sqrt{\frac{mkT}{2\pi\hbar^2}} \exp\left\{-\frac{mkT(x_b - x_a)^2}{2\hbar^2}\right\} \tag{10.46}$$

If we are interested only in the partition function, we set $x_b = x_a$ and find

$$\rho(x_a, x_a) = \sqrt{\frac{mkT}{2\pi\hbar^2}}\, e^{-\beta V(x_a)} \tag{10.47}$$

Then, the partition function is the integral of this expression over all possible initial configurations x_a. Thus

$$Z = \sqrt{\frac{mkT}{2\pi\hbar^2}} \int e^{-\beta V(x_a)}\, dx_a \tag{10.48}$$

This is a formula for the partition function valid in the limit of classical mechanics. It was originally derived, within an uncertain multiplying constant, as a consequence of classical mechanics by Boltzmann. In more complicated cases (e.g., more variables) the classical partition function is simply the product of two factors. The first of these is the path integral which one would get by considering all particles of the system to be free. The second factor is called the configuration integral, and it is just the integral of $e^{-\beta V}$, where the potential energy V of the system depends upon all of the N variables describing the system. For example, for N particles interacting by a potential $V(\mathbf{x}_1, \mathbf{x}_2, \ldots, \mathbf{x}_N)$, where $\mathbf{x}_i$ is the position vector of particle i, the integral required is

$$\int \cdots \int\int \exp\{-\beta V(\mathbf{x}_1, \mathbf{x}_2, \ldots, \mathbf{x}_N)\}\, d^3\mathbf{x}_1\, d^3\mathbf{x}_2 \cdots d^3\mathbf{x}_N$$

This simple form for the partition function is only an approximation valid if the particles of the system cannot wander very far from their initial positions in the "time" $\beta\hbar$. The limit on the distance which the particles can wander before the approximation breaks down can be estimated from Eq. (10.46). We see that if the final point differs from the initial point by as much as

$$\Delta x = \frac{\hbar}{\sqrt{mkT}} \tag{10.49}$$

the exponential function of Eq. (10.46) becomes very greatly reduced. From this, we can infer that intermediate points farther than Δx away from the initial and final points can be reached only on paths which do not contribute greatly to the path integral of Eq. (10.43). If the potential $V(x)$ does not alter very much as x moves over this distance, then the classical statistical mechanics is valid.

For a typical solid or liquid at room temperature, with an atomic mass of about 20, for example, Δx is about 0.1 Å, while the interatomic distances and forces range over one or two angstroms. Thus motions greater than 0.1 Å will not contribute to the density matrix, while the potential function will remain unchanged until motion of about one or two angstroms has been achieved. It is clear that classical statistical mechanics is adequate for such materials.

All of the mysterious transformations between solids, liquids, and gases ordinarily lie in a range where classical statistical mechanics is valid. The mathematical interpretation of all these processes is contained in the problem of evaluating the integral of $e^{-\beta V}$ over the coordinates of all the atoms. That this amazing variety and peculiarity of phenomena comes from just a simple integral is at first surprising, until it is realized that the integral is a multiple integral over a stupendous number of variables. Our usual experience with integrals which involve one or, at most, a few variables of integration does not prepare us for the almost qualitative differences that can arise when the number of variables approaches infinity.

The fascination of the problems in the theory of the solid states, or of liquids and of the condensation of gases, lies, like the behavior of this multiple integral, in the way in which simple descriptions of simple systems when joined together in enormous multiplicity yield such rich phenomena. It is a challenge to the imagination to see how the cooperation between systems can lead to such results. A rough qualitative explanation is readily forthcoming for many of these effects, but the problem of quantitative detail also holds fascination for the theoretical physicist.

There are important statistical phenomena which occur when the classical approximation is not valid. In this case the multiplicity of variables is compounded with the conceptual complexity of quantum mechanics to raise even greater challenges.

Strictly speaking, Eq. (10.48) does contain a little more information than was available to the purely classical statistical mechanics. This is evidenced by the appearance of $\hbar$ in the coefficient in front of the integral. Classical mechanics could not determine the partition function absolutely, but only within an unidentified constant factor. Thus the logarithm of the partition function was determined only to within an additive constant. That meant that a term proportional to T appeared in the expression for the free energy, or an additive constant appeared in the entropy. This constant, which was sometimes called the chemical constant, could be completely evaluated only after the quantum-mechanical solution was worked out.

0-3 QUANTUM-MECHANICAL EFFECTS

There are some cases in which the classical approach is not adequate. For these cases it is necessary to include changes in the potential function which result from the motion along the "path." In this section we shall calculate the first-order effect of the potential when the motion of the particle is taken into account.

Instead of approximating $V(x)$ by the constant value $V(x_a)$ in the expression for the density matrix, Eq. (10.43), we might try a Taylor series expansion for $V(x)$ around the point x_a. However, we would find that we could save effort and increase our accuracy if we chose to expand about the mean position given by

$$\bar{x} = \frac{1}{\beta\hbar} \int_0^{\beta\hbar} x(u)\, du \tag{10.50}$$

which is defined for any particular path. We can characterize each path by its mean position and carry out integrations over all such positions instead of integrations over initial positions x_a, as was done in Eq. (10.48). In this way the partition function becomes

$$Z = \iint_{x_a}^{x_a} \exp\left\{-\frac{1}{\hbar}\int_0^{\beta\hbar}\left[\frac{m}{2}\dot{x}^2(u) + V(x(u))\right] du\right\} \mathcal{D}'x(u)\, d\bar{x} \quad (10.51)$$

In this expression the paths are chosen to satisfy two conditions: (1) that $\bar{x}$ given by Eq. (10.50) is fixed and (2) that the initial and final points are identical. This implies that the integral over all paths must also include an integration over all end points x_a, and that is the meaning of the notation $\mathcal{D}'$ in the differential.

Using a Taylor series expansion for $V(x)$ about the point $\bar{x}$, we find

$$\int_0^{\beta\hbar} V(x(u))\, du = \quad (10.52)$$
$$\beta\hbar V(\bar{x}) + \int_0^{\beta\hbar}(x(u)-\bar{x})V'(\bar{x})\, du + \frac{1}{2}\int_0^{\beta\hbar}(x(u)-\bar{x})^2 V''(\bar{x})\, du + \cdots$$

By virtue of Eq. (10.50) the second term on the right-hand side of this last equation is zero. Thus, by expanding about the mean position, we arrive at an expression for which the first nonzero correction term is of second order. Using this expansion and including no terms of higher order than the second, we have for the partition function

$$Z \approx \int e^{-\beta V(\bar{x})} \quad (10.53)$$
$$\times \int_{x_a}^{x_a} \exp\left\{-\frac{1}{\hbar}\int_0^{\beta\hbar}\left[\frac{m}{2}\dot{x}^2(u) + \frac{1}{2}(x(u)-\bar{x})^2 V''(\bar{x})\right] du\right\} \mathcal{D}'x(u)\, d\bar{x}$$

The path integral in this expression differs from those of our previous experience in one particular way. The paths over which the integral is to be evaluated are constrained by Eq. (10.50), which can be rewritten for present purposes as

$$\frac{1}{\beta\hbar}\int_0^{\beta\hbar}(x(u)-\bar{x})\, du = 0$$

The substitution $y(u) = x(u) - \bar{x}$ as the path coordinate then gives the constraint in the form

$$\frac{1}{\beta\hbar}\int_0^{\beta\hbar} y(u)\, du = 0$$

and the path integral itself is

$$\int_{x_a-\bar{x}}^{x_a-\bar{x}} \exp\left\{-\frac{1}{\hbar}\int_0^{\beta\hbar}\left[\frac{m}{2}\dot{y}^2(u) + \frac{1}{2}y^2(u)V''(0)\right] du\right\} \mathcal{D}'y(u) \quad (10.54)$$

The integrand of this path integral is the same as that for a harmonic oscillator, with the frequency given by $\omega^2 = V''(0)/m$.

We now apply the constraint to this path integral in the following way. We multiply the whole path integral by the Dirac delta function

$$\delta\left(\frac{1}{\beta\hbar}\int_0^{\beta\hbar} y(u)\,du\right)$$

In order to manipulate the delta function within the path integral, we express it by its Fourier transform

$$\delta(x) = \int_{-\infty}^{\infty} \exp\{ikx\}\,\frac{dk}{2\pi}$$

and write Eq. (10.54) as

$$\int_{-\infty}^{\infty}\int_{x_a-\bar{x}}^{x_a-\bar{x}} \exp\left\{-\frac{1}{\hbar}\int_0^{\beta\hbar}\left[\frac{m}{2}\dot{y}^2 + \frac{1}{2}V''(0)y^2 - \frac{iky}{\beta}\right]du\right\}\mathcal{D}y(u)\,\frac{dk}{2\pi} \tag{10.55}$$

In this form, the path integral contains the constraint of Eq. (10.50), and we can drop the prime on $\mathcal{D}'$ and proceed directly with standard path integral techniques to obtain the desired solution. We note that the integrand of the path integral now has the same form as the path integral for a forced harmonic oscillator if we interpret both m and $V''(0)$ to be imaginary. However, we are interested only in the case $V''(0)$ small, and the approximation of including only the first-order term $V''(0)$ can be made at any convenient stage.

Problem 10-2 Use the methods of Chap. 3 and, in particular, Eq. (3.66) to solve this path integral. Remember that paths of interest in this problem have the same initial and final points and that completion of the path integral requires an integration over all values of this point. Finally, carry out the integration over k to get as a solution

$$\text{const}\ \frac{\beta\hbar\omega/2}{\sinh(\beta\hbar\omega/2)} = \text{const}\ \left[1 - \frac{\beta^2\hbar^2}{24m}V''(\bar{x}) + \cdots\right] \tag{10.56}$$

The partition function which results from the solution obtained in Prob. 10-2 is written best in the form (valid to first order in V'')

$$Z = \sqrt{\frac{mkT}{2\pi\hbar^2}}\int \exp\left\{-\beta\left[V(\bar{x}) + \frac{\beta\hbar^2}{24m}V''(\bar{x})\right]\right\}d\bar{x} \tag{10.57}$$

Here the unknown constant has been evaluated simply by comparison with the classical result of Eq. (10.48). We see from this result that the partition function has the same form which we derived under classical assumptions. The only difference is that a temperature-dependent corrective term has been added to the potential. This corrective term, $(\beta\hbar^2/24m)V''(\bar{x})$, is clearly quantum-mechanical in nature, as can be seen from the inclusion of Planck's constant $\hbar$.

Problem 10-3 Show that for many particles (which we identify by subscripts so that the mass of the ith particle is m_i) moving in three dimensions, the correction to the potential is

$$\frac{\beta^2\hbar^2}{24}\sum_i \frac{1}{m_i}\nabla_i^2 V \tag{10.58}$$

In practice, the results of this calculation are not very useful. In most problems, e.g., in a gas of colliding molecules, the potential rises very sharply so that there is a violent repulsion at small distances. In such a case the second derivative is very large. When this is not so, the formula may be of some use. It has one advantage in that it may be easily extended to another order of accuracy.

Problem 10-4 Show that the correction to the partition function up to the order of $\hbar^4$ contains the factor

$$\left[1 - \frac{\beta^2\hbar^2}{24m}V''(\bar{x}) + \frac{7\beta^4\hbar^4}{8\times 720m^2}[V''(\bar{x})]^2 - \frac{\beta^3\hbar^4}{24\times 48m^2}V''''(\bar{x}) + \cdots\right]$$

The Effective Potential Method. We have seen above that quantum-mechanical effects might be represented by calculating the partition function exactly as in the classical formula of Eq. (10.48), but instead of using the correct potential $V(x)$, we use a modified potential $V(x) + (\beta\hbar^2/24m)V''(x)$. This suggests that we try to go further and seek some possibly better effective potential $U(x)$ which, when substituted for the true $V(x)$ in the classical equation (10.48), would represent an even better approximation to the correct quantum-mechanical partition function.

We start out with the exact expression

$$Z = \int e^{-\beta V(\bar{x})} \tag{10.59}$$
$$\times \int \exp\left\{-\frac{m}{2\hbar}\int_0^{\beta\hbar}\dot{x}^2\,du - \frac{1}{\hbar}\int_0^{\beta\hbar}[V(x(u)) - V(\bar{x})]\,du\right\}\mathcal{D}'x(u)\,d\bar{x}$$

The path integral within this expression is related to an average over paths $x(u)$. To be specific, for any functional g of $x(u)$, the weighed average of g over all paths starting and ending at the same point, and with average value $\bar{x}$, weighed by $\exp\{-(m/2\hbar)\int \dot{x}^2\,dx\}$, is

$$\langle g[x(u)]\rangle_{\bar{x}} = \frac{\displaystyle\int \exp\left\{-\frac{m}{2\hbar}\int_0^{\beta\hbar} \dot{x}^2\,du\right\} g[x(u)]\,\mathcal{D}'x(u)}{\displaystyle\int \exp\left\{-\frac{m}{2\hbar}\int_0^{\beta\hbar} \dot{x}^2\,du\right\} \mathcal{D}'x(u)}$$

We call the denominator $B(\bar{x})$, so the path integral within Eq. (10.59) is $B(\bar{x})\langle e^{f[x(u)]}\rangle_{\bar{x}}$, where

$$f[x(u)] = -\frac{1}{\hbar}\int_0^{\beta\hbar} [V(x(u)) - V(\bar{x})]\,du \tag{10.60}$$

If we were to replace this average of an exponential with the exponential of an average, thus

$$\langle e^f\rangle \to e^{\langle f\rangle} \tag{10.61}$$

we know we would make an error of the second order in f or, better, of the order of the difference between $\langle f\rangle^2$ and $\langle f^2\rangle$. We shall see at Eq. (11.6) that we can determine the sign of this error, i.e., the left-hand side is greater than the right. The exact and approximate partition functions are then

$$Z = \int e^{-\beta V(\bar{x})} B(\bar{x})\langle e^f\rangle_{\bar{x}}\,d\bar{x} \quad \text{and} \quad Z' = \int e^{-\beta V(\bar{x})} B(\bar{x}) e^{\langle f\rangle_{\bar{x}}}\,d\bar{x}$$

To evaluate the path integral

$$\langle f\rangle_{\bar{x}} = \frac{1}{B(\bar{x})}\int \exp\left\{-\frac{m}{2\hbar}\int_0^{\beta\hbar} \dot{x}^2\,du\right\} \times \left[-\frac{1}{\hbar}\int_0^{\beta\hbar} [V(x(u)) - V(\bar{x})]\,du\right] \mathcal{D}'x(u) \tag{10.62}$$

we first change variables to the paths $y(u) = x(u) - \bar{x}$ where

$$y(0) = y(\beta\hbar) = Y \qquad \frac{1}{\beta\hbar}\int_0^{\beta\hbar} y(u)\,du = 0 \tag{10.63}$$

so that

$$\langle f\rangle_{\bar{x}} = \frac{1}{B(\bar{x})}\left[-\frac{1}{\hbar}\int_0^{\beta\hbar}\int \exp\left\{-\frac{m}{2\hbar}\int_0^{\beta\hbar} \dot{y}^2\,du\right\} \times [V(\bar{x} + y(t)) - V(\bar{x})]\,\mathcal{D}'y(u)\,dt\right]$$

Second, we define the related path integral

$$I(\bar{x}) = \int \exp\left\{-\frac{m}{2\hbar}\int_0^{\beta\hbar} \dot{y}^2\,du\right\}[V(\bar{x}+y(t)) - V(\bar{x})]\,\mathcal{D}'y(u) \tag{10.64}$$

where t is some specific value of u between 0 and $\beta\hbar$. It is clear that

$$\langle f\rangle_{\bar{x}} = \frac{1}{B(\bar{x})}\left[-\frac{1}{\hbar}\int_0^{\beta\hbar} I(\bar{x})\,dt\right]$$

and that $B(\bar{x})$ is just $I(\bar{x})$ in the special case $[V(\bar{x}+y(t)) - V(\bar{x})] = 1$.

At first glance it appears that $I(\bar{x})$ is a function of t, but the following argument shows that $I(\bar{x})$ is in fact independent of t. Suppose each path in the integral is not of finite length, but is really a $\beta\hbar$-length segment of a periodic path whose period is $\beta\hbar$, as shown in Fig. 10-1. Consider two of the family of all such paths, one $y(u)$ and the other $y(u+t_1) = y_1(u)$, as shown in Fig. 10-2. The value which the first attains at $u = t_1$, namely, $y(t_1)$, is reached by the second when its argument is 0, that is, $y(t_1) = y_1(0)$. Furthermore, for any other point t_i there exists in the family the analogous function $y_i(u)$ for which $y(t_i) = y_i(0)$, and all such paths give the same contribution to

$$\int_0^{\beta\hbar} \dot{y}^2(u)\,du$$

Of course, all these statements apply to each path included in the path integral. Thus we see that we lose nothing by arbitrarily setting $t = 0$ in the path integral over all paths $y(u)$, which is the same as saying that the integral $I(\bar{x})$ is independent of t.

Problem 10-5 Using the method outlined above Prob. 10-2, and Eq. (3.62), show that

$$I(\bar{x}) = \frac{\sqrt{3}}{\pi}\frac{m}{\beta\hbar^2}\int_{-\infty}^{\infty} e^{-Y^2(6m/\beta\hbar^2)}[V(\bar{x}+Y) - V(\bar{x})]\,dY \tag{10.65}$$

$$\langle f\rangle_{\bar{x}} = -\beta\sqrt{\frac{6m}{\pi\beta\hbar^2}}\int_{-\infty}^{\infty} e^{-Y^2(6m/\beta\hbar^2)}V(\bar{x}+Y)\,dY + \beta V(\bar{x})$$

Suppose we call our approximation to the partition function Z' and the corresponding Helmholtz free energy F', so that $Z' = e^{-\beta F'}$. We then have

$$Z' = \int e^{-\beta V(\bar{x}) + \langle f\rangle_{\bar{x}}} B(\bar{x})\,d\bar{x} \tag{10.66}$$

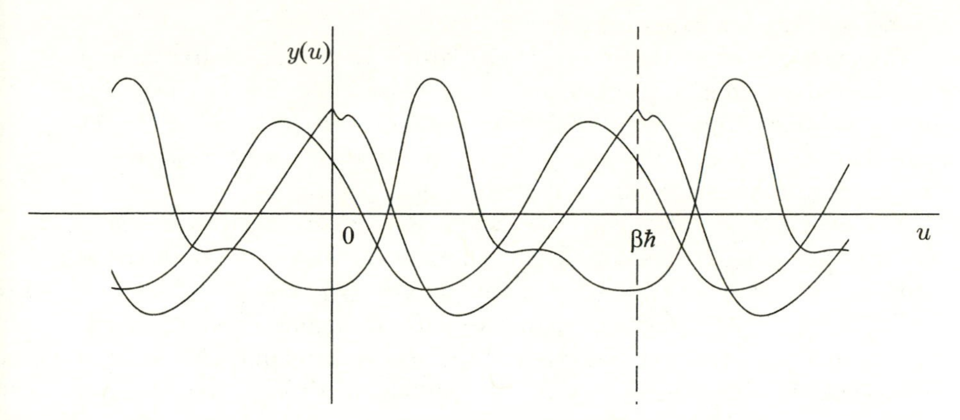

Fig. 10-1 All paths which return at $u = \beta\hbar$ to their initial value (at $u = 0$) can be considered as $\beta\hbar$-length segments of periodic paths where the period is $\beta\hbar$.

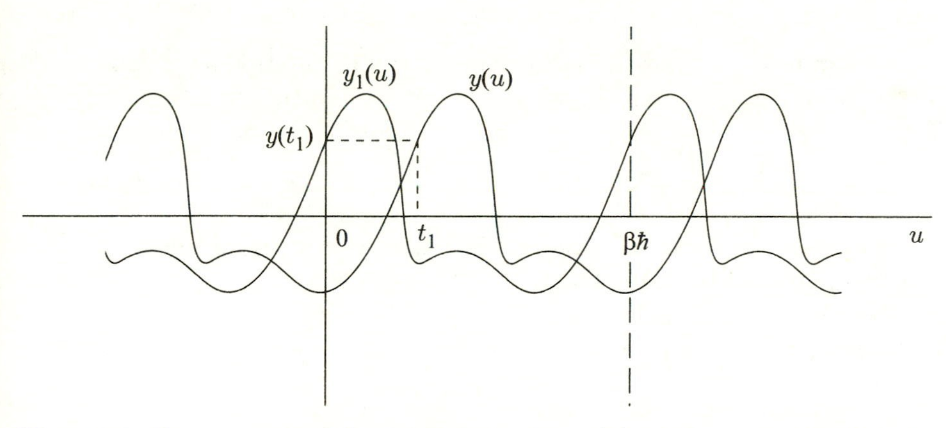

Fig. 10-2 Suppose one of the "periodic" paths $y(u)$, as shown in Fig. 10-1, has the value $y(t_1)$ at $u = t_1$. Then the collection of all "periodic" paths must contain this same path slipped left a distance t_1, that is, $y(u + t_1)$, which will have this same value at $u = 0$. The result of a path integral average over all such paths must then be independent of the selection of the initial point on the u axis.

The factor $B(\bar{x})$ was evaluated in Eq. (10.46), and we have

$$Z' = \sqrt{\frac{mkT}{2\pi\hbar^2}} \int e^{-\beta U(\bar{x})}\, d\bar{x} \tag{10.67}$$

where

$$U(\bar{x}) = \sqrt{\frac{6m}{\pi\beta\hbar^2}} \int_{-\infty}^{\infty} e^{-y^2(6m/\beta\hbar^2)} V(\bar{x} + y)\, dy \tag{10.68}$$

The term $V(\bar{x})$ has cancelled out.

These results mean that we can calculate an approximate free energy F' in a classical manner (i.e., using an expression like Eq. (10.48)) and get a good approximate result if we use an effective potential $U(\bar{x})$, as defined by Eq. (10.68), in place of $V(\bar{x})$. Incidentally, we note that the effective potential is temperature-dependent.

The effective potential is a mean value of $V(x)$ averaged over points near $\bar{x}$ in a gaussian fashion where the root-mean-square spread (or standard deviation) of the gaussian weighing function is $(\beta\hbar^2/12m)^{1/2}$. Furthermore, if we follow through the various inequalities which are involved in our approximation, we find that the approximate free energy F' exceeds the true free energy F. The details of this are discussed in the next chapter, at Eq. (11.9) and following.

Problem 10-6 Show that the relation of Eq. (10.68) becomes the "corrected" potential of Eq. (10.57) (that is, the argument of the exponent in that equation) if V is expanded as a Taylor series.

Problem 10-7 Test the validity of the approximation as it applies to the harmonic oscillator, for which the exact value of the free energy is

$$F_{\text{exact}} = kT \ln\left[2\sinh\frac{\hbar\omega}{2kT}\right] \tag{10.69}$$

Evaluate the approximate value for the free energy by means of the effective potential U. Show that

$$U(x) = \frac{m\omega^2}{2}\left(x^2 + \frac{\beta\hbar^2}{12m}\right) \tag{10.70}$$

and that

$$F_{\text{approx}} = kT\left[\ln\frac{\hbar\omega}{kT} + \frac{1}{24}\left(\frac{\hbar\omega}{kT}\right)^2\right] \tag{10.71}$$

Determine the free energy or, better, the ratio of the free energy to kT, for various values of the frequency. It is suggested that the values of 1.0, 2.0, and 4.0 be used for the ratio $\hbar\omega/kT$. Show that F' is greater than F, as expected, and that the errors grow as the temperature falls. Note that if we are even very far from the classical region (e.g., where the ratio $\hbar\omega/kT = 2.0$, so that the system has an 85 per cent probability of being in the ground state) the approximate results are still surprisingly close to the true results.

Compare these results with those obtained through the classical approximation in which the free energy is given by $kT\ln(\hbar\omega/kT)$. Your results should show the values given in the accompanying table.

$\hbar\omega/kT$	1.0	2.0	4.0
$2F_{\text{exact}}/\hbar\omega$	0.08265	0.8546	0.9908
$2F_{\text{approx}}/\hbar\omega$	0.08333	0.8598	1.0265
$2F_{\text{classical}}/\hbar\omega$	0.00000	0.6931	0.6931

10-4 SYSTEMS OF SEVERAL VARIABLES

If a system has several variables, the formulas describing them are obtained by direct extension of the methods we have already studied, except for some special problems which arise from consideration of symmetry properties.

Liquid Helium. As an example consider the problem of finding the partition function for liquid helium. Suppose we have N identical atoms, each of mass m, confined in some volume. Suppose further that atoms interact in pairs through a potential $V(r_{1,2})$. This potential is a weak attraction at large distances and a very strong repulsion at short distances. Just to orient our thinking, we might imagine $V(r)$ as the potential for hard spheres. That is

$$V(r) = \begin{cases} 0 & r > a \\ \infty & r < a \qquad a = 2.7\ \text{Å} \end{cases} \tag{10.72}$$

The lagrangian for such a system has the form

$$L = \frac{m}{2}\sum_i |\dot{\mathbf{x}}_i|^2 - \frac{1}{2}\sum_{i,j} V(r_{i,j}) \tag{10.73}$$

which means that the partition function is

$$Z = \iint_{\mathbf{x}_i(0)}^{\mathbf{x}_i(0)} \exp\left\{-\frac{1}{\hbar}\left[\frac{m}{2}\sum_i \int_0^{\beta\hbar} |\dot{\mathbf{x}}_i(u)|^2\,du + \frac{1}{2}\sum_{i,j}\int_0^{\beta\hbar} V(|\mathbf{x}_i(u) - \mathbf{x}_j(u)|)\,du\right]\right\} \mathcal{D}^{3N}\mathbf{x}(u)\,d^{3N}\mathbf{x}(0) \tag{10.74}$$

Here the symbol $\mathcal{D}^{3N}\mathbf{x}(u)$ stands for $\mathcal{D}^3\mathbf{x}_1(u)\,\mathcal{D}^3\mathbf{x}_2(u)\cdots\mathcal{D}^3\mathbf{x}_N(u)$ and similarly $d^{3N}\mathbf{x}(0)$ means $d^3\mathbf{x}_1(0)\,d^3\mathbf{x}_2(0)\cdots d^3\mathbf{x}_N(0)$. The path integral is performed over paths taken between initial points $\mathbf{x}_i(0)$ and final points $\mathbf{x}_i(\beta\hbar)$ such that $\mathbf{x}_i(\beta\hbar) = \mathbf{x}_i(0)$.

The form which we have written down in Eq. (10.74) is actually not correct. The symmetry properties which we mentioned above will affect this result. This characteristic is one of the interesting features of the quantum mechanics of identical particles. In Sec. 1-3 we mentioned that if an event occurs in two indistinguishable ways, then the amplitudes for the two ways will add. In particular, when we are dealing with indistinguishable particles, one alternative way for accomplishing any event always exists; namely, the interchange of two particles. In such a case the amplitudes for the particles (1) as interchanged and (2) as not interchanged must be added. (This addition rule applies to Bose particles. For Fermi particles the contributions for amplitudes which arise from odd permutations of particles will subtract from each other.) Ordinary helium atoms are of isotopic mass 4 and contain 6 particles: 2 protons, 2 neutrons, and 2 electrons. This means that helium atoms are Bose particles and the amplitudes for interchange of particles add. (For instance, we say that Bose particles follow symmetrical statistics, whereas Fermi particles follow antisymmetrical statistics.)

To see how this addition of amplitudes comes about, at least for helium atoms, we can follow this line of argument: In the final state the atoms cannot be distinguished from each other. Thus, although the appearance of the configuration of atoms may be the same finally as it was initially, the identity of some of the atoms may have been exchanged.

For example, an atom which we shall designate as 1 starts at position $x_1(0)$. We have assumed that some atom at least will be in this same position at the close. Thus, for some atom $x(\beta\hbar)$ is equal to $x_1(0)$. However, it may not be atom 1 which ends up in this particular place. Instead, atom 1 may go to the initial position of atom 2, say $x_2(0)$, while at the same time atom 2 has moved into the initial position of atom 1. That is, it is possible that atoms 1 and 2 exchange places in the final configuration.

To describe this situation in the most general terms, let Px_i stand for some permutation among the atoms which are initially at x_i. Thus, for example, in the situation in which atoms 1 and 2 were exchanged and all others remained where they were, we would have

$$Px_1 = x_2 \qquad Px_2 = x_1 \qquad Px_3 = x_3 \qquad \cdots \qquad Px_N = x_N \tag{10.75}$$

In general, the final state can be any permutation of the initial state:

$$x_j(\beta\hbar) = Px_i(0) \tag{10.76}$$

Thus in order to construct the complete amplitude, we must sum amplitudes over all the $N!$ possible permutations, since each permutation represents an alternate possibility. The normalization is correct if we average over all the permutations. The resulting rules for symmetrical statistics mean that Eq. (10.74) must be replaced by

$$Z = \frac{1}{N!}\sum_P \iint_{\mathbf{x}_i(0)}^{P\mathbf{x}_i(0)} \exp\left\{-\frac{1}{\hbar}\left[\frac{m}{2}\sum_i \int_0^{\beta\hbar} |\dot{\mathbf{x}}_i(u)|^2\,du \right.\right. \tag{10.77}$$

$$\left.\left. + \frac{1}{2}\sum_{i,j}\int_0^{\beta\hbar} V(|\mathbf{x}_i(u) - \mathbf{x}_j(u)|)\,du\right]\right\} \mathcal{D}^{3N}\mathbf{x}(u)\,d^{3N}\mathbf{x}(0)$$

where $\sum_P$ means a sum over all permutations P.

If we were dealing with Fermi particles, e.g., the isotope of helium which has three nucleons, we would have to include an extra factor of ± 1, positive for even permutations and negative for odd permutations. There would also be some extra features which depend upon the spin of the atom in our result.

It is possible to give a more detailed derivation of Eq. (10.77) in the following manner. For helium-4 atoms the quantum-mechanical amplitude for two atoms which start at positions a and b to get to positions c and d is

$$K(c, a; d, b) + K(d, a; c, b) \tag{10.78}$$

(Amplitudes for alternative final conditions add, since these conditions cannot be distinguished from each other.) In this expression $K(c, a; d, b)$ is the complex amplitude for one particle to go from a to c and for one particle to go from b to d.

Since the particles are indistinguishable, their symmetry properties imply that the amplitude to find the two particles finally at the points c and d must be a symmetric function of c and d. That is, the wave function $\psi(c, d)$ must be a symmetric function of the variables $\mathbf{x}_c$, $\mathbf{x}_d$. That is

$$\psi(c, d) = \psi(d, c) \tag{10.79}$$

If the particles were Fermi, the wave function would have to be an antisymmetric function of these positions.

If many particles are involved, the rule is simply extended, that is,

$$\begin{aligned}\phi(1, 2, 3, \ldots, N) &= \phi(1, 3, 2, \ldots, N)\\ &= \phi(1, 2, 4, \ldots, N)\\ &= \text{etc.}\end{aligned} \tag{10.80}$$

The simplest statement of the general rule is that the wave function must be symmetric (antisymmetric for Fermi particles). Although other solutions of Schrödinger's wave equation exist, only symmetric and antisymmetric ones appear in nature. Hence in the sum defining the partition function in Eq. (10.2), we do not wish the sum over *all* energy eigenstates of the hamiltonian H which can be obtained from solution of $H\phi_n = E_n\phi_n$, but only over those for which the wave function ϕ_n is a symmetric function. For example, the density matrix $\rho(x', x)$ is defined by Eq. (10.28) with a disregard for the statistics of the N atoms involved. How can we reduce this sum to include only symmetric wave functions?

To accomplish this reduction, we use the following trick. First we notice that for any function a symmetric function can be produced simply by permuting all variables and adding together the resulting functions. Thus for any function $f(x_1, x_2)$ the combination $f(x_1, x_2) + f(x_2, x_1)$ is a symmetric function. It follows that for any wave function $\phi(x_1, x_2, \ldots, x_N)$ the function

$$\phi'(x_i) = \sum_P \phi(Px_i) \tag{10.81}$$

is symmetrical. Now if $\phi_n(x_i)$ is a solution to the Schrödinger equation, then $\phi'_n(x_i)$ as defined by Eq. (10.81) is also a solution, since the hamiltonian H is symmetric for an interchange of coordinates. Therefore, each interchanged form $\phi_n(Px)$ is a solution, as is the sum.

Some of the energy eigenvalues E_n have eigenfunctions ϕ_n which are symmetric, and some do not. Suppose E_k is an energy eigenvalue for which the Schrödinger equation does not have a symmetric solution. Then the sum $\sum_P \phi_k(Px)$ must vanish, since if it existed it would be a symmetric solution for E_k. This result implies that the operation defined by Eq. (10.81) selects just those solutions to the wave equation which are symmetric. All other solutions vanish. If $\phi_n(x)$ is symmetric, then it is equal to $\phi_n(Px)$; and since there are $N!$ ways of permuting the N atoms, we have

$$\sum_P \phi_n(Px) = \begin{cases} N!\,\phi_n(x) & \text{if } \phi_n \text{ is symmetric} \\ 0 & \text{if } \phi_n \text{ is of any other symmetry} \end{cases} \tag{10.82}$$

These results give us an answer to our question. We can now select out of the sum defining the density matrix those particular elements which apply to symmetric states. Thus

$$\begin{aligned}
\sum_P \rho(Px', x) &= \sum_n^{\text{all}} \sum_P \phi_n(Px')\phi_n^*(x)e^{-\beta E_n} \\
&= N! \sum_n^{\text{sym}} \phi_n(x')\phi_n^*(x)e^{-\beta E_n} \\
&= N!\, \rho_{\text{sym}}(x', x)
\end{aligned} \tag{10.83}$$

This is the reason why in Eq. (10.77) defining the partition function for symmetric statistics we permute all the particles and divide by $N!$. The resulting partition function corresponds to

$$\int \rho_{\text{sym}}(x_0, x_0)\, d^{3N}x_0 = Z_{\text{sym}} = \sum_n^{\text{sym}} e^{-\beta E_n} \tag{10.84}$$

We note some of the features of Eq. (10.77). At high temperatures, we should expect a classical solution for the partition function with no quantum-mechanical effects in evidence. Suppose we disregard the effects of the potential for the moment and consider the effect of the motion of an atom from its initial point to some other point a distance d away. In the path integral of Eq. (10.77), this is a motion from the initial point $\mathbf{x}_i(0)$ to the permuted position $P\mathbf{x}_i(0)$, and the contribution of that particular permutation to the sum over all permutations is proportional to $\exp\{-mkTd^2/2\hbar^2\}$, thus decreasing with increasing temperature or increasing spacing between atoms. Hence, unless the atoms are extremely close together, no permutation in the sum is important — even the simplest interchange between two atoms — in comparison with the identity permutation which leaves all atoms in their original locations. If we include the effects of the potential which increases steeply at a radius of 2.7 Å from the center of an atom in liquid helium, then no configurations in which the atomic spacing is less than this value are important.

Since only the identity permutation makes a significant contribution to the summations, all that remains for our consideration is the factor $1/N!$. In the early days of *classical* statistical mechanics it was realized that such a factor was necessary when dealing with identical particles, but its significance was not clearly understood. Its effect on the chemical constant is called the *entropy of mixing* when systems of several different kinds of atoms are studied.

As the temperature falls, the exponential factor $\exp\{-mkTd^2/2\hbar^2\}$ prejudicing against migrations to new final positions becomes smaller and smaller. This means that at extremely low temperatures new terms become important in the summation over permutations. Of course, the quantum modifications must be included; and we saw they could be included as a first approximation by replacing the potential V with an effective potential U. As the temperature falls, the specific heat of liquid helium begins to rise slightly near about 2.3 or 2.4 K.

Problem 10-8 The density of liquid helium is 0.17 g/cm^3. Give an order-of-magnitude estimate of the temperature at which permutation terms should begin to play an important role in the description of liquid helium.

At first sight, one would not expect very elaborate exchanges of atoms to ever be important. An exponential factor involving the spacing must be included each time an atom moves to its neighboring location. If we call this factor y, then for r atoms to move to neighboring spots the factor y^r must be included, and since y is certainly less than 1 at any temperature, y^r could become quite small for large r. We certainly would think that as r approaches any reasonable fraction of the approximately 10^{22} atoms in a cubic centimeter of liquid helium, contributions from factors like y^r must be infinitesimal. However, this first sight does not take into account the fact that with r atoms permuting, there is an enormous number of possible permutations, namely $r! \approx e^{r(\ln r - 1)}$. Thus the small weight of one particular permutation is offset by the large number involved.

Another question which arises in the description of liquid helium concerns the type of permutations which are involved. Any permutation can be described by cycles; thus 1–4, 4–7, 7–6, 6–1 is a cycle. Are the important cycles long or short? A careful estimate shows that at moderate temperatures, only simple exchanges of two atoms are important. Then as the temperature falls, cycles of three atoms become important, then four, and so on. But then suddenly, at a certain critical temperature, cycles of much greater length L offset by their great number the small value of y^L. At this temperature cycles of importance become very long, involving nearly all of the atoms inside a container. At this point the curve of specific heat vs. temperature shows a discontinuity. Below this temperature the behavior of the liquid is very strange. It flows through very thin tubes without resistance for low velocities. It simulates infinite heat conductivity in bulk, etc. These odd characteristics are manifestations of quantum mechanics, particularly the constructive interference between amplitudes for replacing one atom with another. Quantitatively, the details of the behavior of the specific heat just at the transition temperature are not on a very firm foundation. But the qualitative reason for the transition is clear.[1]

The expression analogous to Eq. (10.77) for Fermi particles, such as ^{3}He, is also easily written down. However, in the case of liquid ^{3}He, the effect of the potential is very hard to evaluate quantitatively in an accurate manner. The reason for this is that the contribution of a cycle to the sum over permutations is either positive or negative depending on whether the cycle has an odd or even number of atoms in its length L. At low temperature, the contributions of cycles such as $L = 51$ and $L = 52$ are very nearly equal but opposite in sign, and therefore they very nearly cancel. It is necessary to compute the difference between such terms, and this requires very careful calculation of each term separately. It is very difficult to sum an alternating series of large terms which are decreasing slowly in magnitude when a precise analytic formula for each term is not available.

Progress could be made in this problem if it were possible to arrange the mathematics describing a Fermi system in a way that corresponds to a sum of positive terms. Some such schemes have been tried, but the resulting terms appear to be much too hard to evaluate even qualitatively.

For molecules which are separated by distances in the neighborhood of 1 Å we have seen that the effects of exchange (the nonidentical permutations) are important only when the temperature is down to a few degrees absolute. In contrast to this, consider the behavior of electrons in a solid metal. The mass of the electron is so much smaller than that of a molecule that the critical temperature is much higher. At room temperature, electrons in a metal are described accurately only by equations which include the exchange effects of these cyclic permutations. From this point of view, room temperature is very cold for electrons. The exchange effects are of dominant importance, or, to put it another way, the electron gas is degenerate. Of course, the electrons interact by Coulomb's law, which is quite strong. But since the effects of the Coulomb attraction are of long range, they tend to average out. To a fair approximation, the electrons act as if they are independent, although, of course, each moves in the same periodically varying potential produced by the arrangement of the nuclei and the average of the positions of neighboring electrons. From the study of the ideal Fermi gas neglecting interactions, we can learn a lot about the behavior of electrons in metals.

However, it is apparent that we cannot learn quite enough, for the su-

[1] A more detailed discussion of the partition function of liquid helium from this point of view may be found in R.P. Feynman, Atomic Theory of the λ Transition in Helium, *Phys. Rev.*, vol. 91, pp. 1291–1301, 1953.

perconductivity of metals occurring below a few degrees absolute would remain a mystery. This phenomenon, in some metals at least, involves an interaction in which the slow vibratory motion of the atoms is involved. We conclude this because the transition temperatures for two different isotopes of the same metal depend on the atomic mass. This value of the isotopic mass would not be important if the transition were simply a matter of mutual interaction between electrons, or interaction of the electrons with an idealized array of *fixed* atoms. The idealization that the atoms are fixed must be incorrect. But how does the motion of the atoms produce a sudden jump in specific heat in some metals and permit electrical conductivity without resistance below this temperature? This question was first answered in a convincing way by Bardeen, Cooper, and Schrieffer.[1] The path integral approach played no part in their analysis, and in fact it has never proved useful for degenerate Fermi systems.

The Planck Blackbody Radiation Law. The partition function for any system of interacting oscillators is easily worked out. Such a system is equivalent to a set of independent oscillators of frequencies ω_i. However, the value of the free energy F for independent systems is the sum of the values of F for each of the separate systems, which we find directly from the sum of Eq. (10.2) to be

$$kT\ln\left(2\sinh\frac{\hbar\omega}{2kT}\right)$$

This gives the free energy of a linear system as

$$\begin{aligned} F &= kT\sum_i \ln\left(2\sinh\frac{\hbar\omega_i}{2kT}\right) \\ &= kT\sum_i \ln(1-e^{-\hbar\omega_i/kT}) + \sum_i \frac{1}{2}\hbar\omega_i \end{aligned} \tag{10.85}$$

The last term in this expression is a ground-state energy of the system.

For an electromagnetic field in a box of volume V, the modes are specified by the vector wave number $\mathbf{K}$, two for each $\mathbf{K}$. The zero-point energy is omitted. Thus the free energy of the electromagnetic field per unit volume is

$$\frac{F}{V} = kT\int 2\ln(1-e^{-\hbar Kc/kT})\,\frac{d^3\mathbf{K}}{(2\pi)^3} \tag{10.86}$$

[1] J. Bardeen, L.N. Cooper, and J.R. Schrieffer, Theory of Superconductivity, *Phys. Rev.*, vol. 106, pp. 162–164, 1957 and vol. 108, pp. 1175–1204, 1957.

The internal energy U is the partial derivative of βF with respect to β which becomes (putting $\omega = Kc$)

$$\frac{U}{V} = 2\int \frac{\hbar\omega}{e^{\hbar\omega/kT} - 1}\frac{d^3\mathbf{K}}{(2\pi)^3} \tag{10.87}$$

The volume element in $\mathbf{K}$ space can be written as

$$d^3\mathbf{K} = 4\pi K^2\,dK = 4\pi\frac{\omega^2}{c^3}\,d\omega \tag{10.88}$$

This means that the energy density in the electromagnetic field in the range of frequencies from ω to $\omega + d\omega$ is

$$\frac{2\cdot 4\pi}{(2\pi c)^3}\frac{\hbar\omega^3}{e^{\hbar\omega/kT} - 1}\,d\omega \tag{10.89}$$

This is the famous blackbody-radiation law discovered by Planck. It was the first real quantitative quantum-mechanical result discovered and was the first step in the discovery of the new laws.

Another early quantum-mechanical triumph was the explanation of the temperature dependence of the specific heat of solids by Einstein and by Debye. This also comes from Eq. (10.85), but the oscillators are now the normal modes of the crystal, as described in Chap. 8. For example, the thermal energy per unit volume of such a crystal is, like Eq. (10.87) (leaving out the zero-point energy), just

$$\frac{U}{V} = \sum_{3p\text{ modes}}\int \frac{\hbar\omega(\mathbf{K})}{e^{\hbar\omega(\mathbf{K})/kT} - 1}\frac{d^3\mathbf{K}}{(2\pi)^3} \tag{10.90}$$

where $\omega(\mathbf{K})$ is the frequency of a phonon of wave vector $\mathbf{K}$. In a crystal, this is a multiple-valued function (there are $3p$ values for each $\mathbf{K}$ if there are p atoms in a unit cell), and we must sum over each of the possible values of ω for each $\mathbf{K}$. The $\mathbf{K}$ integral extends only over the finite range proper for the crystal. For light there are two modes for each $\mathbf{K}$, each of frequency $\omega = Kc$, so the sum gives a factor 2 and Eq. (10.87) results, the integral on $\mathbf{K}$ now going to infinity.

The result of Eq. (10.90), studied in various approximations by Einstein and Debye, gave a good accounting of the main features of the specific-heat curve, particularly the behavior at low temperatures, which had been in direct contradiction to the classical expectations. Today, putting a more complete knowledge of the phonon spectrum $\omega(\mathbf{K})$ into Eq. (10.90) yields a completely satisfactory description of that part of the specific heat of solids due to internal vibration of the atoms.

10-5 REMARKS ON METHODS OF DERIVATION

The presentation of statistical mechanics given in the early part of this chapter leaves much to be desired. The fundamental law which shows that the probability for finding a system in an energy state with energy E is proportional to $e^{-E/kT}$ is usually derived by considering the interaction of complex systems over long periods of time. But an entertaining problem presents itself.

We started our discussion of physics in this book by expressing the laws of quantum mechanics in terms of path integrals (Chap. 2). Just as a question of curiosity, let us take the point of view that this is the fundamental law. Then ultimately these statistical properties of a system whose quantum-mechanical properties are defined by such a path integral are found to be expressible in terms of the partition function Z. This function can be defined by a path integral of an obviously very similar and closely related form, as shown by Eq. (10.77). Yet the derivation of this result requires noting the wave equation, the existence of stationary states and eigenvalues, and the argument about interaction over long periods of time to which we referred, all of which leads to the expression (10.2) for the partition function in terms of the energy levels E_n. Finally, we proceed to the reverse argument producing the path integral formulation for Z. Is there any way to derive the path integral expression for Z for a system in equilibrium directly from the path integral description for the time-dependent motion? Can we find a short cut which avoids the mention of energy levels altogether? If it is possible, we do not yet know how to do it.

One might ask: Why try it at all? It is like showing that you can swim with your hands tied behind your back. After all, you know there are energy levels. The only excuse for trying to avoid their mention would be that in so doing a deeper understanding of physical processes might result or possibly more powerful methods of statistical mechanics might be evolved. At any rate, it would be interesting to solve the problem.

It was the promptings of a similar quest, to get the well-known variational principle for the lowest energy level directly from the path integral formulation (instead of indirectly via the Schrödinger equation), which resulted in the methods described in Chap. 11. Thus the results of this apparently academic problem were of some use as well as of some interest.

Nevertheless, if we prefer, we can suppose our desire for one particular course in achieving a solution is prompted simply by an academic interest in the methods of classical physics. Suppose that we have a system obeying the principle of least action, with the action defined by

$$S = \frac{m}{2}\int \dot{x}^2(t)\,dt - \frac{k}{2}\int x(t)x(t+a)\,dt \tag{10.91}$$

so that the equation of motion is

$$m\ddot{x}(t) = -\frac{k}{2}[x(t+a) + x(t-a)] \tag{10.92}$$

Here we have created the curious situation in which a particle is driven by a force depending on the average value of coordinates that were and that will be. There are exponentially exploding solutions of Eq. (10.92), but let us say that only motions for which x remains finite both in the distant past and in the distant future will be allowed. (Incidentally, it is likely that solutions which we wish to ignore are excluded anyway if the action law is stated as $\delta S = 0$ for all variations of path δx subject to the restraint $\delta x \to 0$ for $t \to \pm\infty$.)

For such a system it is possible to define an expression for energy which is conserved; for the equations of motion of the system do not depend on time. (No simple hamiltonian gives the equations of motion.) Presumably, such a system could possess properties which allow it, for example, to be perturbed by molecules of a gas and thus achieve thermal equilibrium. We might ask: What are the averages of various quantities describing a system obeying the equations of motion of Eq. (10.92) and appropriate boundary conditions at infinity when it is in equilibrium at the temperature T? Perhaps such a problem is not definable, or perhaps it is easily solved only in this special case because the equations of motion are linear. But the aim of these remarks is to ask whether the existence of a hamiltonian and momentum variables is indeed necessary to the formulation of classical statistical mechanics — or whether a wider class of mechanical systems can be analyzed, a system in which the equation of motion comes most simply from the principle of least action, even though that action involves more than the instantaneous positions and velocities of the particles in the system.

This question is the classical analogue of our more interesting question, namely, how do we proceed directly from the path integral formulation of quantum-mechanical laws of a mechanical system to the path integral formulation of statistical mechanical laws for the same system in equilibrium?

Problem 10-9 Show that the expression

$$E(t) = \frac{m}{2}\dot{x}^2(t) + \frac{k}{2}x(t)x(t+a) - \frac{k}{2}\int_t^{t+a} x(t'-a)\dot{x}(t')\,dt' \tag{10.93}$$

defines a conserved energy for the equation of motion (10.92).

In general, for any action functional, like S, that does not involve the time explicitly (i.e., is invariant for the transformation $t \to t + \text{const}$) there is an expression $E(T)$ for the energy at time T which is conserved. It can be found by asking for the first-order change in the action S when all paths are changed from $x(t)$ to $x(t + \eta(t))$, where $\eta(t) = +\epsilon/2$ for $t < T$ and $\eta(t) = -\epsilon/2$ for $t > T$, with constant ϵ. δS is then $\epsilon E(T)$ for infinitesimal ϵ.

Problem 10-10 Discuss the problem of the path integral formulation of statistical mechanics for a particle in a time-constant magnetic field.

11

The Variational Method

IN this chapter we discuss a method based on a variational principle for the approximate evaluation of certain path integrals. First, we shall illustrate the method by some examples. Later, we consider those problems for which the method may be useful.

11-1 A MINIMUM PRINCIPLE

Suppose we wish to evaluate the free energy F of a system. This problem can be expressed in terms of path integrals by starting with the partition function for the system defined in Eq. (10.4) as

$$Z = e^{-\beta F} \tag{11.1}$$

In Eq. (10.30) the partition function was expressed as an integral of the density matrix $\rho(x, x)$. Then, in Sec. 10-2, a kernel expression for $\rho(x, x)$ was developed. It allows us to write

$$Z = \int_{-\infty}^{\infty} \int_{x_a}^{x_a} e^{S/\hbar} \, \mathcal{D}x(u) \, dx_a \tag{11.2}$$

so long as we use the "time" variable u in the way described below Eq. (10.43). (Also, S here is the negative of the functional S used in Chap. 10.)

In Sec. 10-3 we developed a perturbation technique for the evaluation of the path integral defining the partition function for certain special cases. We shall now describe another technique, applicable in those cases where S *is real.* For ordinary cases without a magnetic field (and no spin) S is real.

Throughout the remainder of this chapter, we choose units in which the value of $\hbar$ is 1. Whenever it is necessary to include $\hbar$ symbolically in order to visualize the quantum-mechanical character of a result, it can be so included by a straightforward dimensional inspection.

Let us suppose that some other S' can be found which satisfies two conditions: First, S' is simple enough that expressions such as $\int e^{S'} \, \mathcal{D}x(u)$ or $\int G e^{S'} \, \mathcal{D}x(u)$, for simple functionals G, can be evaluated. Second, the important paths in the integral $\int e^{S} \, \mathcal{D}x(u)$ and those in the integral $\int e^{S'} \, \mathcal{D}x(u)$ are similar, that is, S' and S are similar when they are both large. Now suppose F' is the free energy associated with S'. That is,

$$e^{-\beta F'} = \int_{-\infty}^{\infty} \int_{x_a}^{x_a} e^{S'} \, \mathcal{D}x(u) \, dx_a \tag{11.3}$$

so that

$$\frac{\iint e^{S} \, \mathcal{D}x(u) \, dx_a}{\iint e^{S'} \, \mathcal{D}x(u) \, dx_a} = e^{-\beta(F-F')} \tag{11.4}$$

Then since $e^S = e^{S-S'}e^{S'}$, we can write Eq. (11.4) as

$$\frac{\iint e^{S-S'}e^{S'}\,\mathcal{D}x(u)\,dx_a}{\iint e^{S'}\,\mathcal{D}x(u)\,dx_a} = e^{-\beta(F-F')} \tag{11.5}$$

This says simply that $e^{-\beta(F-F')}$ is the average value of $e^{S-S'}$ where this average is taken over all paths with the same initial and final point and the weight of each path is $e^{S'}$. All possible values of x_a are included in the averaging process.

One way to proceed now would be to suppose that $S-S'$ is small and that $F-F'$ is small and then expand both sides up to the first power in their respective exponents. This method appears dubious because $\beta(F-F')$ is not small if β is large. However, comparison of higher-power terms shows that this is nevertheless a legitimate approximation to $F-F'$.

The argument can be made much more rigorous and powerful in the following way. The average value of e^x when x is a random variable always exceeds or equals the exponential of the average value of x, as long as x is real and the weights used in the averaging process are positive. That is,

$$\langle e^x\rangle \geq e^{\langle x\rangle} \tag{11.6}$$

where $\langle x\rangle$ means the weighted average of x. This follows because he curve of e^x is concave upward, as shown in Fig. 11-1, so that if a number of masses (weights) lie along this curve, the center of gravity of these masses — the point with coordinates $(\langle x\rangle, \langle e^x\rangle)$ — lies above the curve. The vertical height of this center of gravity is the average vertical

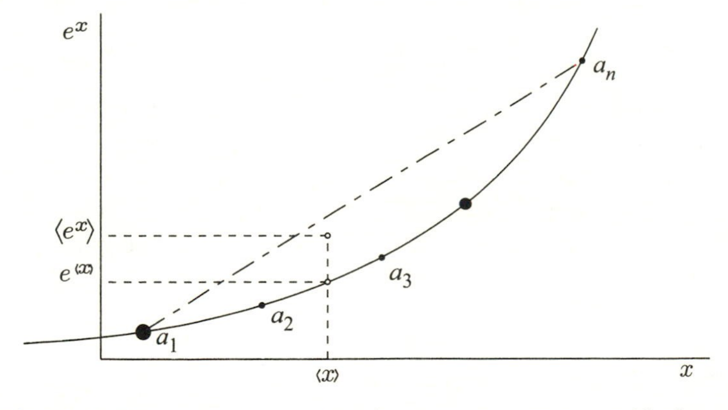

Fig. 11-1 We assume the weighing factors a_i are positive and look on them as different masses positioned along a string. Then the exponential of the weighted average of x, that is, $e^{\langle x\rangle}$, must lie below the weighted average of the exponentials $\langle e^x\rangle$ because of the concave nature of the curve e^x. The value of $e^{\langle x\rangle}$ must lie on the curve, but $\langle e^x\rangle$, the center of gravity of the several points, must lie under the dot-dashed line and above the curve.

position $\langle e^x \rangle$ of the points. It exceeds $e^{\langle x \rangle}$, the ordinate of the curve e^x at the abscissa position of the center of gravity, which is the average value $\langle x \rangle$.

On the left-hand side of Eq. (11.5) we take the average value of $e^{S-S'}$ over paths with the positive weight $e^{S'}$, where S' and S are real. Hence, by Eq. (11.6), this exceeds the quantity $e^{\langle S-S' \rangle}$, where $\langle S - S' \rangle$ is the average of $S - S'$ with this same weighting scheme, namely, with the weight $e^{S'}$. That is,

$$\langle S - S' \rangle = \frac{\iint (S - S') e^{S'} \, \mathcal{D}x(u) \, dx_a}{\iint e^{S'} \, \mathcal{D}x(u) \, dx_a} \tag{11.7}$$

We have then

$$e^{-\beta(F-F')} \geq e^{\langle S-S' \rangle} \tag{11.8}$$

This result implies that

$$F \leq F' - \frac{1}{\beta} \langle S - S' \rangle \tag{11.9}$$

Our final result is then

$$F \leq F' - \delta \tag{11.10}$$

where

$$\delta = \frac{1}{\beta} \frac{\iint (S - S') e^{S'} \, \mathcal{D}x(u) \, dx_a}{\iint e^{S'} \, \mathcal{D}x(u) \, dx_a} \tag{11.11}$$

It is very fortunate that we have a minimum principle here. It says that, if we calculate $F' - \delta$ for various "actions" S', that calculation which gives the smallest result is nearest to the true free energy F.[†] The energy F is actually obtained, of course, if $S' = S$; but we can guess that if S and S' differ in some sense to a first order of smallness, then the deviation of $F' - \delta$ from F must be of second order.

If only a reasonable general form of S' can be guessed but certain parameters still remain uncertain, the calculation of $F' - \delta$ can be made leaving these parameters undetermined. Then the nearest approximation to F will be the lowest $F' - \delta$ available. That is, the "best" values of the parameters are those which minimize $F' - \delta$, "best" in the sense that the resultant $F' - \delta$ differs least from the true F.

[†]It is worth emphasizing again that neither S nor S' is an action functional in the proper physical meaning of the term, since both are defined with the variable u used as the "time" variable. However, operations with path integrals are the same for these functionals as for proper physical actions used previously.

This same minimum principle can be used to find an approximate value for the lowest energy state of the system, E_0. Recall that

$$Z = e^{-\beta F} = \sum_{n=0}^{\infty} e^{-\beta E_n} \tag{11.12}$$

As the temperature of the system becomes lower and lower, that is, as β grows larger and larger, terms involving higher values of energy become less and less important in this series. Eventually the series for Z is dominated by the term of smallest energy, $e^{-\beta E_0}$. That is,

$$\lim_{\beta \to \infty} Z = e^{-\beta E_0} \tag{11.13}$$

Now following the line of argument developed in the preceding paragraphs, we can simply replace F with E_0. We define E_0' as the result of the path integral involving the new action S' and finally derive

$$E_0 \leq E_0' - \delta \tag{11.14}$$

as an approximation in the limit of large β.

In approximating E_0 by this technique, our task is somewhat simpler than it was for the free energy F. Specifically, we can disregard the specification that the initial and final points of the paths be the same. To understand this, we refer back to Eq. (10.28) and note that as β becomes large the density matrix $\rho(x', x)$ is also dominated by the zero-order term and approaches $\phi_0(x')\phi_0^*(x)e^{-\beta E_0}$. Thus the dependence on x' and x enters into a multiplying factor but does not affect the nature of the exponential behavior of the function. It is this exponential behavior which is fundamental in the evaluation of E_0 by this technique.

1-2 AN APPLICATION OF THE VARIATIONAL METHOD

As an example of the evaluation of a partition function using this variational principle, consider the example of a single particle constrained to move in one dimension. Using the approach developed in Chap. 10, we write the action for such a particle as

$$S = -\int_0^\beta \left[\frac{m}{2}\dot{x}^2(u) + V(x(u))\right] du \tag{11.15}$$

So the partition function is

$$Z = \int_{-\infty}^{\infty} \int_{x_a}^{x_a} \exp\left\{-\int_0^\beta \left[\frac{m}{2}\dot{x}^2(u) + V(x(u))\right] du\right\} \mathcal{D}x(u)\, dx_a \tag{11.16}$$

This path integral is over paths which return to the initial starting points; and after the path integral has been evaluated, a further integration over all possible starting points is carried out.

In Sec. 10-2 we considered this same problem and pointed out how the classical approximation may be derived by inspection. In the classical limit of high temperatures, or high values of kT compared with $\hbar$, the value of $\beta\hbar$ is so small that paths which get very far away from x_a do not contribute. Thus, the potential can be replaced by the constant value $V(x_a)$, and the path integral contributes only a constant, giving

$$Z_{\text{classical}} = e^{-\beta F_{\text{classical}}} = \sqrt{\frac{m}{2\pi\beta}} \int_{-\infty}^{\infty} e^{-\beta V(x)}\, dx \tag{11.17}$$

as shown in Eq. (10.48).

In Sec. 10-3, one quantum-mechanical improvement was made on the classical result by expanding the potential about the average position of the path and using terms up through the second order in this expansion. Then a still greater improvement was achieved by using an effective potential U, developed through a particular averaging process. From the point of view of this chapter, we see that that approach was a special application of the variational method. To clarify this point, we shall review the key steps using the notation and concepts of this chapter.

Thus we wish to derive a suitable trial function $W(\bar{x})$, a substitute for the potential, where $\bar{x}$ is the average position of a path defined by

$$\bar{x} = \frac{1}{\beta} \int_0^\beta x(u)\, du \tag{11.18}$$

Along any particular path, this $W(\bar{x})$ is a constant, so that the new form of the action along that path becomes

$$S'[x(u)] = -\frac{m}{2} \int_0^\beta \dot{x}^2(u)\, du - \beta W(\bar{x}) \tag{11.19}$$

With this more general form, it is possible to calculate both F' and $\langle S - S' \rangle$.

Proceeding along this course, we use Eq. (11.11). Substituting into this expression, we have

$$\delta = \frac{-\displaystyle\iint \left[\frac{1}{\beta}\int_0^\beta V(x(u'))\, du' - W(\bar{x})\right] e^{S'[x(u)]}\, \mathcal{D}x(u)\, dx_a}{\displaystyle\iint e^{S'[x(u)]}\, \mathcal{D}x(u)\, dx_a} \tag{11.20}$$

where

$$e^{S'[x(u)]} = \exp\left\{-\frac{m}{2}\int_0^\beta \dot{x}^2(u)\,du\right\}\exp\{-\beta W(\bar{x})\}$$

It is to be remembered that the paths to be used in the path integrals of Eq. (11.20) are those which have the same initial and final points, and, as in Eq. (11.16), a further integration over all end points x_a is to be carried out.

Note that the numerator of δ is quite similar to the term $I(\bar{x})$ introduced in Eq. (10.64), if we restrict ourselves to paths that have a specific average value $\bar{x}$ and count on integrating over all possible values of $\bar{x}$ at a later stage of the calculation. By the same arguments as were used in the discussion of $I(\bar{x})$, we see that the numerator of δ is independent of u'. We can evaluate the path integrals in both numerator and denominator by the methods used in Chap. 10 and take the answer from Eq. (10.65), remembering that

$$Y = x_a - \bar{x} \tag{11.21}$$

Since the denominator is simply a special form of the expression appearing in the numerator, the result is

$$\delta = \frac{-\displaystyle\int_{-\infty}^{\infty}\int_{-\infty}^{\infty}[V(x_a) - W(\bar{x})]e^{\sigma(x_a,\bar{x})}\,d\bar{x}\,dx_a}{\displaystyle\int_{-\infty}^{\infty}\int_{-\infty}^{\infty} e^{\sigma(x_a,\bar{x})}\,d\bar{x}\,dx_a} \tag{11.22}$$

where

$$e^{\sigma(x_a,\bar{x})} = \exp\left\{-\frac{6m}{\beta}(x_a - \bar{x})^2\right\}\exp\{-\beta W(\bar{x})\}$$

The integral over x_a in the denominator in Eq. (11.22) can be easily evaluated to give $(\pi\beta/6m)^{1/2}$. Furthermore, the integral over the term in the numerator containing the factor $W(\bar{x})$ results in this same multiplying constant. It will be more convenient for our future work if we carry out that particular integration in the numerator and further simplify the resulting expression by defining the function $\overline{V(\bar{x})}$ as

$$\overline{V(\bar{x})} = \sqrt{\frac{6m}{\pi\beta}}\int_{-\infty}^{\infty} V(x_a)\exp\left\{-\frac{6m}{\beta}(x_a - \bar{x})^2\right\}dx_a \tag{11.23}$$

The form of $\overline{V(\bar{x})}$ reveals the quantum-mechanical effect we have introduced. This function is a weighted average of $V(x_a)$ with a gaussian weighting function just like the function $U(x_a)$ defined in Eq. (10.68), and the gaussian spread is again $(\beta\hbar^2/12m)^{1/2}$. For a helium atom at a temperature of 2 K, this spread amounts to about 0.7 Å. At room temperature, however, it is only about 2 per cent of the 2.7-Å diameter of the atom. The value of δ can now be written as

$$\delta = \frac{\displaystyle\int_{-\infty}^{\infty} \left[W(\bar{x}) - \overline{V(\bar{x})}\right] e^{-\beta W(\bar{x})}\, d\bar{x}}{\displaystyle\int_{-\infty}^{\infty} e^{-\beta W(\bar{x})}\, d\bar{x}} \tag{11.24}$$

The next step is to evaluate $W(\bar{x})$ by the requirement that we obtain the minimum value for $F' - \delta$, as shown in Eq. (11.10). F' is given by

$$\begin{aligned} e^{-\beta F'} &= \int_{-\infty}^{\infty}\int_{x_a}^{x_a} e^{S'}\, \mathcal{D}x(u)\, dx_a \qquad (11.25) \\ &= \int_{-\infty}^{\infty}\int_{x_a}^{x_a} \exp\left\{-\frac{m}{2}\int_0^\beta \dot{x}^2\, du - \beta W(\bar{x})\right\} \mathcal{D}x(u)\, dx_a \\ &= \int_{-\infty}^{\infty}\int_{-\infty}^{\infty} e^{-\beta W(\bar{x})} \int_{\bar{x}\text{ fixed}} \exp\left\{-\frac{m}{2}\int_0^\beta \dot{x}^2\, du\right\} \mathcal{D}'x(u)\, d\bar{x}\, dx_a \end{aligned}$$

The path integral is a simple one (see Eq. 11.17) whose value we know to be $\sqrt{m/2\pi\beta}$, so that we obtain

$$e^{-\beta F'} = \sqrt{\frac{m}{2\pi\beta}} \int_{-\infty}^{\infty} e^{-\beta W(\bar{x})}\, d\bar{x} \tag{11.26}$$

The next step, finding the optimum choice for $W(\bar{x})$, requires us to determine the effect of a small variation in the function $W(\bar{x})$ on the value of $F' - \delta$ and set this effect equal to 0. Thus, imagining W to be replaced by

$$W \to W(\bar{x}) + \eta(\bar{x}) \tag{11.27}$$

we find from Eq. (11.26) that the variation in F' is

$$\partial F' = \frac{\int \eta(\bar{x}) e^{-\beta W(\bar{x})}\, d\bar{x}}{\int e^{-\beta W(\bar{x})}\, d\bar{x}} \tag{11.28}$$

and from Eq. (11.24) that the variation in δ is

$$\begin{aligned} \partial\delta = {} & \frac{\displaystyle\int \left\{\eta(\bar{x}) - \beta\eta(\bar{x})\left[W(\bar{x}) - \overline{V(\bar{x})}\right]\right\} e^{-\beta W(\bar{x})}\, d\bar{x}}{\displaystyle\int e^{-\beta W(\bar{x})}\, d\bar{x}} \qquad (11.29) \\ & + \frac{\displaystyle\int \left[W(\bar{x}) - \overline{V(\bar{x})}\right] e^{-\beta W(\bar{x})}\, d\bar{x} \int \beta\eta(\bar{x}) e^{-\beta W(\bar{x})}\, d\bar{x}}{\left(\displaystyle\int e^{-\beta W(\bar{x})}\, d\bar{x}\right)^2} \end{aligned}$$

Finding a stationary value for the right-hand side of Eq. (11.10) requires simply that

$$\partial F' - \partial\delta = 0 \tag{11.30}$$

which will be true if we take

$$W(\bar{x}) = \overline{V(\bar{x})} \tag{11.31}$$

This, in turn, implies that δ is zero and that the upper bound on F has the same form as the classical free energy of Eq. (11.17). However, the potential, in this upper bound, has been replaced by $\overline{V(\bar{x})}$. That is,

$$e^{-\beta F} \geq \sqrt{\frac{m}{2\pi\beta}} \int_{-\infty}^{\infty} e^{-\beta\overline{V(\bar{x})}}\, d\bar{x} \tag{11.32}$$

where $\overline{V(\bar{x})}$, the effective classical potential, is given by Eq. (11.23). For large values of β, the free energy is essentially the same as the lowest energy level E_0; thus we can interpret Eq. (11.32) as providing an approximation to E_0. This means that the variational approach has produced the same result as that obtained in Chap. 10 and shown in Eqs. (10.67) and (10.68).

11-3 THE STANDARD VARIATIONAL PRINCIPLE

There is in quantum mechanics a standard variational principle, called the Rayleigh-Ritz method, which is this: If H is the hamiltonian of the system, whose lowest energy level is E_0, then with f representing any arbitrary function in configuration space Γ,

$$E_0 \leq \frac{\int f^* H f\, d\Gamma}{\int f^* f\, d\Gamma} \tag{11.33}$$

This has wide application and is very easily demonstrated. If the function f is expanded as a series in the eigenfunctions ϕ_n belonging to the hamiltonian, i.e., if $f = \sum_n a_n \phi_n$, it is evident that

$$\frac{\int f^* H f\, d\Gamma}{\int f^* f\, d\Gamma} = \frac{\sum_n |a_n|^2 E_n}{\sum_n |a_n|^2} \tag{11.34}$$

This latter expression is an average of the energy values (with positive weights $|a_n|^2$) which therefore exceeds (or equals) the least value E_0. The principle expressed in Eq. (11.33) has characteristics similar to the principle of Eq. (11.14). In fact, Eq. (11.33) is a special case of Eq. (11.14). (To be more precise, we should restrict this conclusion to those cases for which the hamiltonian H is derived from a lagrangian which does not contain any magnetic field. Under this restriction, then, the conclusion holds.) To see the relation between these two equations, we shall consider the following example:

Suppose the action S is connected with a lagrangian such as

$$L = \frac{m}{2}\dot{x}^2 - V(x) \tag{11.35}$$

where $V(x)$ is independent of t. (Otherwise, of course, there are no fixed energy levels to seek!) We shall limit ourselves to the case of a single variable x, but the general case follows directly. We note here that if the lagrangian contains the term $\dot{x}A$ — for example, if the lagrangian represents a particle in a magnetic field — then Eq. (11.33) is still correct. However, the action S is complex. In this case we suspect that Eq. (11.14) (or some simple modification of this equation) is still valid. However, this has not been proved. So, for the present we shall limit our discussion to a case in which no magnetic field is present. Then in the limit for large values of β we have

$$e^{-\beta E_0} \approx \int \exp\left\{-\frac{m}{2}\int_0^\beta \dot{x}^2(u)\,du - \int_0^\beta V(x(u))\,du\right\} \mathcal{D}x(u) \tag{11.36}$$

Now suppose we use for our trial action S' the form

$$S' = -\frac{m}{2}\int_0^\beta \dot{x}^2(u)\,du - \int_0^\beta V'(x(u))\,du \tag{11.37}$$

which involves some other potential $V'(x)$. This means that

$$S - S' = \int_0^\beta [V'(x(u)) - V(x(u))]\,du \tag{11.38}$$

or

$$\delta = \frac{\displaystyle\int \frac{1}{\beta}\int_0^\beta [V'(x(u)) - V(x(u))]\,du\, e^{S'}\,\mathcal{D}x(u)}{\displaystyle\int e^{S'}\,\mathcal{D}x(u)} \tag{11.39}$$

If we were to define the mean value of any function which depends on the path $x(u)$ in such a manner as this, we would find that the value is nearly independent of u so long as u was not too close to either 0 or β. Therefore, to a sufficient approximation, we can write

$$\delta = \frac{\int [V'(x(\breve{u})) - V(x(\breve{u}))]e^{S'}\,\mathcal{D}x(u)}{\int e^{S'}\,\mathcal{D}x(u)} = \langle V'(x(\breve{u})) - V(x(\breve{u}))\rangle \tag{11.40}$$

where $\breve{u}$ is a "representative" value of u between 0 and β. Following the methods developed earlier, we can evaluate this path integral if we assume that the energy values E'_n and the energy functions $\phi'_n(x)$ belonging to the S' are known. If our path goes from x_a to x_b, for example, we obtain°

$$\langle f[x(\breve{u})]\rangle = \frac{\displaystyle\sum_{m,n} e^{-(\beta-\breve{u})E'_m}\phi'_m(x_b) f_{mn} e^{-\breve{u}E'_n}\phi'^*_n(x_a)}{\displaystyle\sum_n e^{-\beta E'_n}\phi'_n(x_b)\phi'^*_n(x_a)} \tag{11.41}$$

where

$$f_{mn} = \int_{-\infty}^{\infty} \phi'^*_m(x) f(x)\phi'_n(x)\,dx \tag{11.42}$$

But, if β approaches infinity and $\breve{u}$ is likewise large (for example, $\breve{u} = \beta/2$), all the higher exponentials are negligible compared to the exponential involving the lowest energy term E'_0. Thus in the limit

$$\lim_{\beta\to\infty} \langle f\rangle = f_{00} \tag{11.43}$$

This result can be written as

$$\delta = \int_{-\infty}^{\infty} \phi'^*_0(x)V'(x)\phi'_0(x)\,dx - \int_{-\infty}^{\infty} \phi'^*_0(x)V(x)\phi'_0(x)\,dx \tag{11.44}$$

Of course, to use Eq. (11.14) we must subtract this value from E'_0. If H' is the hamiltonian associated with S', that is, if

$$H' = \frac{p^2}{2m} + V'(x) \tag{11.45}$$

then

$$H'\phi'_0(x) = E'_0\phi'_0(x) \tag{11.46}$$

so that

$$E'_0 - \delta = \int \phi'^*_0 H'\phi'_0\,dx + \int \phi'^*_0 V\phi'_0\,dx - \int \phi'^*_0 V'\phi'_0\,dx \tag{11.47}$$

But the true Hamiltonian can be written as

$$H = \frac{p^2}{2m} + V = \frac{p^2}{2m} + V' + V - V' = H' + V - V' \tag{11.48}$$

and this means that

$$E_0 \le \int_{-\infty}^{\infty} \phi'^*_0(x) H\phi'_0(x)\,dx \tag{11.49}$$

where $\phi'_0(x)$ is normalized and is the wave function corresponding to the lowest energy state of the hamiltonian H'. The estimate of the lowest energy level given in Eq. (11.49) involves the arbitrary potential $V'(x)$ only through the wave function $\phi'_0(x)$. Since this potential was arbitrary, so is the wave function $\phi'_0(x)$. Therefore, instead of choosing an arbitrary potential and finding from it the resulting wave function and then proceeding to evaluate Eq. (11.49), we could instead pick the wave function itself and then evaluate Eq. (11.49) without ever bothering about the potential to which this arbitrary wave function belongs. The variable function in this process is then the wave function $\phi'_0(x)$ rather than the potential function $V'(x)$. We find, then, that this result is simply another way of stating the Rayleigh-Ritz method Eq. (11.33)

If the problems such as the one given in this example were the only ones in which the concept expressed in Eq. (11.14) were useful, then there would not be much point to this long discussion. But there are much more complicated integrals for which Eq. (11.14) can be used in a way that, at least as far as we can tell, is not so easily transformable into Eq. (11.33). We shall describe such an example in the next section.

11-4 SLOW ELECTRONS IN A POLAR CRYSTAL[1]

We imagine an electron moving in a polar crystal, such as sodium chloride. The electron interacts with ions, which are not rigidly fixed. Thus, the electron creates in its neighborhood a distortion of the crystal lattice, and if the electron moves about, the region of distortion moves with it. This electron, together with its distorted environment, has been called a *polaron*.

One consequence of the lattice distortion is that the energy of the electron is lowered. Furthermore, since as the electron moves the ions must move to adjust the distortions, the effective inertia of the electron (or, to use the currently accepted term, the mass of the polaron) is higher than simply the mass the electron would obtain if the lattice were composed of rigidly fixed points. The precise motion of such a polaron analyzed quantum-mechanically is exceedingly complicated. We shall, however, make a number of approximations whose justification in the real case may be quite difficult. Nevertheless, we shall arrive at an idealized problem which has been studied by a number of physicists.[2]

[1] R.P. Feynman, Slow Electrons in a Polar Crystal, *Phys. Rev.*, vol. 97, pp. 660–665, 1955.

[2] For example, H. Fröhlich, Electrons in Lattice Fields, *Advanc. Phys.*, vol. 3, pp. 325–361, 1954. References to other works are given in this article.

It has been studied not only because of its possible connection with the real behavior of an electron and a crystal, but also because it represents one of the simplest examples of the interaction of a particle and a field. The path integral variational method has been very successful in the solution of this idealized problem.

First, we note that even if the ions were rigidly fixed in the crystal, the electron would still move in a very complicated potential. In such a case, one can show that there are solutions of the Schrödinger equation for the electron with characteristic wave numbers $\mathbf{k}$. The energy levels of these solutions are generally very complicated functions of the wave number. Nevertheless, we assume that the relation between the energy E and the wave number $\mathbf{k}$ is still a quadratic form, such as

$$E = \frac{\hbar^2 k^2}{2m} \tag{11.50}$$

where m is a constant (not necessarily the mass of an electron in a vacuum). Next, we note that the force which the electron exerts on the lattice is such as to push away the negative ions and attract the positive ions. The motion of these ions will be analyzed by considering them as a set of harmonic oscillators and employing the methods of Chap. 8. However, we shall assume that the only harmonic modes which we need are those with high frequency, in which ions of opposite sign of charge move in opposite directions. The frequency $\omega_{\mathbf{k}}$ of each mode then depends on the wave number $\mathbf{k}$ of the mode. However, we shall neglect this dependence and assume that ω is a constant.

Our object is to find the electrical force generated by a distortion characterized by the wave number $\mathbf{k}$ and find the interaction of the electron in this force. Here, we neglect the atomic structure and treat the material of our crystal as simply a continuous dielectric which carries waves of polarization. If $\mathbf{P}$ is the polarization, written in the form of a longitudinal wave

$$\mathbf{P} = \frac{\mathbf{k}}{k} a_{\mathbf{k}} e^{i\mathbf{k}\cdot\mathbf{r}} \tag{11.51}$$

then the charge density from the ions is

$$\rho(\mathbf{r}) = -\boldsymbol{\nabla}\cdot\mathbf{P} = -ika_{\mathbf{k}}e^{i\mathbf{k}\cdot\mathbf{r}} \tag{11.52}$$

If the potential is $\phi(\mathbf{r})$, we have

$$\nabla^2\phi = -4\pi\rho(\mathbf{r}) \tag{11.53}$$

Thus if $q_{\mathbf{k}}$ is the amplitude of the $\mathbf{k}$th longitudinal running wave, the polarization amplitude $a_{\mathbf{k}}$ is proportional to $q_{\mathbf{k}}$, and the energy of interaction between the wave of polarization and the electron at $\mathbf{x}$ is proportional to the sum over all values of $\mathbf{k}$ of the terms $(q_{\mathbf{k}}/k)e^{i\mathbf{k}\cdot\mathbf{x}}$.

Since the energy and the momentum of the electron are related through $E = p^2/2m$, we can write the lagrangian of the entire system as

$$L = \frac{1}{2}|\dot{\mathbf{x}}|^2 + \frac{1}{2}\sum_{\mathbf{k}}(\dot{q}_{\mathbf{k}}^2 - q_{\mathbf{k}}^2) - \left(\frac{2\sqrt{2}\pi\alpha}{V}\right)^{1/2}\sum_{\mathbf{k}}\frac{1}{k}q_{\mathbf{k}}e^{i\mathbf{k}\cdot\mathbf{x}} \tag{11.54}$$

The first term of this expression is the energy of the electron in a rigid lattice, where $\mathbf{x}$ is its position. The second term is the lagrangian of the oscillations of polarizations taken alone, where it is assumed that all waves of polarization have the same frequency and the coordinate of the $\mathbf{k}$th mode is $q_{\mathbf{k}}$. The last term is the lagrangian of the interaction between the electron and the lattice vibrations, where V represents the volume of the crystal and α is a constant. To simplify writing of all our subsequent formulas, we have written this in dimensionless form. That is, the scales of energy, length, and time are so chosen that not only $\hbar$ but also the common frequency ω of the oscillators and the mass m of the electron are all unity. The coupling constant α is then the dimensionless ratio

$$\alpha = \frac{1}{\sqrt{2}}\left(\frac{1}{\epsilon_\infty} - \frac{1}{\epsilon}\right)\mathrm{e}^2 \tag{11.55}$$

where ϵ and ϵ_∞ are the static and high-frequency dielectric constants, respectively. In a typical case, such as the crystal of NaCl, the value of α is about 5. The values of the energy which we shall calculate are in units of $\hbar\omega$.

Now we can study the quantum-mechanical motion of the electron, solving the motion of the harmonic oscillators completely. For example, the amplitude that the electron starts at $\mathbf{x}_a$ with the oscillators in the ground state and ends a time T later at $\mathbf{x}_b$ with the oscillators still in the ground state is

$$G_{0,0}(b,a) = \int e^{iS}\,\mathcal{D}\mathbf{x}(t) \tag{11.56}$$

where, using Eq. (8.138),

$$S = \frac{1}{2}\int_0^T |\dot{\mathbf{x}}|^2\,dt + \iint\int_0^T\int_0^T \frac{\sqrt{2}\pi\alpha}{k^2}e^{i\mathbf{k}\cdot\mathbf{x}(t)}e^{-i\mathbf{k}\cdot\mathbf{x}(s)}e^{-i|t-s|}\,ds\,dt\,\frac{d^3\mathbf{k}}{(2\pi)^3} \tag{11.57}$$

Performing the integral over wave numbers $\mathbf{k}$ gives

$$S = \frac{1}{2}\int_0^T |\dot{\mathbf{x}}|^2\,dt + \frac{\alpha}{\sqrt{8}}\int_0^T\int_0^T \frac{e^{-i|t-s|}}{|\mathbf{x}(t) - \mathbf{x}(s)|}\,ds\,dt \tag{11.58}$$

The quantity $G_{0,0}(b,a)$ depends upon the initial and final positions of the electron, $\mathbf{x}_a$ and $\mathbf{x}_b$, and upon the time interval we are considering, T. Since this function is a kernel, it is a solution of the Schrödinger wave equation, considered as a function of the time interval T. Therefore, we realize that it will contain frequencies in its exponentials which are proportional to the energy levels E_n. It is the lowest one of these energy levels which we now seek.

In developing our variational principle, as we have explained, we are not interested in the kernel for real time intervals T. Instead, we want quantities such as those which appear in Eq. (11.8) for large values of β. By following all the steps leading to Eq. (11.58), it can be readily shown for imaginary values of the time variable that the resulting kernel has the form

$$K(b,a) = \int e^{S}\,\mathcal{D}\mathbf{x}(t) \tag{11.59}$$

where the variable t (previously called u) goes from 0 to β and

$$S = -\frac{1}{2}\int_0^\beta |\dot{\mathbf{x}}|^2\,dt + \frac{\alpha}{\sqrt{8}}\int_0^\beta\int_0^\beta \frac{e^{-|t-s|}}{|\mathbf{x}(t)-\mathbf{x}(s)|}\,ds\,dt \tag{11.60}$$

This result is just that which one might expect from the replacement of t in Eq. (11.58) by the imaginary time variable $-it$ (previously called $-iu$). Asymptotically, for large values of β, this kernel becomes proportional to $e^{-\beta E_0}$.

We now have a relatively complicated path integral on which to try our variational principle. Next, we shall have to choose some simple action S', which roughly approximates the true action S, and then find E_0' and δ.

We note that in Eq. (11.60) the particle considered at any particular time[1] "interacts" with its position at a past time by a reaction which is inversely proportional to the distance traveled between these two times, and which dies out exponentially with the time difference. The reason for this is that the disturbance set up by the electron in the crystal lattice in the past takes some time to die out. That is, it takes some time for the ions to relax, and during this relaxation period the electron still "feels" the old disturbance.

We shall try an action S' which has this same property, except that instead of involving the inverse distance as a coupling law, the attraction will have the geometric form of a parabolic well. This would be a

[1] Although t in Eq. (11.60) is not really a time, but an integration variable instead, it is useful to think about t as a time, just as we thought of u as a time below Eq. (10.43).

poor approximation if the distance $|\mathbf{x}(t)-\mathbf{x}(s)|$ could very often become exceedingly large. However, since there is a limited time available before the exponential time factor forces the interaction to die out, large values of this difference will not make any important contributions to the integral. Thus, we shall try

$$S' = -\frac{1}{2}\int_0^\beta |\dot{\mathbf{x}}|^2\,dt - \frac{C}{2}\int_0^\beta\int_0^\beta |\mathbf{x}(t)-\mathbf{x}(s)|^2 e^{-w|t-s|}\,ds\,dt \tag{11.61}$$

The constant C is a measure of the strength of the attraction between the electron and the previously created disturbance. We take this as an adjustable parameter. Furthermore, we can with no extra difficulty permit the exponential cutoff law to contain the adjustable parameter w, which may differ from unity. With this extra parameter we can partly compensate for the imperfection which we have introduced by replacing the inverted distance effect by a parabolic effect. (We also note in this regard that adding an extra constant to the parabolic term $|\mathbf{x}(t)-\mathbf{x}(s)|^2$ leads to no further freedom, since such a term would drop out in evaluating a formula for E_0'.) We shall adjust variable parameters C and w later in the evaluation in order to make E_0' a minimum.

Since the action S' we have picked is quadratic, all of the path integrals which result are easily worked out by the methods described in Sec. 3-5.

By comparing Eqs. (11.60) and (11.61), we find that

$$\begin{aligned}
\delta &= \frac{1}{\beta}\langle S - S'\rangle \\
&= \frac{\alpha}{\sqrt{8}\beta}\int_0^\beta\int_0^\beta \left\langle \frac{1}{|\mathbf{x}(t)-\mathbf{x}(s)|}\right\rangle e^{-|t-s|}\,ds\,dt \\
&\qquad + \frac{C}{2\beta}\int_0^\beta\int_0^\beta \langle|\mathbf{x}(t)-\mathbf{x}(s)|^2\rangle e^{-w|t-s|}\,ds\,dt \\
&\equiv A + B
\end{aligned} \tag{11.62}$$

We shall concentrate our attention on the first term on the right-hand side of this equation, A. In this term we can express $|\mathbf{x}(t)-\mathbf{x}(s)|^{-1}$ by a Fourier transform. As a matter of fact, this term is the result of the Fourier transform involved in the step between Eqs. (11.57) and (11.58). So we have

$$\frac{1}{|\mathbf{x}(t)-\mathbf{x}(s)|} = \int \frac{4\pi}{k^2}\exp\{i\mathbf{k}\cdot[\mathbf{x}(t)-\mathbf{x}(s)]\}\,\frac{d^3\mathbf{k}}{(2\pi)^3} \tag{11.63}$$

For this reason we need to study

$$\langle\exp\{i\mathbf{k}\cdot[\mathbf{x}(\tau)-\mathbf{x}(\sigma)]\}\rangle = \frac{\int \exp\{i\mathbf{k}\cdot[\mathbf{x}(\tau)-\mathbf{x}(\sigma)]\}e^{S'}\,\mathcal{D}\mathbf{x}(t)}{\int e^{S'}\,\mathcal{D}\mathbf{x}(t)} \tag{11.64}$$

The integral in the numerator is of the form

$$I = \int \exp\left\{-\frac{1}{2}\int_0^\beta |\dot{\mathbf{x}}|^2\,dt - \frac{C}{2}\int_0^\beta\int_0^\beta |\mathbf{x}(t)-\mathbf{x}(s)|^2 e^{-w|t-s|}\,ds\,dt \right.$$
$$\left. + \int_0^\beta \mathbf{f}(t)\cdot\mathbf{x}(t)\,dt\right\}\,\mathcal{D}\mathbf{x}(t) \tag{11.65}$$

where specifically

$$\mathbf{f}(t) = i\mathbf{k}\delta(t-\tau) - i\mathbf{k}\delta(t-\sigma) \tag{11.66}$$

Now we shall evaluate Eq. (11.65) in so far as it depends on $\mathbf{f}$ or $\mathbf{k}$ aside from a normalization factor which drops out in Eq. (11.64). Incidentally, let us notice that the three rectangular components separate in Eq. (11.65) and we need consider only a scalar case. The method of integration is the same as that introduced in Sec. 3-5 for the evaluation of gaussian path integrals. Thus we substitute $X(t) = \bar{X}(t) + Y(t)$ where $\bar{X}(t)$ is that special function for which the exponent is maximum. The variable of integration is now $Y(t)$. Since the exponent is quadratic in $X(t)$ and $\bar{X}(t)$ renders it an extremum, it can contain $Y(t)$ only quadratically; so $Y(t)$ then separates off as a factor not containing f, which may be integrated to give an unimportant constant (depending on β only). Therefore, within such a constant

$$I = \exp\left\{-\frac{1}{2}\int_0^\beta \dot{\bar{X}}^2(t)\,dt - \frac{C}{2}\int_0^\beta\int_0^\beta [\bar{X}(t)-\bar{X}(s)]^2 e^{-w|t-s|}\,ds\,dt \right.$$
$$\left. + \int_0^\beta f(t)\bar{X}(t)\,dt\right\} \tag{11.67}$$

where $\bar{X}(t)$ is that function which minimizes the expression (subject for convenience to the boundary condition $\bar{X}(0) = \bar{X}(\beta) = 0$). The variational problem gives the integral equation

$$\frac{d^2\bar{X}(t)}{dt^2} = 2C\int_0^\beta [\bar{X}(t)-\bar{X}(s)]e^{-w|t-s|}\,ds - f(t) \tag{11.68}$$

Using Eq. (11.68), Eq. (11.67) can be simplified to

$$I = \exp\left\{\frac{1}{2}\int_0^\beta f(t)\bar{X}(t)\,dt\right\} \tag{11.69}$$

We need merely solve Eq. (11.68) and substitute into Eq. (11.69). To do this, we define

$$Z(t) = \frac{w}{2}\int_0^\beta \bar{X}(s)e^{-w|t-s|}\,ds \tag{11.70}$$

so that

$$\frac{d^2 Z(t)}{dt^2} = w^2[Z(t) - \bar{X}(t)] \tag{11.71}$$

while Eq. (11.68) is

$$\frac{d^2 \bar{X}(t)}{dt^2} = \frac{4C}{w}[\bar{X}(t) - Z(t)] - f(t) \tag{11.72}$$

The equations are readily separated and solved. The solution for $\bar{X}(t)$ substituted into Eq. (11.69) gives, for the case of Eq. (11.66),

$$\begin{aligned} I &= \exp\left\{ik[\bar{X}(\tau) - \bar{X}(\sigma)]\right\} \\ &= \exp\left\{-\frac{2C}{v^3 w}k^2(1 - e^{-v|\tau-\sigma|}) - \frac{w^2}{2v^2}k^2|\tau - \sigma|\right\} \end{aligned} \tag{11.73}$$

where we have defined

$$v^2 = w^2 + \frac{4C}{w} \tag{11.74}$$

The result is correctly normalized, since it is valid for $k = 0$. Upon substitution from Eq. (11.73) into Eq. (11.63) there results an integral over $\mathbf{k}$ which is a simple gaussian, so that substitution into A gives, in the limit $\beta \to \infty$,

$$A = \alpha\frac{v}{\pi^{1/2}} \int_0^\infty \left[w^2\tau + \frac{v^2 - w^2}{v}(1 - e^{-v\tau})\right]^{-1/2} e^{-\tau}\, d\tau \tag{11.75}$$

To find B, we need $\langle|\mathbf{x}(t) - \mathbf{x}(s)|^2\rangle$ This can be obtained by expanding both sides of Eq. (11.73) with respect to $\mathbf{k}$ up to order k^2. Therefore

$$\frac{1}{3}\langle|\mathbf{x}(\tau) - \mathbf{x}(\sigma)|^2\rangle = \frac{4C}{v^3 w}(1 - e^{-v|\tau-\sigma|}) + \frac{w^2}{v^2}|\tau - \sigma| \tag{11.76}$$

The integral in B is now easily performed and, in the $\beta \to \infty$ limit, the expression simplifies to

$$B = \frac{3C}{vw} = \frac{3}{4}\frac{v^2 - w^2}{v} \tag{11.77}$$

In addition we need E_0', the ground-state energy associated with our action S'. This is most easily obtained by noting that, in parallel with Eqs. (11.2) and (11.13),

$$e^{-\beta E_0'} = \lim_{\beta\to\infty} \int_{-\infty}^{\infty} \int_{x_a}^{x_a} e^{S'}\, \mathcal{D}x(u)\, dx_a$$

Differentiating both sides with respect to C, one finds immediately

$$\frac{dE_0'}{dC} = \frac{B}{C} \tag{11.78}$$

so that, in view of Eqs. (11.77) and (11.74), integration gives

$$E_0' = \frac{3}{2}(v - w) \tag{11.79}$$

since $E_0' = 0$ when $C = 0$ (the free particle). Finally, using Eqs. (11.14) and (11.62), we obtain for the true ground state energy

$$E_0 \leq \frac{3}{2}(v - w) - A - B = \frac{3}{4}\frac{(v - w)^2}{v} - A \tag{11.80}$$

with A given in Eq. (11.75). The quantities v and w are two parameters which may be varied separately to obtain a minimum.

The integral in A, unfortunately, cannot be performed in closed form, so that a complete determination of E_0 requires numerical integration. It is, however, possible to obtain approximate expressions in various limiting cases. The choice $w = 0$, corresponding to a fixed harmonic binding potential in Eq. (11.61), leads to

$$A = \alpha\left(\frac{v}{\pi}\right)^{1/2}\int_0^\infty (1 - e^{-v\tau})^{-1/2}e^{-\tau}\,d\tau = \alpha\frac{\Gamma(1/v)}{v^{1/2}\Gamma(1/2 + 1/v)} \tag{11.81}$$

and to $E_0' = 3v/4$. The case of large α corresponds to large v, in which case $e^{-v\tau}$ can be neglected, so that $A = \alpha(v/\pi)^{1/2}$. For α less than 5.8 and $w = 0$, Eq. (11.80) does not give a minimum unless $v = 0$, so that the $w = 0$ case does not give a single expression for all ranges of α. In spite of this disadvantage, the result with Eq. (11.81) is relatively simple and fairly accurate. For $\alpha > 6$, only fairly large values of v are important, and the asymptotic formula (good to 1 per cent for $v > 4$)

$$A = \alpha\left(\frac{v}{\pi}\right)^{1/2}\left(1 + \frac{2\ln 2}{v}\right) \tag{11.82}$$

is convenient. Fröhlich, however, considers the discontinuity at $\alpha = 6$ as a serious disadvantage — a disadvantage which can be avoided in our present approach by choosing w different from zero.

Let us study Eq. (11.80) in case w is not zero. For small α, the minimum will occur for v near w. Therefore, we write $v = (1 + \epsilon)w$, consider ϵ small, and expand the root in Eq. (11.75). This gives

$$A = \alpha\frac{v}{w}\left[1 - \epsilon\int_0^\infty \tau^{-3/2}(1 - e^{-w\tau})e^{-\tau}\,\frac{d\tau}{w\pi^{1/2}} + \cdots\right] \tag{11.83}$$

The integral is

$$(2/w)[(1+w)^{1/2}-1] \equiv P \tag{11.84}$$

The problem of Eq. (11.80) then corresponds, in this order, to minimizing

$$E_0 = \tfrac{3}{4}w\epsilon^2 - \alpha - \alpha\epsilon(1-P) \tag{11.85}$$

That is,

$$\epsilon = \frac{2\alpha(1-P)}{3w} \tag{11.86}$$

which is valid for small α only, because ϵ was assumed small. The resulting energy is

$$E_0 = -\alpha - \frac{\alpha^2(1-P)^2}{3w} \tag{11.87}$$

Our method therefore gives a correction even for small α. It is least for $w = 3$, in which case it gives

$$E_0 = -\alpha - \frac{\alpha^2}{81} = -\alpha - 1.23\left(\frac{\alpha}{10}\right)^2 \tag{11.88}$$

It is not sensitive to the choice of w. For example, for $w = 1$ the 1.23 falls only to 0.98. The method of Lee and Pines[1] gives exactly the result of Eq. (11.88) to this order. The perturbation expansion has been carried out to second order by Haga,[2] who shows that the exact coefficient of the $(\alpha/10)^2$ term should be 1.26, so that our variational method is remarkably accurate for small α.

The opposite extreme of a large α corresponds to large v and, as we shall see, to w near 1. Since $v \gg w$, the integral Eq. (11.75) reduces in the first approximation to Eq. (11.81), which we can use in its asymptotic form. The next approximation in w can be obtained by expanding the radical in Eq. (11.75), considering $w/v \ll 1$. Furthermore, $e^{-v\tau}$ is negligible. In this way we get

$$E_0 = \frac{3}{4}\frac{(v-w)^2}{v} - \alpha\left(\frac{v}{\pi}\right)^{1/2}\left(1 + \frac{2\ln 2}{v} - \frac{w^2}{2v}\right) \tag{11.89}$$

[1] T.-D. Lee and D. Pines, Interaction of a Nonrelativistic Particle with a Scalar Field with Application to Slow Electrons in Polar Crystals, *Phys. Rev.*, vol. 92, pp. 883–889, 1953.

[2] E. Haga, Note on the Slow Electrons in a Polar Crystal, *Prog. Theoret. Phys. (Kyoto)*, vol. 11, pp. 449–460, 1954.

This is minimum, within our approximation of large v, when $w = 1$ and $v = (4\alpha^2/9\pi) - (4\ln 2 - 1)$. Then we find[1]

$$E_0 = -\frac{\alpha^2}{3\pi} - 3\ln 2 - \frac{3}{4} = -0.1061\alpha^2 - 2.829 \tag{11.90}$$

The approximations do not keep E_0 as an upper limit because, unfortunately, the further terms, of order $1/\alpha^2$, are probably positive.

Detailed and numerical work based on this approach has been carried out by T.D. Schultz.[2] Using a digital computer, Schultz worked out values of v and w which would give a minimum for several different values of α. He also evaluated E_0 and compared it with the values which would be obtained from several alternative theories. In particular, he worked out the self-energy from the theories of Lee, Low, and Pines[3] (E_{llp}), Lee and Pines[4] (E_{lp}), Gross[5] (E_g), and Pekar,[6] Bogoliubov,[7] and Tyablikov[8] (E_{pbt}).

The results for α, v, and w and also for the energies given by the Feynman theory (E_f) compared with energies derived from the other theories are given below, in a table reproduced from the paper of Schultz. In this table, both $\hbar$ and ω are assumed to have the value 1. Note that for all values of α, the value of E_f is less than all others.

α	3.00	5.00	7.00	9.00	11.00
v	3.44	4.02	5.81	9.85	15.5
w	2.55	2.13	1.60	1.28	1.15
E_f	−3.1333	−5.4401	−8.1127	−11.486	−15.710
E_{llp}	−3.0000	−5.0000	−7.0000	−9.000	
E_{lp}	−3.10	−5.30	−7.58	−9.95	−12.41
E_g	−3.09	−5.24	−7.43	−9.65	−11.88
E_{pbt}			−6.83	−10.31	−14.7

[1] S.I. Pekar in Theory of Polarons, *Zh. Eksperim. i Teor. Fiz.*, vol. 19, pp. 796–806, 1949, has shown that E_0 goes as $-0.1088\alpha^2$ for the case of large α.

[2] T.D. Schultz, Slow Electrons in Polar Crystals: Self-Energy, Mass, and Mobility, *Phys. Rev.*, vol. 116, pp. 526–543, 1959.

[3] T.-D. Lee, F.E. Low, and D. Pines, The Motion of Slow Electrons in a Polar Crystal, *Phys. Rev.*, vol. 90, pp. 297–302, 1953.

[4] *Op. cit.*

[5] E.P. Gross, Small Oscillation Theory of the Interaction of a Particle and Scalar Field, *Phys. Rev.*, vol. 100, pp. 1571–1578, 1955.

[6] S.I. Pekar, "Untersuchungen über die Elektronentheorie der Kristalle," Akademie-Verlag, Berlin, 1954.

[7] N.N. Bogoliubov, On a New Form of the Adiabatic Theory of Disturbances in the Problem of Interaction of Particles with a Quantum Field, *Ukrainskii Matematicheskii Zhurnal*, vol. 2, no. 2, pp. 3–24, 1950.

[8] S.V. Tyablikov, An Adiabatic Form of Excitation Theory in the Problem of Exchange Effects of a Particle with the Quantum Field, *Zh. Eksperim. i Teor. Fiz.*, vol. 21, pp. 377–388, 1951.

12

Other Problems in Probability

In the preceding chapters we have seen how to use path integrals to treat a number of quantum-mechanical problems which are, by their very physical nature, probabilistic problems. We have also used the path integral method to analyze some aspects of statistical mechanics wherein the probabilistic nature of the functions permitted the path integral technique to be particularly effective. We can continue this line of development into a wide variety of probability problems where this approach has special and valuable applications.

It is the purpose of this chapter to explore a number of these probability problems. They will be of two kinds. First we shall discuss direct applications of path integral ideas to classical probability problems (Sec. 12-1 through 12-6). This is quite different from all preceding chapters, in which all applications were to quantum mechanics. Following that, we shall deal with problems in which both probability and quantum mechanics are involved (Sec. 12-7 through 12-10). We cannot, in this chapter, deal with these matters in any detail. We shall only outline by some examples how certain problems may be set up and thereby suggest to the reader other applications of the path integral approach.

The main direct application of path integrals to probability problems is due to the ability of path integrals to deal directly with the notion of the probability of a path or a function. To make this idea clear, we proceed in steps from the well-known[1] ideas of probability applied to discrete events and to continuous variables.

12-1 RANDOM PULSES

To start with, suppose we consider a typical probability problem for a discrete variable. We are given a situation in which a series of discrete events is taking place at random times, e.g., cosmic rays striking a detector or raindrops falling on a specifically demarked area of ground. We know that the particles fall at random times, but in any long period of time T we expect $\bar{n} = \mu T$ particles will be observed. In other words, μ is the mean counting rate.

Of course, in any actual measurement the exact number of particles n recorded will not, in general, correspond to the expected number. But we can ask directly: "What is the probability of observing a particular number n of particles during a period when the expected number of

[1] Harold Cramér, "Mathematical Methods of Statistics," Princeton University Press, Princeton, N.J., 1951. We assume knowledge of usual probability theory.

particles is $\bar{n}$?" It is given by the Poisson distribution

$$P_n = \frac{\bar{n}^n}{n!} e^{-\bar{n}} \tag{12.1}$$

On the other hand, we might ask a probability question of a different kind. We might, for example, ask: "What is the probability that the interval from one particle impact to the next will be some particular time t?" Actually, there is no correct answer to the question phrased this way. If we were to ask the probability that the time interval will be equal to or greater than t, then we could give an answer (it is $e^{-\mu t}$). That is, we can get an answer to a question about t falling within a certain range. Thus, if we are interested in a particular value, we must allow ourselves an infinitesimal range and ask the question: "What is the (infinitesimal) probability that the time interval will fall within the range dt centered on t?" The answer is written as

$$P(t)\,dt = \mu e^{-\mu t}\,dt \tag{12.2}$$

So we create a concept of a probability distribution of a continuous variable: $P(t)$ is the probability per unit range of t that the interval is t. We write the probability distribution of x as $P(x)$ if $P(x)\,dx$ is the probability that the variable lies in the range dx about x. We can easily extend this to two variables and write the probability distribution of x and y as $P(x,y)\,dx\,dy$. By this we mean that the probability of finding the variables x and y in the region R of the xy plane is given by

$$\int_R P(x,y)\,dx\,dy$$

We wish to expand the concepts of probability still further. We want to consider the distribution not of single variables but of complete curves; i.e., we want to construct probability functions, or rather functionals, which will permit us to answer the question: "What is the probability of obtaining a particular time history of a physical phenomenon, such as the voltage in a resistor or the price of a commodity, or, in two variables, the probability of a certain shape of the surface of the sea as a function of latitude and longitude?" Thus, we are led to consider the *probability of a function.*

We shall write it down this way. The probability of observing the function $f(t)$ is a functional $P[f(t)]$. But we must be careful to remember that questions relating to such a probability have meaning only if we define the range within which we are looking for a specific curve. Just as in the example above we had to ask the question: "What is the probability of finding the time interval within the range dt?" so now we must ask: "What is the probability of finding the function within some more or less restricted class of functions, for example, those curves which are bounded between values a and b for the complete time history in which we are interested?" If we call such a subset of curves the class A, then we ask: "What is the probability of finding $f(t)$ in the class A?" and we write the answer as the path integral

$$\int_A P[f(t)]\,\mathcal{D}f(t) \tag{12.3}$$

where the integral extends over all functions of class A.

Actually, this expression can be thought of as similar to the probability function for a number of different variables. If we imagine time to be divided into discrete intervals (as we imagined it when we were first defining path integrals in Chap. 2) taking on the values of $t_1, t_2, \ldots$, then the values of the function at those particular times $f(t_1), f(t_2), \ldots = f_1, f_2, \ldots$ are analogous to the variables of a multivariable distribution function. The probability of observing a particular curve can then be thought of as the probability of obtaining a particular set of values $f_1, f_2, \ldots$ in the range $df_1, df_2, \ldots$, that is, $P(f_1, f_2, \ldots)\,df_1\,df_2 \cdots$.

If we then proceed to the limit as the number of discrete intervals in time becomes infinite, with suitable normalization, we obtain the probability of observing the continuous curve $f(t)$ in the range $\mathcal{D}f(t)$ as the integrand in the path integral of Eq. (12.3). It is this probability concept and this probability functional with which we shall be working in the remainder of this chapter.

12-2 CHARACTERISTIC FUNCTIONS

It is helpful to continue using the analogy between the probability functional of a path and the more traditional probability function of a variable. A number of concepts, such as the concept of a mean value, are common to the two approaches. With usual probability distributions for quantities which have discrete values, so that the probability of observing the specific number n is P_n, the mean is

$$\sum_{n=1}^{\infty} nP_n = \bar{n} \tag{12.4}$$

For a continuously distributed variable, it is

$$\int_{-\infty}^{\infty} xP(x)\,dx = \bar{x} \tag{12.5}$$

and in an analogous fashion, the mean value of the functional $Q[f(t)]$ is written

$$\frac{\int Q[f(t)]P[f(t)]\,\mathcal{D}f(t)}{\int P[f(t)]\,\mathcal{D}f(t)} = \langle Q \rangle \tag{12.6}$$

In this last equation, as in Sec. 11-1, we have included a path integral in the denominator to remind ourselves that we are always faced with normalizing problems. In principle, it would be possible to work out the path integral of the distribution function, set it equal to 1, and so evaluate the normalizing constant to begin with. However, in many practical cases it is more convenient to leave the function unnormalized and simply cancel out factors on the top and bottom of the expression which might, in actuality, be extremely difficult to evaluate.

Just as the mean value of the function can be expressed in the path integral notation, so can the mean-square value of the function at a particular time, say $t = a$. Thus,

$$\langle f^2(a) \rangle = \frac{\int f^2(a)P[f(t)]\,\mathcal{D}f(t)}{\int P[f(t)]\,\mathcal{D}f(t)} \tag{12.7}$$

for this is only a special functional.

One of the most important mean values of a function, as evaluated with Eq. (12.5), is the mean of e^{ikx}. It is called the characteristic function, and it is

$$\phi(k) = \langle e^{ikx} \rangle = \int_{-\infty}^{\infty} e^{ikx} P(x)\,dx \tag{12.8}$$

This is sometimes also called the moment-generating function. It is simply the Fourier transform of $P(x)$, and it is an extremely useful function for evaluating various characteristics of the distribution, since it is equivalent to a knowledge of the distribution function itself. This last fact is the result of the possibility of performing the inverse transform as

$$P(x) = \int_{-\infty}^{\infty} e^{-ikx} \phi(k)\,\frac{dk}{2\pi} \tag{12.9}$$

A number of important parameters of the distribution can be determined by taking the derivatives of the characteristic function. Thus, for example, the mean value of x is

$$\langle x \rangle = -i\,\frac{d\phi(k)}{dk}\bigg|_{k=0} \tag{12.10}$$

as is readily demonstrated by differentiating each expression in Eq. (12.8) with respect to k and then setting $k = 0$. In fact, a series of such relations exists:

$$\phi(0) = 1 \qquad \phi'(0) = i\langle x\rangle \qquad \phi''(0) = -\langle x^2\rangle \qquad \cdots \tag{12.11}$$

Of course, our next step is to generalize the concept of the characteristic function to the functional distribution case. We construct a mathematical definition of such a generalization by returning to our picture of discrete time intervals. We then wish to perform the Fourier transform on the probability function of a large number of variables, using the kernel $e^{ik_1f_1}e^{ik_2f_2}\cdots$. As we go to the limit of an infinite number of time intervals, this becomes simply $e^{i\int k(t)f(t)\,dt}$. This, then, is the functional whose mean value we wish to take in order to develop the *characteristic functional.* By using Eq. (12.6), we obtain

$$\Phi[k(t)] = \frac{\int e^{i\int k(t)f(t)\,dt}P[f(t)]\,\mathcal{D}f(t)}{\int P[f(t)]\,\mathcal{D}f(t)} \tag{12.12}$$

This characteristic functional also has important special properties. For example, $\Phi[0] = 1$, and the mean value of the function $f(t)$ evaluated at the particular time $t = a$ is

$$\langle f(a)\rangle = -i\left.\frac{\delta\Phi[k(t)]}{\delta k(a)}\right|_{k(t)=0} \tag{12.13}$$

where we have used the technique of the functional derivative as described in Sec. 7-2.

In principle, we can invert our path integral Fourier transform and write the probability functional as

$$P[f(t)] = \int e^{-i\int k(t)f(t)\,dt}\Phi[k(t)]\,\mathcal{D}k(t) \tag{12.14}$$

where now, of course, the path integral is carried out in the space of the $k(t)$ functions.

We may remark, for use in interpretation later on, that if the function $f(t)$ is not uncertain but is definitely known to be some particular function $F(t)$, that is, $P[f(t)]$ is zero for all $f(t)$ except $f(t) = F(t)$, then the characteristic functional is

$$\Phi[k(t)] = e^{i\int k(t)F(t)\,dt} \tag{12.15}$$

12-3 NOISE

Suppose we apply the ideas so far developed to a particular example and in the process develop a few more concepts. Let us consider the situation in which we are counting some sort of pulses, perhaps pulses generated by the impact of cosmic rays on a Geiger counter or perhaps thermal-noise pulses in a resistor. In such cases the pulses are not simply discrete spikes of energy but are represented by a rising and falling voltage. Thus, careful inspection of the actual voltage history associated with such a pulse would show that it has the form $g(t)$ for a pulse occurring at $t = 0$. So, if the pulse occurred at t_0, the shape of the voltage curve would be $g(t - t_0)$.

Now, suppose we conduct our counting experiment for the time interval of length T (much longer than the length of a single pulse) during which a number of pulses centered on the times $t_1, t_2, \ldots, t_n$ occurred. The complete voltage history over this experiment would be

$$f(t) = \sum_{j=1}^{n} g(t - t_j)$$

Since we know when all the events occurred, our probability function would simply be the representation of certainty, and by use of Eq. (12.15) the corresponding characteristic functional becomes

$$\Phi[k(t)] = \exp\left\{ i \sum_{j=1}^{n} \int k(t) g(t - t_j)\, dt \right\} \tag{12.16}$$

But now suppose that we wish to determine the probability of finding a particular time history of the voltage before conducting the experiment. In that case we permit the n events to be randomly distributed with uniform probability over the complete time interval. That is, the probability of an event happening within the time interval dt is dt/T. In this case the characteristic functional becomes

$$\begin{aligned} \Phi[k(t)] &= \int_0^T \cdots \int_0^T \int_0^T \exp\left\{ i \sum_{j=1}^{n} \int k(t) g(t - t_j)\, dt \right\} \frac{dt_1}{T} \frac{dt_2}{T} \cdots \frac{dt_n}{T} \\ &= \left(\int_0^T e^{i \int k(t+s) g(t)\, dt} \frac{ds}{T} \right)^n \end{aligned} \tag{12.17}$$

We call the expression in parentheses A and write this result as A^n.

If the number of events in the time interval is distributed in such a way that the Poisson distribution applies, i.e., the occurrence of each event is independent of the time of occurrence of any other event and there is a constant rate μ for the expected number of events per unit time, then the expected number of events in the time interval T is $\bar{n} = \mu T$. The characteristic functional is

$$\Phi[k(t)] = \sum_n A^n \frac{\bar{n}^n}{n!} e^{-\bar{n}} \tag{12.18}$$

The sum on the right-hand side of this equation is the expansion of an exponential function, so that we can write the characteristic functional as

$$\begin{aligned}\Phi[k(t)] = e^{-(1-A)\bar{n}} &= \exp\left\{-\mu T\left(1 - \int_0^T e^{i\int k(t+s)g(t)\,dt}\frac{ds}{T}\right)\right\} \\ &= \exp\left\{-\mu \int_0^T \left(1 - e^{i\int k(t+s)g(t)\,dt}\right) ds\right\}\end{aligned} \tag{12.19}$$

Thus, we may determine the characteristic functional for many different situations. We next go on to discuss this result under various approximate circumstances.

Suppose we imagine that the pulses get very weak while the expected number of pulses per unit time, that is μ, becomes large. In that case $g(t)$ is small, so we can expand $e^{i\int k(t+s)g(t)\,dt}$ in a power series and we can approximate the characteristic functional as

$$\exp\left\{i\mu \int_0^T \int_0^T k(t+s)g(t)\,dt\,ds\right\} = \exp\left\{i\mu G \int_0^T k(t)\,dt\right\} \tag{12.20}$$

where we have defined $G = \int g(t)\,dt$, the area of the pulse. This means that $\Phi[k(t)]$ is in the form of Eq. (12.15) with $F(t) = \mu G$ (a constant independent of t). This is equivalent to saying that $f(t)$ is certainly μG or, in other words, that there is unit probability for observing the function $f(t) = \mu G$ and zero probability for observing any other $f(t)$. That is to say, the pile-up of a large number of small pulses generates a nearly steady direct voltage of value equal to the number of pulses per second μ times the average voltage G supplied by each. Next, we go to one higher approximation and study the fluctuations or irregularities of this nearly constant voltage.

Equation (12.20) is a first-order approximation to the exponential $e^{i\int k(t+s)g(t)\,dt}$ in the description of the characteristic functional of

Eq. (12.19). Suppose now that we go on to the next-order approximation and include the second-order term. This is

$$-\frac{\mu}{2}\iint k(t)g(t+s)\,dt \int k(t')g(t'+s)\,dt'\,ds \tag{12.21}$$

To simplify this expression, we define a function which measures the overlap between two nearby pulses as

$$\lambda(\tau) = \int g(t)g(t+\tau)\,dt \tag{12.22}$$

By use of this substitution, the second-order term is reduced to

$$-\frac{\mu}{2}\int_0^T\int_0^T k(t)k(t')\lambda(t-t')\,dt\,dt' \tag{12.23}$$

Including both first- and second-order terms, the characteristic functional is

$$\Phi[k(t)] = e^{i\mu G\int k(t)\,dt}\,e^{-(\mu/2)\iint k(t)k(t')\lambda(t-t')\,dt\,dt'} \tag{12.24}$$

The first factor in this expression is the constant average level, which we might call the DC level if we are thinking about voltage pulses. We can, if we wish, neglect this level and concentrate only on the variations around it by shifting the origin of $f(t)$. That is, we can always take out a factor $e^{i\int k(t)F(t)\,dt}$ by shifting the origin of $f(t)$ (i.e., by writing $f(t) = F(t) + f'(t)$ and studying the probability distribution of $f'(t)$ and its characteristic functional). If we make this change of origin, we are in a position to study the fluctuations of voltage around the DC level.

We note one special approximation to Eq. (12.24) which is often adequate. Generally, $\lambda(\tau)$ is a narrow function of τ. The pulse shape $g(t)$ rises and falls with a finite width, so if two pulses are spaced a very great distance apart, their overlapping area vanishes. This is another way of saying that $\lambda(\tau)$ approaches 0 rapidly as τ becomes large. As a result of this, if $\lambda(\tau)$ is narrow enough, the second factor in Eq. (12.24) can be approximated by

$$e^{-(q/2)\int k^2(t)\,dt} \tag{12.25}$$

where $q = \mu\displaystyle\int_{-\infty}^{\infty}\lambda(\tau)\,d\tau$. This is equivalent to the probability distribution

$$P[f(t)] = e^{-(1/2\sigma^2)\int f^2(t)\,dt} \tag{12.26}$$

Such fluctuations are often called *gaussian noise.*

Characteristics of distribution functionals describing noise functions have been studied extensively in recent years in the theory of communications. A number of characteristics of noise spectra have been defined and evaluated, and we shall carry through similar discussions here and in the next section, where we treat gaussian noise.

Now we shall continue to show, by giving one further example, how characteristic functionals are set up. We shall consider pulses which come at random times all with a given characteristic shape, say $u(t)$, but each with a different scale height, so a typical pulse is written $au(t)$. We might allow the height a to be either plus or minus. So now we suppose the timings of the pulses are randomly spaced instants t_j and the heights take on random positive and negative values a_j. The resulting function is

$$f(t) = \sum_{j=1}^{n} a_j u(t - t_j) \tag{12.27}$$

If first we set aside the random nature of the events, we obtain a characteristic functional equivalent to that of Eq. (12.16) as

$$\Phi[k(t)] = \exp\left\{ i \sum_{j=1}^{n} a_j \int k(t) u(t - t_j)\, dt \right\} \tag{12.28}$$

Next, if we include the presumed random nature of the scale heights of the pulses and say that the probability of obtaining for the jth pulse a particular scale height of a_j in the region da_j is $p(a_j)\, da_j$, then the characteristic functional becomes

$$\Phi[k(t)] = \int \cdots \int\int \exp\left\{ i \sum_{j=1}^{n} a_j \int k(t) u(t - t_j)\, dt \right\} \times p(a_1)\, da_1\, p(a_2)\, da_2 \cdots p(a_n)\, da_n \tag{12.29}$$

Of course, each of these probability functions for the values of a_j has associated with it a characteristic function (also called a moment-generating function). We call this function

$$W[\omega] = \int_{-\infty}^{\infty} e^{i\omega a} p(a)\, da \tag{12.30}$$

Then the expression for $\Phi[k(t)]$ is

$$\Phi[k(t)] = \prod_j W\left[\int k(t) u(t - t_j)\, dt\right] \tag{12.31}$$

Now we can proceed as we did in the derivation of Eq. (12.17) and introduce the notion that the exact time at which a pulse occurs is randomly and uniformly distributed over the interval $0 \le t \le T$. If we suppose that there are precisely n pulses in this interval, the characteristic functional becomes

$$\Phi[k(t)] = \left(\frac{\tau[k(t)]}{T}\right)^n \tag{12.32}$$

where

$$\tau[k(t)] = \int W\left[\int k(t)u(t-s)\,dt\right]\,ds \tag{12.33}$$

If again we assume, as we did in the derivation of Eq. (12.18), that the pulse distribution satisfies the Poisson distribution, then we must multiply Eq. (12.32) by $(\bar{n}^n/n!)e^{-\bar{n}}$ and sum over n to get

$$\Phi[k(t)] = e^{-\mu(T-\tau[k(t)])} = \exp\left\{-\mu\int\left(1 - W\left[\int k(t)u(t-s)\,dt\right]\right)\,ds\right\} \tag{12.34}$$

As a special example of this result, we assume that that pulse shape is extremely narrow. In fact, we assume that we can approximate the shape function by a Dirac delta function, that is, $u(t) = \delta(t)$. Then the characteristic functional is

$$\Phi[k(t)] = \exp\left\{-\mu\int(1 - W[k(s)])\,ds\right\} \tag{12.35}$$

Next, we assume that the distribution of scale heights is gaussian with zero mean and a root-mean-square value of σ; in other words, the ordinary normal distribution is given by

$$p(a)\,da = \frac{1}{\sqrt{2\pi}\,\sigma}e^{-a^2/2\sigma^2}\,da \tag{12.36}$$

In that case, the characteristic function is

$$W[\omega] = e^{-(\sigma^2/2)\omega^2} \tag{12.37}$$

and for Φ there results

$$\Phi[k(t)] = \exp\left\{-\mu\int\left(1 - e^{-(\sigma^2/2)k^2(s)}\right)\,ds\right\} \tag{12.38}$$

So we find again that a characteristic functional $\Phi[k(t)]$ can be derived to fit our assumed conditions. At any stage in this derivation, approximations that would reduce this to a quadratic form may be valid.

For example, in the case just described a small value of the root-mean-square scale height σ corresponds to weak signals. If, at the same time, the expected number of signals arriving in a time interval is not small, then Eq. (12.38) can be approximated quite well by

$$\Phi[k(t)] = \exp\left\{-\mu\frac{\sigma^2}{2}\int k^2(t)\,dt\right\} \tag{12.39}$$

A distribution like this is called *white noise.*

12-4 GAUSSIAN NOISE

The type of distribution whose characteristic functional is gaussian comes up in many situations, and we shall discuss it here.

We have been working with probability distributions which are gaussian, i.e., exponentials of second order in the defining functions. Although we arrived at this gaussian functional by making a second-order approximation to the exponential term introduced by our assumption of a Poisson distribution of random pulses, it is worth remarking that a number of physical processes actually seem so distributed by their nature. In traditional probability theory the normal, or gaussian, distribution fits physical phenomena which are the result of the combination of a large number of independent events occurring randomly. This is the conclusion of the central-limit theorem of probability theory.[1] The same conclusion applies to distribution functionals and results in the fact that many important cases for study of physical phenomena have gaussian distributions. For further reference, we write here the most general form of a gaussian characteristic functional as

$$\Phi[k(t)] = e^{i\int k(t)F(t)\,dt}e^{-(1/2)\int\int k(t)k(t')A(t,t')\,dt\,dt'} \tag{12.40}$$

The first factor in this expression can be removed by a shift of the origin defining $f(t)$, as we discussed in deriving the distribution of fluctuations of voltage around a DC level. Thus, we could define $f'(t) = f(t) - F(t)$. Next we note that, if the system we are describing behaves in a manner independent of the absolute value of time, then the kernel $A(t,t')$ must have the form $A(t-t')$.

In actual physical situations this function A may be defined by mechanisms in some sort of experimental situation or by approximating a particular piece of reality in such a way that it behaves nearly like the distribution function we are studying. We have an example of such an

[1] *Ibid.*, pp. 213ff.

approximation in the derivations given above on the noise spectrum. For it, $A(t,t') = \mu\lambda(t-t')$. In either case theorems of the behavior of the system which result from the use of this function will be the same so long as the characteristic functional Φ can be suitably approximated by the quadratic or gaussian form of Eq. (12.40).

Of course, by now we know how to deal with gaussian functionals, since we have spend quite a bit of time in the preceding chapters manipulating them in one way or another. In this particular case the appearance of the factor i is different from that in typical quantum-mechanical cases. This means that functions which were real in Sec. 7-4, for example, are imaginary here. However, this does not require any review of the mathematical aspects of the subject; it simply is an awareness of and preparation for certain differences in detail in the results.

The probability distribution which corresponds to the characteristic functional of Eq. (12.40) is

$$P[f(t)] = \exp\left\{-\tfrac{1}{2}\iint [f(t)-F(t)][f(t')-F(t')]B(t,t')\,dt\,dt'\right\} \tag{12.41}$$

where the function $B(t,t')$ is a kernel reciprocal to $A(t,t')$. That is, the functions A and B are related by

$$\int A(t,\tau)B(\tau,s)\,d\tau = \delta(t-s) \tag{12.42}$$

Problem 12-1 Prove this.

All the parameters of the distribution can be calculated from the characteristic functional by the methods introduced in Chap. 7.

We shall now study in more detail some of the physical characteristics of gaussian noise that is time-independent; i.e., we shall study distributions whose characteristic functional is

$$\Phi[k(t)] = e^{-(1/2)\iint k(t)k(t')A(t-t')\,dt\,dt'} \tag{12.43}$$

This function $A(\tau)$ is called the correlation function. Eq. (12.43) means that the probability of observing a particular noise function $f(t)$ is

$$P[f(t)] = e^{-(1/2)\iint f(t)f(t')B(t-t')\,dt\,dt'} \tag{12.44}$$

The function B appearing in this last expression is the *inverse* of the correlation function A. That is, $\int A(t-s)B(s)\,ds = \delta(t)$, or, if

$$\mathcal{P}(\omega) = \int A(\tau)e^{i\omega\tau}\,d\tau \tag{12.45}$$

is the Fourier transform of $A(\tau)$, the Fourier transform of $B(\tau)$ is $1/\mathcal{P}(\omega)$.

We shall begin by calculating some of the properties of this distributional. We first show that the average value of the noise signal vanishes. This is because the average value of the noise function at a particular time $t = a$ is, as in Eq. (12.13),

$$\langle f(a) \rangle = -i \frac{\delta \Phi}{\delta k(a)} \tag{12.46}$$

In this expression, the functional derivative of Φ in Eq. (12.43) is given by (see Sec. 7-2)

$$\frac{\delta \Phi}{\delta k(a)} = \left[- \int k(t) A(t-a)\, dt \right] \Phi \tag{12.47}$$

and, if it is evaluated for the particular function $k(t) = 0$, then it becomes 0.

Next we calculate the average of the square of the noise function or, better, the expected value of the product of two noise functions at times a and b. This is called the correlation function of the noise. It is (by differentiating both sides of Eq. (12.12) twice)

$$\begin{aligned} \langle f(a) f(b) \rangle &= - \frac{\delta^2 \Phi}{\delta k(a) \delta k(b)} \qquad (12.48) \\ &= A(b-a)\Phi - \left[\int k(t) A(t-a)\, dt \right] \left[\int k(t') A(t'-b)\, dt' \right] \Phi \end{aligned}$$

and, if this is evaluated for the function $k(t) = 0$, it is simply $A(b-a)$. That is why A is called the correlation function.

12-5 NOISE SPECTRUM

A most useful characteristic of the noise distribution is the power spectrum of the noise (see Prob. 6-26), which is defined as the mean value of the square of the Fourier transform of the noise function, that is, the mean square of

$$\phi(\omega) = \int f(t) e^{i\omega t}\, dt \tag{12.49}$$

By using our previous results, we can evaluate this as

$$\begin{aligned} \langle |\phi(\omega)|^2 \rangle &= \langle \int f(a) e^{i\omega a}\, da \int f(b) e^{-i\omega b}\, db \rangle \\ &= \int\int \langle f(a) f(b) \rangle e^{i\omega(a-b)}\, da\, db \\ &= \int\int A(b-a) e^{i\omega(a-b)}\, da\, db \\ &= \int \mathcal{P}(\omega)\, da \end{aligned} \tag{12.50}$$

Here we have made use of the function $\mathcal{P}(\omega)$, the Fourier transform of the correlation function A (see Eq. 12.45).

If we carried out the integration shown in the last step of Eq. (12.50) we would, of course, get an infinite result. Therefore, the mean-square value which we are attempting to work out can be defined only for some finite time interval. If we take a unit time interval, then we say that the mean power per second is

$$\text{Mean of } |\phi(\omega)|^2 \text{ per second} = \mathcal{P}(\omega) \tag{12.51}$$

We can apply some of these general results to our special example of noise produced by a multitude of small pulses. The correlation function for our problem is $\mu\lambda(\tau)$ introduced in Eq. (12.22). That is,

$$A(\tau) = \mu \int g(t)g(t+\tau)\,d\tau \tag{12.52}$$

This means that the power spectrum is

$$\mathcal{P}(\omega) = \mu \iint g(t)g(t+\tau)e^{i\omega\tau}\,d\tau\,dt = \mu|\gamma(\omega)|^2 \tag{12.53}$$

where $\gamma(\omega)$ is the Fourier transform of our pulse function $g(t)$. We can explain this simple result more directly for our problem as follows. If the pulses occur at times t_i so that $f(t) = \sum_i g(t - t_i)$, the Fourier transform of $f(t)$ is $\phi(\omega) = \sum_i \gamma(\omega)e^{i\omega t_i}$. Thus the square of $\phi(\omega)$ has the average

$$\langle|\phi(\omega)|^2\rangle = \left\langle |\gamma(\omega)|^2 \sum_{i,j} e^{i\omega(t_i - t_j)} \right\rangle \tag{12.54}$$

But, since the times t_i are random, and independent of t_j for $i \neq j$, all the terms with $i \neq j$ average out, because the average of $e^{i\omega(t_i-t_j)}$ is zero. Only the terms with $i = j$ remain. Each is $|\gamma(\omega)|^2$, and they are μT in number; so the mean of $|\phi(\omega)|^2$ per second is $\mu|\gamma(\omega)|^2$.

In the special case that the characteristic function can be approximated by the white-noise characteristic of Eq. (12.39), the function $A(t - t') = \text{const}\ \delta(t - t')$. This means that $\mathcal{P}(\omega)$ is independent of ω and there is the same "power" per unit frequency range (mean $|\phi(\omega)|^2$ per second) at all frequencies.

The distributions we are describing can very conveniently be described by giving the probability distribution not for $f(t)$ but for its

Fourier transform $\phi(\omega)$ directly, and the characteristic functional not in terms of $k(t)$ but its Fourier transform

$$\mathrm{K}(\omega) = \int k(t) e^{i\omega t}\, dt \tag{12.55}$$

Using these functions, the characteristic functional for the noise distribution corresponding to Eq. (12.43) is

$$\Phi = e^{-(1/2)\int |\mathrm{K}(\omega)|^2 \mathcal{P}(\omega)\, d\omega/2\pi} \tag{12.56}$$

by direct substitution of the inverse of Eq. (12.55) into Eq. (12.43). The corresponding probability functional is

$$P = e^{-(1/2)\int [|\phi(\omega)|^2/\mathcal{P}(\omega)]\, d\omega/2\pi} \tag{12.57}$$

We deduce Eq. (12.57) from Eq. (12.56) as follows. Note that

$$\int k(t) f(t)\, dt = \int \mathrm{K}^*(\omega)\phi(\omega)\, d\omega/2\pi \tag{12.58}$$

so that Eq. (12.14) implies

$$P = \int \Phi e^{-i\int \mathrm{K}^*(\omega)\phi(\omega)\, d\omega/2\pi} \mathcal{D}\mathrm{K}(\omega) \tag{12.59}$$

If we now imagine the possible values of ω to be discrete and separated by an infinitesimal spacing $2\pi\Delta$, the integrals in the exponent in Eqs. (12.56) and (12.57) can be replaced by Riemann sums, and our path integral becomes

$$P = \prod_{\omega} \int e^{-(1/2)|\mathrm{K}(\omega)|^2 \mathcal{P}(\omega)\Delta} e^{-i\mathrm{K}^*(\omega)\phi(\omega)\Delta}\, d\mathrm{K}(\omega) \tag{12.60}$$

The integral for each value of ω can be done separately (by completing the square), and we get

$$P = \prod_{\omega} e^{-(1/2)[|\phi(\omega)|^2/\mathcal{P}(\omega)]\Delta} \tag{12.61}$$

Putting the product together gives Eq. (12.57). It is clear that what happens at one frequency is independent of what happens at another, and that the signal strength $\phi(\omega)$ at frequency ω is distributed as a gaussian with a mean square proportional to $\mathcal{P}(\omega)$.

12-6 BROWNIAN MOTION

It is usually true that the path integral method does not really help to get the solution to problems that cannot be solved in some other manner. Nevertheless, someone who has followed us this far and who is now familiar with path integrals will find its mode of expression and logic very simple and direct when applied to probability problems.

For example, in the theory of brownian motion we might have a linear system — say, a damped harmonic oscillator being driven by a fluctuating force $f(t)$. Assume the mass of the oscillator equal to 1, and we must solve

$$\ddot{x}(t) + \gamma\dot{x}(t) + \omega_0^2 x(t) = f(t) \tag{12.62}$$

where $x(t)$ is the coordinate of the oscillator. If the function $f(t)$ is not known but is given by a known probability distribution $P_f[f(t)]$, what is the probability distribution $P_x[x(t)]$ for the various responses $x(t)$? Equation (12.62) relates $x(t)$ to $f(t)$; that is, for each $f(t)$ there is an $x(t)$. Hence the probability of given x's is the same as that for the corresponding f's, or

$$P_x[x(t)]\,\mathcal{D}x(t) = P_f[f(t)]\,\mathcal{D}f(t) \tag{12.63}$$

where $x(t)$ is related to $f(t)$ via Eq. (12.62). In general, we must be very careful in relating path differentials like $\mathcal{D}x(t)$ to $\mathcal{D}f(t)$, there being an analogue of a "jacobian" between the "volume" elements. But if $f(t)$ and $x(t)$ are linearly related (as above), this jacobian is a constant; so if, as is usual with path integrals, we can trust ourselves to normalize our answer in the end, we have

$$P_x[x(t)] = \text{const}\ P_f[\ddot{x}(t) + \gamma\dot{x}(t) + \omega_0^2 x(t)] \tag{12.64}$$

which gives us a formal solution. If P_f is gaussian, then P_x is and the problem may be worked out in many ways, the most evident being by the method of Fourier series if ω_0^2 and γ are independent of time.

At any rate, many problems can be set up and solved or partly solved by using Eq. (12.64) as a starting point. We shall look at a specific example. A fast particle goes through matter in which it receives small, sharp alterations in velocity as a result of passage by nuclei. After going through a thickness T, what is the probability it will emerge a distance D from the origin (the extension of its original straight-line path) moving with deflection angle θ as in Fig. 12-1?

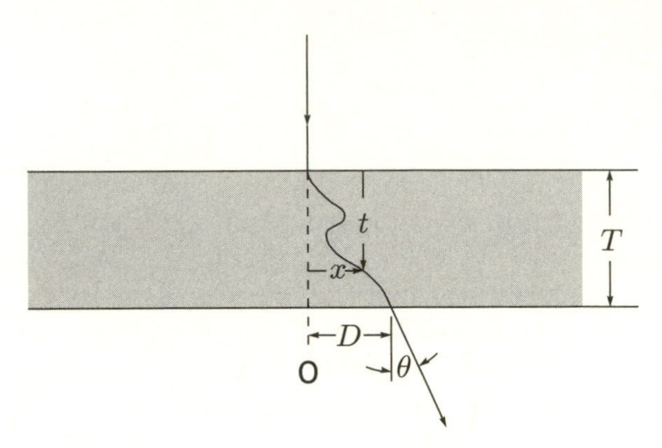

Fig. 12-1 A fast particle impinges perpendicularly on a slab of matter of thickness T. After traveling through a thickness t measured parallel to its original line of flight, it is deflected away from its original trajectory (as extended) by a distance x owing to a number of interactions with the nuclei in the material. Eventually, it emerges from the slab a distance D from the point O, at which it would have emerged with no deflection, and is traveling in a direction that makes an angle θ with its original direction.

We assume that the interactions cause no measurable loss in the longitudinal velocity of the particle and that the matter through which the particle passes is homogeneous. Further, we assume that θ is always small and that the motion is the result of a large number of collisions each of which has a small effect. We assume that the expected number of collisions in the infinitesimal thickness dt is $\mu\, dt$ and that the deflection suffered in each collision is given by the angle Δ, which is governed by the probability distribution $p(\Delta)\, d\Delta$. We further assume that this probability distribution results in a mean-square value of Δ given by

$$\int_{-\infty}^{\infty} \Delta^2 p(\Delta)\, d\Delta = \sigma^2 \tag{12.65}$$

and we shall use the definition $R = \mu\sigma^2$.

We shall confine our attention to the motion as projected onto a two-dimensional plane containing the original path of the particle. Motion in a plane normal to this will follow similar rules, and the motion in either plane can be considered independently of the other. We shall use t to measure the depth of penetration into the slab, θ to represent the instantaneous direction of motion in the plane we are considering, and x to measure the position of the particle away from an extension of its original path of motion, as shown in Fig. 12-1. These parameters are related by $dx = \theta\, dt$, or $\dot{x} = \theta$.

We assume that the deflections of θ occur suddenly, so that $\dot{\theta} = f(t)$, where the functions $f(t)$ are a set of randomly spaced delta functions having random scale heights. This means that $\ddot{x}(t) = f(t)$ and $P_f[f(t)]$ has the characteristic functional (see Eq. 12.34)

$$\Phi[k(t)] = e^{-\mu \int (1-W[k(s)])\, ds} \tag{12.66}$$

where

$$W[\omega] = \int p(\Delta) e^{i\omega\Delta}\, d\Delta \tag{12.67}$$

We note that the mean value of Δ is assumed to be 0, and these deflections themselves are assumed small. Now if we expand $W[\omega]$ as

$$W[\omega] = \int p(\Delta) \left(1 + i\omega\Delta - \frac{\omega^2}{2}\Delta^2 + \cdots\right) d\Delta \tag{12.68}$$

and use terms only through second order in Δ to get $W[\omega] = 1 - \omega^2\sigma^2/2$, then

$$\Phi[k(t)] = e^{-(1/2)R \int k^2(s)\, ds} \tag{12.69}$$

This in turn implies (Eq. 12.44) that

$$P_f[f(t)] = e^{-(1/2R) \int f^2(t)\, dt} \tag{12.70}$$

Hence

$$P_x[x(t)] = \text{const}\ \exp\left\{-\frac{1}{2R}\int_0^T \ddot{x}^2(t)\, dt\right\} \tag{12.71}$$

We wish to evaluate the probability distribution $P(D, \theta)$, which gives the probability that the particle will exit with displacement D and angle of motion θ when it enters with initial conditions $x(0) = 0$ and $\dot{x}(0) = 0$. We are concerned not with the exact path that the particle takes in the material, but only that the particle exits with $x(T) = D$ and $\dot{x}(T) = \theta$. Thus, we express this probability distribution by an integral over all paths as

$$P(D, \theta) = \int \exp\left\{-\frac{1}{2R}\int_0^T \ddot{x}^2(t)\, dt\right\} \mathcal{D}x(t) \tag{12.72}$$

where the paths included in the integral satisfy the assumed end-point conditions. This integral can be carried out by the methods of Sec. 3-5. The integral is a gaussian and becomes an extremum for the path

$$\frac{d^4\bar{x}}{dt^4} = 0 \tag{12.73}$$

The solution of this equation, which satisfies our assumed boundary conditions, is

$$\bar{x}(t) = (3D - \theta T)\left(\frac{t}{T}\right)^2 - (2D - \theta T)\left(\frac{t}{T}\right)^3 \tag{12.74}$$

By using this path in the integrand of the exponential function in Eq. (12.72), we find

$$\frac{1}{2R}\int_0^T \ddot{\bar{x}}^2(t)\,dt = \frac{6}{RT^3}\left(D - \frac{\theta T}{2}\right)^2 + \frac{\theta^2}{2RT} \tag{12.75}$$

which means that our required probability distribution is

$$P(D,\theta) = \text{const}\ \exp\left\{-\frac{6}{RT^3}\left(D - \frac{\theta T}{2}\right)^2 - \frac{\theta^2}{2RT}\right\} \tag{12.76}$$

In some practical cases we may really be concerned not with the exact linear spacing of the particle away from our assumed origin point but, rather, with the deflection angle at which it leaves the slab. Given the overall distribution function of Eq. (12.76), it is simple to evaluate the distribution function in angle alone by integrating over all values of D. The result is $e^{-\theta^2/2RT}$. This is an expected result, because we have already assumed that the mean-square value of the deflection angle which would be acquired in a unit thickness is R, so this value in a total thickness T should be RT.

Suppose next we look only at particles which emerge traveling in a specific angle θ and consider the distribution function of the emerging positions D of those particles. We find that the probability distribution has a maximum at $D = \theta T/2$. This would be the position we would expect if the final deflection angle θ were acquired in a smooth manner as a linear function of thickness starting from 0 and building up to its final value. In that case its average value during the passage through the slab would be $\theta/2$.

Problem 12-2 Show that the constant required to normalize the probability function $P(D,\theta)\,dD\,d\theta$ is

$$\text{const} = \sqrt{\frac{6}{\pi RT^3}}\sqrt{\frac{1}{2\pi RT}} \tag{12.77}$$

12-7 QUANTUM MECHANICS

In this and the following sections we should like to see how to formulate statistical problems in quantum mechanics. In quantum mechanics there are probabilities involved in an intrinsic way, because even a known state implies probabilities to be found in other states. But in addition there may be extrinsic uncertainties. The state, for example, may not be known — we may know only that the state is such and such with a certain probability. This situation is analogous to the classical-mechanics situation in which the initial conditions are not known and only a probability distribution for such conditions is available. We have already dealt with such a situation in statistical mechanics (see Chap. 10), but that is a very special case in which the state of energy E has the probability $e^{-E/kT}$. Here we shall be more general.

Again, under a given external force, say $f(t)$, the behavior of a quantum-mechanical system can be worked out, but what can we say if that force is uncertain and has a probability distribution $P[f(t)]\,\mathcal{D}f(t)$? Need we actually solve the problem for each $f(t)$ and then average, or is there some way to formulate the problem *after* the average of $f(t)$ is taken? (We hope so, because it often occurs that the solution of a statistical problem after an average is taken is, in fact, much easier than finding the general solution of the original problem for a wide range of conditions.) We shall find such a formulation in this section. Then we shall go on to discuss situations in which a quantum-mechanical system is disturbed not just by a classical system but by another quantum-mechanical system about which there are statistical uncertainties.

Our main purpose in this chapter is to show how these and other questions may be formulated. We shall not deal in detail with solving the special problems mentioned; they are brought up only to help us understand the more general formulations we shall arrive at.

We wish first to discuss the analogue of brownian motion for a quantum-mechanical system. That is, we shall suppose that a quantum-mechanical system whose unperturbed action is $S[x(t)]$ is under the influence of an external force $f(t)$ such that its action is[1]

$$S_f[x(t)] = S[x(t)] + \int x(t)f(t)\,dt \tag{12.78}$$

[1] We shall do everything as though there were only one coordinate x. One can immediately generalize to several coordinates x_i (so that a set of forces f_i act) and to cases in which the coefficient in front of $f(t)$ in the action is not simply x but some more complex operator.

Suppose we ask: "What is the probability that, starting at some time t_a with coordinate $x(t_a) = x_a$, we shall arrive at a final time t_b at coordinate x_b?" It is the square of an amplitude: $|K(x_b, t_b; x_a, t_a)|^2$. Or again, if we specify that initially a system is in the state of wave function $\psi(x)$ and finally in the state of wave function $\chi(x)$, the probability of transition from ψ to χ is

$$\begin{aligned} P[\chi(x); \psi(x)] & \qquad (12.79) \\ = \left| \iint \chi^*(x_b) K(x_b, t_b; x_a, t_a) \psi(x_a)\, dx_b\, dx_a \right|^2 & \\ = \iiiint \chi^*(x_b)\chi(x_b') K(x_b, t_b; x_a, t_a) K^*(x_b', t_b; x_a', t_a) \psi(x_a)\psi^*(x_a') & \\ \times dx_b\, dx_b'\, dx_a\, dx_a' & \end{aligned}$$

It is evident that all such problems can be solved if we can evaluate

$$K(x_b, t_b; x_a, t_a) K^*(x_b', t_b; x_a', t_a) \tag{12.80}$$

The first factor is the path integral $\int e^{iS[x(t)]}\, \mathcal{D}x(t)$, whereas the second factor is its complex conjugate[1] $\int e^{-iS[x(t)]}\, \mathcal{D}x(t)$. Each integral is over paths with appropriate end points. In writing the product of Eq. (12.80), we shall call the path variable in the second integral $x'(t)$ and we can then express Eq. (12.80) as the *double path integral*

$$\iint e^{iS[x(t)] - iS[x'(t)]}\, \mathcal{D}x(t)\, \mathcal{D}x'(t) \tag{12.81}$$

The summing of such integrals over various end points gives the required probability.

If the force $f(t)$ is acting, we should replace $S[x(t)]$ in Eq. (12.81) by $S_f[x(t)]$, and the expression becomes

$$\iint e^{i\{S[x(t)] - S[x'(t)] + \int x(t)f(t)\, dt - \int x'(t)f(t)\, dt\}}\, \mathcal{D}x(t)\, \mathcal{D}x'(t) \tag{12.82}$$

But now suppose the force is known only in a probabilistic sense; i.e., we know that there is a probability $P_f[f(t)]\, \mathcal{D}f(t)$ that the force is $f(t)$. Then the probability to go from ψ to χ is given by Eq. (12.79) calculated for each $f(t)$ and then averaged over all $f(t)$ each with the weight $P_f[f(t)]\, \mathcal{D}f(t)$. This is then

$$\begin{aligned} P[\chi(x); \psi(x)] = & \qquad (12.83) \\ \iiiint \chi^*(x_b)\chi(x_b') J(x_b, x_b'; x_a, x_a') \psi(x_a)\psi^*(x_a')\, dx_b\, dx_b'\, dx_a\, dx_a' & \end{aligned}$$

[1] We suppose that $S[x(t)]$ is real and that our units are so defined that $\hbar = 1$, as in Chap. 11.

where J is the average of Eq. (12.82) over all $f(t)$ with weight $P_f[f(t)]\,\mathcal{D}f(t)$; thus

$$J(x_b, x'_b; x_a, x'_a) = \iiint e^{i\{S[x(t)]-S[x'(t)]\}} e^{i\int [x(t)-x'(t)]f(t)\,dt} P_f[f(t)]\,\mathcal{D}x(t)\,\mathcal{D}x'(t)\,\mathcal{D}f(t) \tag{12.84}$$

with the integrals taken between appropriate end points $x(t_a) = x_a$, $x'(t_a) = x'_a$, $x(t_b) = x_b$, $x'(t_b) = x'_b$. Actually, this choosing of end points and then integrating over various values with wave-function distributions depending on the problem (as in Eq. 12.83) is simply a sum of J's for different end conditions, and we shall hereafter simply forget this and speak as though with J we already have our probability — it being left to the reader to remember that a bit more has yet to be done. This is so that we can concentrate on the main feature, the evaluation of the double path integral needed to calculate J.

In this form we can do the integral over $f(t)$ explicitly and see that, to find the probabilities after averaging, we must evaluate a double path integral

$$J = \iint e^{i\{S[x(t)]-S[x'(t)]\}} \Phi[x(t) - x'(t)]\,\mathcal{D}x(t)\,\mathcal{D}x'(t) \tag{12.85}$$

where $\Phi[k(t)]$ is the characteristic functional belonging to the probability distribution P_f, so

$$\Phi[k(t)] = \int e^{i\int k(t)f(t)\,dt} P_f[f(t)]\,\mathcal{D}f(t) \tag{12.86}$$

Equation (12.85) then answers our challenge to express the answer in a form valid after the averaging. It involves evaluation of the double path integral. How to evaluate it is, of course, another question, but the methods discussed in this book may be useful. In these sections we are discussing only how various problems may be formulated.

As an example of the application of Eq. (12.85), suppose $f(t)$ is gaussian noise with zero mean and correlation function $A(t,t')$ as in Eq. (12.40). We must evaluate

$$J = \iint e^{i\{S[x(t)]-S[x'(t)]\}} \times \exp\left\{-\tfrac{1}{2}\iint [x(t)-x'(t)][x(t')-x'(t')]A(t,t')\,dt\,dt'\right\} \mathcal{D}x(t)\,\mathcal{D}x'(t) \tag{12.87}$$

Because in the new factor at least the x and x' appear only quadratically, some of the methods previously discussed for quadratic forms may be useful. Of course, if $S[x]$ is itself quadratic, corresponding to a harmonic oscillator, the path integrals can be evaluated exactly by using the methods of Sec. 3-5.

12-8 INFLUENCE FUNCTIONALS

Now we wish to discuss the behavior of a quantum-mechanical system whose general coordinate we call x in interaction with another quantum-mechanical system whose coordinate we call X.[†] We shall suppose that all measurements which are to be made are on system x only, and no direct measurements of the system X will be made. For example, we may be interested in how an atom makes transitions because it is in the electromagnetic field and can radiate. We contemplate studying only the atom and will not directly measure the light coming from it; then x are the atomic coordinates and X the coordinates of the field. If we study it the other way — that is, if we only observe the light from the atom, emitted, absorbed, or scattered, but never ask for any quantity directly involving the atom's variables — then we may use our present analysis with x being the coordinates of the electromagnetic field and X those of the atom. If, for example, the theory of the index of refraction is wanted, then x are again the field coordinates and X the coordinates of the matter through which the light goes. For one further example, suppose the behavior of an electron in a crystal (or an ion in a liquid) is to be studied: the measurements to be analyzed involve directly only the position of the charge, not the material of the crystal. For example, we might wish the current (electron velocity) generated in some circumstance, but we are not contemplating correlations with the number of phonons produced. Then x can be the coordinates of the electron and X all the other coordinates of the matter of the crystal.

Let $S[x(t)]$ be the action of system x, $S_0[X(t)]$ that of the environmental system alone, and $S_{\text{int}}[x(t), X(t)]$ that of the interaction between the environmental system X and the system of interest x. The action of the combined system is $S[x(t)] + S_0[X(t)] + S_{\text{int}}[x(t), X(t)]$, and the probability of any event involving the combined system can be evaluated from the double path integral, an obvious generalization of Eq. (12.81), and now written as

---

[†] X stands for any number of coordinates — this other system may be, and generally is, very complex. We shall just carry one X variable, but nothing essential will be lost.

$$J = \iiiint \exp\{i(S[x(t)] - S[x'(t)] + S_0[X(t)] - S_0[X'(t)] + S_{\text{int}}[x(t), X(t)] - S_{\text{int}}[x'(t), X'(t)])\} \mathcal{D}x(t)\,\mathcal{D}X(t)\,\mathcal{D}x'(t)\,\mathcal{D}X'(t) \tag{12.88}$$

But, if we need no measurements on system X and if only the dependence on $x(t)$ need ever be studied, then we can write our answer in the form

$$J = \iint e^{i\{S[x(t)] - S[x'(t)]\}} F[x(t), x'(t)]\,\mathcal{D}x(t)\,\mathcal{D}x'(t) \tag{12.89}$$

where we shall call the functional $F[x(t), x'(t)]$ the *influence functional.* It is a functional of the two functions $x(t)$ and $x'(t)$, and for this particular problem it is given by

$$F[x(t), x'(t)] = \sum_{\text{final}} \iint \exp\{i(S_0[X(t)] - S_0[X'(t)] + S_{\text{int}}[x(t), X(t)] - S_{\text{int}}[x'(t), X'(t)])\}\,\mathcal{D}X(t)\,\mathcal{D}X'(t) \tag{12.90}$$

The sum ranges over all possible final states of X. This is because no measurement on X is to be taken, and all final states of the environment are possible. Therefore we must add together the probabilities (i.e., the J functions of Eq. (12.88)) of all. In coordinate representation, for example, $\sum_{\text{final}}$ just means that at the final time t_b after we are no longer interested in the interaction we must take $X(t_b) = X'(t_b) = X_b$ and integrate over all X_b.

To summarize, the behavior of a system in any environment can be discussed in terms of a double path integral like Eq. (12.89), where F is a property of the environment — its "influence" on the system. It summarizes all of the environment that is relevant to $x(t)$. Two different possible surrounding conditions, say, A and B, might physically be very differently constructed; nevertheless, if they happen to lead to the same functional F, they are indistinguishable as far as the behavior of the x system is concerned.

This F is somewhat analogous to the use of "external force" in separating the behavior of interacting systems classically. We can analyze the motion of x alone provided we know what force is produced (as a function of time) by the environment. These newtonian equations of motion of x alone are the rough analogue of Eq. (12.89), whereas Eq. (12.90) corresponds to the calculation of the force produced by a given environment. Two different environments which produce the same force on x are equivalent. Actually, the analogy is only rough; for F contains the entire effect of the environment including the change in behavior of the environment resulting from reaction with x. In the classical analogue, F would correspond to knowing not only what the force *is* as a function of time, but also what it *would be* for every possible motion $x(t)$. The force for a given environmental system depends in general on the motion $x(t)$, of course, since the environmental system is affected by interaction with the system of interest x.

We are therefore led to study the properties of influence functionals. We shall be content to list a few such rules and give some suggestions on how they are arrived at.

Rule I:

$$F[x(t), x'(t)] = F^*[x'(t), x(t)] \tag{12.91}$$

where the asterisk means complex conjugate.

Rule II: If the argument functions $x(t)$ and $x'(t)$ are equal for t exceeding some value t_c, then F does not depend upon the actual values of $x(t)$ for $t > t_c$.

Rule III: If F_i is the influence functional for a particular environment i and we do not know what the environment actually is but know only that the probability of its being i is w_i, then the effective influence functional (for calculating all probabilities) is

$$F = \sum_i w_i F_i \tag{12.92}$$

Rule IV: If the system x is simultaneously in interaction with two external systems A and B, and if A and B do not interact directly with each other, and if there is no correlation between their initial conditions, then

$$F = F_A \cdot F_B \tag{12.93}$$

where F_A is the influence functional if A alone were interacting and F_B is that if B alone were interacting.

Rule V: If the functional F can be adequately approximated by the form

$$F[x(t), x'(t)] = \exp\left\{i \int [x(t) - x'(t)] f(t)\, dt\right\} \tag{12.94}$$

then the system x is acting as though under a classical force $f(t)$ with action of interaction $\int x(t) f(t)\, dt$. If it is of the form

$$F[x(t), x'(t)] = \Phi[x(t) - x'(t)]$$

where $\Phi[k(t)]$ is any functional, then the environment is equivalent to a classical but uncertain force $f(t)$, where Φ is the characteristic functional for the distribution of $f(t)$.

That rule I is true is evident directly from Eq. (12.90).

This expression also explains rule II, but in a much more subtle way. Note that for any given system with any definite action $S_D(X)$ and any given initial state

$$\sum_{\text{final}} \iint e^{i\{S_D[X(t)]-S_D[X'(t)]\}}\, \mathcal{D}X(t)\, \mathcal{D}X'(t) = 1 \tag{12.95}$$

This is because the integrals and the sum over final states are equivalent to

$$\int K(X_b, t_b; X_a, t_a) K^*(X_b, t_b; X'_a, t_a)\, dX_b = \delta(X_a - X'_a) \tag{12.96}$$

by Eq. (4.37). Thus, if the initial wave function were $\psi(X_a)$, we would multiply by $\psi(X_a)\psi^*(X'_a)$ as we did in Eq. (12.79) and integrate to get

$$\int \delta(X_a - X'_a)\psi(X_a)\psi^*(X'_a)\, dX_a\, dX'_a = \int |\psi(X)|^2\, dX = 1 \tag{12.97}$$

Now notice that, if we put $x'(t) = x(t)$ for all time in Eq. (12.90), we have an expression just like Eq. (12.95) where the effective (and definite) action is

$$S_D[X(t)] = S_0[X(t)] + S_{\text{int}}[x(t), X(t)]$$

with

$$S_D[X'(t)] = S_0[X'(t)] + S_{\text{int}}[x(t), X'(t)]$$

as required, as long as $x'(t) = x(t)$. Hence $F[x(t), x(t)] = 1$.

The same argument limited to the time range $t_c \leq t \leq t_b$, using a relation like Eq. (12.96) but with t_a, X_a replaced by t_c, X_c, shows that, if $x'(t) = x(t)$ for $t > t_c$, the dependence of F on $x(t)$ for $t > t_c$ drops away, because the right side of Eq. (12.96) does not depend on $x(t)$ for $t > t_c$.

Rule III is an evident result of the fact that probabilities are determined by adding the value of J over various circumstances.

Rule IV is evident from Eq. (12.90) when it is realized that the conditions of the rule imply that the action that goes into Eq. (12.90) is $S_{0\,A}[X_A(t)] + S_{\text{int}\,A}[x(t), X_A(t)] + S_{0\,B}[X_B(t)] + S_{\text{int}\,B}[x(t), X_B(t)]$ and that the exponential of the sum becomes a product, as does the integral F, if the initial state is itself a product of wave functions.

Rule V is merely a statement of our results shown in Eqs. (12.82) and (12.85).

These are some of the general properties of influence functionals. Calculations with them involve the various methods for doing path integrals applied to Eq. (12.89). We shall conclude this section by discussing certain important influence functionals.

Just as gaussian probability distributions and gaussian noise distributions are simple and important, so influence functionals which depend on $x(t)$, $x'(t)$ as an exponential of a quadratic form — which we shall call gaussian influence functionals — are particularly important.

First, if the environment is a set of harmonic oscillators in their ground states (or at a given temperature) coupled linearly to the system of interest x, evaluation of Eq. (12.90) shows that F is gaussian. But gaussian influence functionals, like gaussian probabilities, occur in good approximation in a much wider class of situations, namely, where the effect is the result of a very large number of influences, each of which by itself has little effect. For example, consider an atom in weak interaction with each of the large number of atoms of an environmental gas. The influence of one atom A is very small, so its influence functional F_A differs only slightly from 1. However, in view of rule IV, the complete F is the product of many such factors, which becomes (nearly) the exponential of the sum of a small contribution from each. This contribution expanded to first and second order in the interaction with each atom leads to influence functionals of the gaussian type.

As an application of this conclusion, a piece of metal placed in a cavity resonator affects the resonator in a simple linear way summarizable by one impedance function, even though the multitude of electrons in the metal behave in such a complex manner. The influence functional of the metal (X) on the cavity oscillator (x) is nearly a gaussian, and to this extent the metal is equivalent to some set of harmonic oscillators which would produce the same influence functional.

The most general exponential functional involving $x(t)$ and $x'(t)$ in linear form is

$$F[x(t), x'(t)] = \exp\left\{i \textstyle\int x(t) f(t)\, dt - i \int x'(t) g(t)\, dt\right\} \tag{12.98}$$

for arbitrary and complex $f(t)$ and $g(t)$. If this is to be an influence functional, however, it must satisfy the conditions of our five rules. Rule I requires $g(t) = f^*(t)$, and rule II implies $g(t) = f(t)$; hence g and f are equal and real. Thus the most general linear functional is that equivalent to the action of a classical external force in accordance with rule V.

We need not discuss this simple case further; for it is completely analyzable just by adding $-x(t)f(t)$ to the hamiltonian of the unperturbed problem. If the exponent has both quadratic and linear terms, the linear term can be factored out, so via rule IV we can say it is a classical force plus the effect of a purely quadratic functional.

The most general exponential functional which involves its arguments purely quadratically is of the form

$$F[x(t), x'(t)] = \exp\left\{-\int_0^T\int_0^t [\alpha(t,t')x(t)x(t') + \beta(t,t')x'(t)x'(t') + \gamma(t,t')x(t)x'(t') + \delta(t,t')x'(t)x(t')]\,dt'\,dt\right\} \tag{12.99}$$

for arbitrary and complex[1] α, β, γ, and δ. The integrals on t', t are over the entire interesting range of time, but we always take $t > t'$. This is no loss of generality, of course, but it is convenient for later analysis. For this to be a satisfactory influence functional, we must have from Rule I

$$\beta(t,t') = \alpha^*(t,t') \tag{12.100}$$

and

$$\gamma(t,t') = \delta^*(t,t') \tag{12.101}$$

Rule II gives us a great deal of information, for putting $x(t) = x'(t)$ for $t > t_c$ and, assuming $t > t_c$, $t' < t_c$, the expression (which is part of the integral in Eq. (12.99))

$$-\int_{t_c}^T\int_0^{t_c} [\alpha(t,t')x(t)x(t') + \beta(t,t')x(t)x'(t') + \gamma(t,t')x(t)x'(t') + \delta(t,t')x(t)x(t')]\,dt'\,dt \tag{12.102}$$

must be independent of $x(t)$ for $t > t_c$ and arbitrary $x(t')$ and of $x'(t')$ for $t' < t_c$. This requires that

$$\begin{aligned}\delta(t,t') &= -\alpha(t,t')\\ \gamma(t,t') &= -\beta(t,t')\end{aligned} \tag{12.103}$$

as long as $t > t_c$, $t' < t_c$. But since t_c is arbitrary, Eqs. (12.103) must hold for all t, t' (under the continuing restriction $t > t'$).

Therefore, the most general gaussian influence functional depends on only one complex function $\alpha(t,t')$ and is of the form

$$F[x(t), x'(t)] = \exp\left\{-\int_0^T\int_0^t [\alpha(t,t')x(t') - \alpha^*(t,t')x'(t')][x(t) - x'(t)]\,dt'\,dt\right\} \tag{12.104}$$

[1] These functions are defined only for $t > t'$.

In the case that $\alpha(t,t')$ is real, say $A(t,t')$, our functional is equivalent to the exponential of Eq. (12.87), and we have the equivalent of a noisy classical perturbation. In quantum-mechanical systems α is generally complex. A special case of importance is when $\alpha(t,t') = \alpha(t-t')$ depends only on the time difference $t - t'$. We are then dealing with an environmental system which has average properties independent of absolute time.

To help understand some of the properties of Eq. (12.104), we shall ask for the probability that the x system makes a transition from energy level n to some other orthogonal level m during an interval of time T in the case that α is very small and we can use perturbation theory. If we expand F in Eq. (12.104), the leading term, 1, gives nothing because the states are orthogonal. The next term, linear in α, has four pieces. One is $-\int_0^T\int_0^t \alpha(t,t')x(t)x(t')\,dt'\,dt$. When this is substituted for F in Eq. (12.89), and the resulting J is used in Eq. (12.83) with $\psi = \phi_n$ and $\chi = \phi_m$, the integral over paths $x(t)$ and $x'(t)$ is seen to be the product of two factors. One, the integral over $x(t)$, involves

$$\int e^{iS[x]}\left[-\int_0^T\int_0^t \alpha(t,t')x(t)x(t')\,dt'\,dt\right]\mathcal{D}x(t)$$

and when passed through Eq. (12.83) results in the transition element (see Chap. 7)

$$\left\langle m\left|-\int_0^T\int_0^t \alpha(t,t')x(t)x(t')\,dt'\,dt\right|n\right\rangle = \\ -\int_0^T\int_0^t \alpha(t,t')\langle m|x(t)x(t')|n\rangle\,dt'\,dt \tag{12.105}$$

The integral over $x'(t)$ is just $\int e^{-iS[x']}\,\mathcal{D}x'$ and results in the complex conjugate of the transition element $\langle m|1|n\rangle$. Analyzing the other three pieces in a similar way, the total transition probability is

$$\begin{aligned} P(n\to m) = \int_0^T\int_0^t & [-\alpha(t,t')\langle m|x(t)x(t')|n\rangle\langle m|1|n\rangle^* \\ & +\alpha(t,t')\langle m|x(t)|n\rangle^*\langle m|x(t')|n\rangle \\ & +\alpha^*(t,t')\langle m|x(t)|n\rangle\langle m|x(t')|n\rangle^* \\ & -\alpha^*(t,t')\langle m|1|n\rangle\langle m|x(t)x(t')|n\rangle^*]\,dt'\,dt \end{aligned} \tag{12.106}$$

If m and n are orthogonal, $\langle m|1|n\rangle = 0$. If $S[x]$ comes from a constant hamiltonian with energy level E_k for state k, then

$$\langle m|x(t)|n\rangle = x_{mn}e^{i(E_m-E_n)t} \tag{12.107}$$

Only the middle two terms of Eq. (12.106) survive, and they are complex conjugates of each other, so that

$$P(n \to m) = 2|x_{mn}|^2 \,\Re e \left\{ \int_0^T \int_0^t \alpha(t,t') e^{-i(E_m - E_n)(t-t')} \, dt' \, dt \right\} \tag{12.108}$$

Problem 12-3 For $m = n$, verify $P(m \to m) = 1 - \sum_n P(m \to n)$ as required by conservation of probability.

In the case of a time-steady environment $\alpha(t,t') = \alpha(t-t')$. Suppose we define the Fourier transform

$$a(\omega) = \int_0^\infty \alpha(\tau) e^{-i\omega\tau} \, d\tau \tag{12.109}$$

($\alpha(t)$ is not defined for $t < 0$.) Then since $P(n \to m)$ in Eq. (12.108) is proportional to the time interval over which the integrals extend, we can define a rate of transition per second and find the probability of transition

$$P(n \to m) \text{ per second} = 2|x_{mn}|^2 a_R(E_m - E_n) \tag{12.110}$$

where we have broken $a(\omega)$ into real and imaginary parts

$$a(\omega) = a_R(\omega) + i a_I(\omega) \tag{12.111}$$

We may note that, for a disturbance by a classical force under gaussian noise, $\alpha(\tau)$ is real (see Eq. 12.87) and the real part of $a(\omega)$ is the power-spectrum function of the noise as defined in Eq. (12.50). So, for such *classical noise* systems

$$a_R(\omega) = a_R(-\omega) \tag{12.112}$$

and in first-order perturbation

$$\text{Rate of transition } n \to m = \text{Rate of transition } m \to n \tag{12.113}$$

and both rates are proportional to the power $\mathcal{P}(\omega)$ at the frequency of the transition. Thus classical forces have equal probability of causing transitions up and down.

Another interesting example is when the environment cannot supply energy with any reasonable probability. For example, it may be initially in the ground state or at zero temperature. We shall call such an environment "cold." For such a situation transitions of the system x going up in energy ($E_m > E_n$) are unlikely. Hence for such *cold environment* systems

$$a_R(\omega) = 0 \qquad \text{for } \omega > 0 \tag{12.114}$$

and for first-order perturbations

$$\text{Rate of transition } n \to m = 0 \qquad \text{if } E_m > E_n \tag{12.115}$$

Since any $a(\omega)$ can be written as the sum of one of the type shown in Eq. (12.112) plus one of the type shown in Eq. (12.114), it is readily apparent that any time-independent gaussian functional is equivalent to a system in some cold environment acted on by a fluctuating classical force described by a gaussian expression. This conclusion follows from the fact that the product of any two gaussian functions is also a gaussian and from rule IV. If the interaction of one environment on the system is represented by $A_1(t, t')$ in the manner of Eq. (12.87) and the interaction of the other environment as $A_2(t, t')$, then the single interaction term in the single resulting gaussian functional is $A_1 + A_2$.

12-9 INFLUENCE FUNCTIONAL FROM A HARMONIC OSCILLATOR

We shall next give an example of how F can be worked out from Eq. (12.90) for an environment consisting of a harmonic oscillator with coordinates X, in the ground state, coupled to x linearly through an interaction $S_{\text{int}}[x(t), X(t)] = C \int x(t)X(t)\,dt$. We suppose the oscillator of unit mass and frequency ω_0, so that

$$S_0[X(t)] = \frac{1}{2}\int [\dot{X}^2(t) - \omega_0^2 X^2(t)]\,dt \tag{12.116}$$

Then

$$F[x(t), x(t')] = \sum_m \iint \exp\left\{i\int [\tfrac{1}{2}\dot{X}^2(t) - \tfrac{1}{2}\omega_0^2 X^2(t) + Cx(t)X(t)]\,dt\right\} \exp\left\{-i\int [\tfrac{1}{2}\dot{X}'^2(t) - \tfrac{1}{2}\omega_0^2 X'^2(t) + Cx'(t)X'(t)]\,dt\right\} \times \mathcal{D}X(t)\,\mathcal{D}X'(t) \tag{12.117}$$

where m is the final state and the initial state is the ground state. The integral over X is clearly gaussian, and in fact we have already done it; for it is exactly the transition amplitude G_{m0} worked out in Sec. 8-9 for a forced harmonic oscillator. The forcing function there called $f(t)$ is here $Cx(t)$.† It is therefore given by Eq. (8.145) with $n = 0$ or

$$G_{m0} = \frac{(i\beta^*)^m}{\sqrt{m!}} G_{00} \tag{12.118}$$

with G_{00} given in Eq. (8.138) and β^* in Eq. (8.143) replacing $f(t)$ by $Cx(t)$. Likewise, the integral over X' is the complex conjugate of a similar expression but with $f(t)$ replaced by $Cx'(t)$ this time. We distinguish values of this substitution with a prime. Then the sum over final states in Eq. (12.117) gives us

$$F[x, x'] = \sum_m G_{m0} G'^*_{m0} = \sum_m \frac{(i\beta^*)^m}{\sqrt{m!}} G_{00} \frac{(-i\beta')^m}{\sqrt{m!}} G'^*_{00} = G_{00} G'^*_{00} e^{\beta^* \beta'} \tag{12.119}$$

Substitution from Eqs. (8.138) and (8.143) produces, as expected, an F of the form of Eq. (12.104) but with

$$\alpha(t, t') = \frac{C^2}{2\omega_0} e^{-i\omega_0(t-t')} \tag{12.120}$$

For example, the terms in xx' in Eq. (12.104) come from the $\beta^*\beta'$ in the exponential; for this product by Eq. (8.143) is

$$\begin{aligned} &\frac{C^2}{2\omega_0} \left[\int_0^T x(t) e^{i\omega_0 t}\, dt \right] \left[\int_0^T x'(t) e^{-i\omega_0 t}\, dt \right] \\ &= \frac{C^2}{2\omega_0} \int_0^T \int_0^t [x(t) x'(t') e^{i\omega_0(t-t')} + x'(t) x(t') e^{-i\omega_0(t-t')}]\, dt'\, dt \end{aligned} \tag{12.121}$$

†The reader may prefer to observe that Eq. (12.117) is

$$F[x(t), x'(t)] = \iiint K(X_b, t_b; X_a, t_a) K'^*(X_b, t_b; X'_a, t_a) \phi_0(X_a) \phi_0^*(X'_a)\, dX_a\, dX'_a\, dX_b$$

where K is the kernel of Eq. (3.66) for a forced harmonic oscillator with $f(t) = Cx(t)$ and K' is that with $f(t) = Cx'(t)$. $\phi_0(X)$ is the wave function of the oscillator in the ground state. All variables X_a, X'_a, and X_b then appear in a simple gaussian way and may be directly integrated. We shall then find it as easy to do the finite-temperature case. For here state n is the initial state with probability proportional to $e^{-\beta E_n}$, so, in view of rule III, the resulting F is obtained by the expression above but with the wave functions $\phi_0(X_a)\phi_0^*(X'_a)$ replaced by

$$\text{const} \sum_n \phi_n(X_a) \phi_n^*(X'_a) e^{-\beta E_n}$$

that is, by the density matrix $\rho(X_a, X'_a)$ worked out in Prob. 10-1. The integrations are again gaussian.

The quantity $a(\omega)$ defined in Eq. (12.109) is therefore (see Eq. (5.17) and the Appendix)

$$a(\omega) = \frac{C^2}{2\omega_0}\int_0^\infty e^{-i\omega_0\tau}e^{-i\omega\tau}\,d\tau = \frac{C^2}{2\omega_0}\left[-i\,\text{P.P.}\frac{1}{\omega_0+\omega} + \pi\delta(\omega_0+\omega)\right] \tag{12.122}$$

so that the real part of $a(\omega)$ is

$$a_R(\omega) = \frac{\pi C^2}{2\omega_0}\delta(\omega_0+\omega) \tag{12.123}$$

This is zero for positive ω. As expected, we have a "cold environment" as specified in Eq. (12.114).

If many independent oscillators at different frequencies are all acting, then by rule IV, their $a_R(\omega)$ functions add; so any cold system (to this gaussian approximation) is equivalent to a continuum of oscillators in their ground state. This follows, since any function $a_R(\omega)$, for negative ω, can be built up of delta functions of the form of Eq. (12.123).

Another interesting example is the interaction with an oscillator at finite temperature. If the temperature is T, the initial state is energy state n with relative probability $e^{-E_n/kT}$. For our case, the absolute probability is

$$w_n = e^{-n\hbar\omega_0/kT}/(1-e^{-\hbar\omega_0/kT}) \tag{12.124}$$

If the initial state were n, the influence functional would be

$$F_n = \sum_m G_{mn}G'^*_{mn} \tag{12.125}$$

instead of the form in Eq. (12.119). Using rule III, we add these with probabilities w_n, so our final F is

$$F = \sum_{m,n} G_{mn}G'^*_{mn}e^{-n\hbar\omega_0/kT}/(1-e^{-\hbar\omega_0/kT}) \tag{12.126}$$

The sum is difficult to work out directly from Eq. (8.145), but it is

$$F = G_{00}G'^*_{00}e^{\beta^*\beta'}\exp\left\{-\frac{(\beta-\beta')(\beta^*-\beta'^*)}{e^{\hbar\omega_0/kT}-1}\right\} \tag{12.127}$$

The $a_R(\omega)$ that results from this in place of Eq. (12.123) is

$$a_R(\omega) = \frac{\pi C^2}{2\omega_0}\left[\frac{e^{\hbar\omega_0/kT}}{e^{\hbar\omega_0/kT}-1}\delta(\omega_0+\omega) + \frac{1}{e^{\hbar\omega_0/kT}-1}\delta(\omega_0-\omega)\right] \tag{12.128}$$

and sums of such expressions of many oscillators constitute the environment. Now transitions can go down in energy ($\omega < 0$) or up in energy.

We note that if $\omega > 0$, the first delta function fails, whereas if $\omega < 0$, the second fails, and that indeed

$$a_R(-|\omega|) = e^{\hbar|\omega|/kT} a_R(+|\omega|) \tag{12.129}$$

This definite relation means that in perturbation theory, if $E_n > E_m$,

$$\frac{\text{probability per second of a transition up } (m \to n)}{\text{probability per second of a transition down } (n \to m)} = e^{-(E_n - E_m)/kT} \tag{12.130}$$

by using Eq. (12.110).

Thus, if the system x occupies states n with relative probabilities $e^{-E_n/kT}$, the net number of up and down transitions will balance out and the system will be in statistical equilibrium for weak perturbation with the environment. This is just what we expect for the laws of statistical equilibrium. Any environment at temperature T producing a quadratic influence functional will have the property of Eq. (12.129).

For an atom as system x in interaction with the electromagnetic field at temperature T as the environment, $a_R(\omega)$ is given by an expression like Eq. (12.128) integrated over all the modes of the field of various frequencies ω_0. It can be split into the cold environment of Eq. (12.123) plus a noisy external force:

$$a_R(\omega) = \frac{\pi C^2}{2\omega_0}\left[\delta(\omega_0 + \omega) + \frac{1}{e^{\hbar\omega_0/kT} - 1}[\delta(\omega_0 + \omega) + \delta(\omega_0 - \omega)]\right] \tag{12.131}$$

The first term produces only transitions down in energy and is called spontaneous emission. The second produces transitions up and down with equal ease and is called induced emission or induced absorption. We say that the transition is induced by an external force or noise whose mean-square strength at frequency ω varies with temperature as $1/(e^{\hbar\omega/kT} - 1)$. This is the way Einstein first discussed the blackbody-radiation laws. As we see here, any environment giving a quadratic influence functional at temperature T (we say it is an environment responding linearly) can be treated in the same way. Many people have extended Einstein's argument to other systems, like the voltage fluctuation noise in a resistor at temperature T. The first term measures the rate at which energy is taken out of our system x in a one-way manner. It measures the amount of "dissipation" produced by the environment (e.g., electrical resistance of a metal or radiation resistance of the electromagnetic field). At temperature T we can then say that things behave as if, in addition to the dissipation, there is a noisy signal generated by the environment whose mean square at each frequency is proportional to the dissipation at that frequency and to $1/(e^{\hbar\omega/kT}-1)$. This is called the *dissipation-fluctuation theorem.*

We cannot pursue this subject further here.[1]

12-10 CONCLUSIONS

In these applications of path integrals to probability theory it is evident that, if the integrands are gaussian, we can make considerable use of the technique. But these problems are precisely those for which other methods, not requiring path integrals, are also available to solve the problem. One may reasonably question the real utility of the path integrals. We can only say that if the problem is not gaussian, it can at least be formulated and studied by using path integrals — and that we might hope that someday, when the techniques of analysis improve, more can be done with it. The only example of a result obtained with path integrals which cannot be obtained in simple manner by more conventional methods is the variational principle discussed in Chap. 11. We hope that further study of these methods may yield more such results.

In the meantime, however, it is worth pointing out that the path integral method does permit a rapid passage from one formulation of a problem to another and often gives a clear or quick suggestion of a relation which can then be more slowly derived in a more ordinary fashion.

With regard to application to quantum mechanics, path integrals suffer most grievously from a serious defect. They do not permit a discussion of spin operators or other such operators in a simple and lucid way. They find their greatest use in systems for which coordinates and their conjugate momenta are adequate. Nevertheless, spin is a simple and vital part of real quantum-mechanical systems. It is a serious limitation that the half-integral spin of the electron does not find a simple and ready representation. It can be handled if the amplitudes and quantities are considered as quaternions instead of ordinary complex numbers, but the lack of commutativity of such numbers is a serious complication.

[1] The subject of influence functionals is discussed in detail by R.P. Feynman and F.L. Vernon, Jr., The Theory of a General Quantum System Interacting with a Linear Dissipative System, *Ann. Phys. (N.Y.)*, vol. 24, pp. 118–173, 1963, and by W.H. Wells, Quantum Formalism Adapted to Radiation in a Coherent Field, *Ann. Phys. (N.Y.)*, vol. 12, pp. 1–40, 1961. An application to calculation of mobility of the polaron is in R.P. Feynman, R.W. Hellwarth, C.K. Iddings, and P.M. Platzmann, Mobility of Slow Electrons in a Polar Crystal, *Phys. Rev.*, vol. 127, pp. 1004–1017, 1962.

Nevertheless, many of the results and formulations of path integrals can be reexpressed by another mathematical system, a kind of ordered operator calculus.[1] In this form many of the results of the preceding chapters find an analogous but more general representation (only for the special problems of Chap. 11 is the generalization not known) involving noncommuting variables. For example, in this chapter discussing influence functionals it must have struck the reader that an environment coupled not to the coordinate x but to a noncommuting operator, such as the spin, would be an important and interesting generalization. Such things cannot be conveniently expressed in the path integral formulation but can be easily expressed in the closely related operator calculus.

An effort to extend the path integral approach beyond its present limits continues to be a worthwhile pursuit; for the greatest value of this technique remains in spite of its limitations, i.e., the assistance which it gives one's intuition in bringing together physical insight and mathematical analysis.

---

[1] R.P. Feynman, An Operator Calculus Having Applications in Quantum Electrodynamics, *Phys. Rev.*, vol. 84, pp. 108–128, 1951.

Appendix

Some Useful Definite Integrals

$$\int_{-\infty}^{\infty} e^{ax^2+bx}\,dx = \sqrt{\frac{\pi}{-a}}e^{-b^2/4a} \tag{A.1}$$

for $\Re e\{a\} \leq 0$ but $a \neq 0$

$$\int_{-\infty}^{\infty} e^{a(x_1-x)^2} e^{b(x_2-x)^2}\,dx = \sqrt{\frac{-\pi}{a+b}} \exp\left\{\frac{ab}{a+b}(x_1-x_2)^2\right\} \tag{A.2}$$

for $\Re e\{a+b\} \leq 0$ but $a+b \neq 0$

$$\int_0^{\infty} \exp\left\{\frac{ia}{x^2} + ibx^2\right\} dx = \sqrt{\frac{i\pi}{4b}} \exp\{i2\sqrt{ab}\} \tag{A.3}$$

for a, b real and positive

$$\int_0^T \exp\left\{\frac{ia}{T-\tau} + \frac{ib}{\tau}\right\} \frac{d\tau}{\sqrt{(T-\tau)\tau^3}} = \sqrt{\frac{i\pi}{bT}} \exp\left\{\frac{i}{T}(\sqrt{a}+\sqrt{b})^2\right\}$$

for a, b real and positive (A.4)

$$\int_0^T \exp\left\{\frac{ia}{T-\tau} + \frac{ib}{\tau}\right\} \frac{d\tau}{\left[\sqrt{(T-\tau)\tau}\right]^3}$$

$$= \sqrt{\frac{i\pi}{T^3}} \frac{\sqrt{a}+\sqrt{b}}{\sqrt{ab}} \exp\left\{\frac{i}{T}(\sqrt{a}+\sqrt{b})^2\right\} \tag{A.5}$$

for a, b real and positive

$$\int_0^{\pi/2} e^{q \sin x} \sin(2x)\,dx = \frac{2}{q^2}[(q-1)e^q + 1] \tag{A.6}$$

$$\left.\begin{aligned}&\int_0^\pi e^{p\cos x}\sin(p\sin x)\sin(ax)\,dx\\&\int_0^\pi e^{p\cos x}\cos(p\sin x)\cos(ax)\,dx\end{aligned}\right\}=\frac{\pi p^a}{2a!} \tag{A.7}$$

$$\int_0^\infty e^{-\lambda x^m}x^k\,dx=\frac{1}{m}\lambda^{-(k+1)/m}\Gamma\left(\frac{k+1}{m}\right) \tag{A.8}$$
$$\text{for } k>-1,\quad \lambda>0,\quad m>0$$

$$\int_{-\infty}^\infty e^{i\omega t}\,dt=2\pi\delta(\omega) \tag{A.9}$$

$$\int_0^\infty e^{i\omega t}\,dt=\pi\delta_+(\omega)=\pi\delta(\omega)+\text{P.P.}\left(\frac{i}{\omega}\right)=\lim_{\epsilon\to 0+}\frac{i}{\omega+i\epsilon} \tag{A.10}$$

$$\int f(\mathbf{k})\frac{d^3\mathbf{k}}{(2\pi)^3}=\frac{1}{\text{Vol}}\sum_{\mathbf{k}}f(\mathbf{k})\qquad\text{[see Sec. 4-3]} \tag{A.11}$$

$$\int_{t_a}^{t_b}\int_{t_a}^{t}f(t,s)\,ds\,dt=\int_{t_a}^{t_b}\int_{t}^{t_b}f(s,t)\,ds\,dt \tag{A.12}$$

Appendix

Notes

These notes were added by the editor to explicate, amplify, or update the book's discussion. A relevant note is signaled in the text through the symbol°.

Throughout: The book is often careless in distinguishing between "probability," "relative (i.e. unnormalized) probability," and "probability density". Similarly for amplitude.

Page 3: In this book a sequence of two events is labeled as a (initial) to b (final); a sequence of three events is labeled as a to c to b; a sequence of four events is labeled as a to d to c to b; and so forth. This scheme for inserting intermediate events proves its value many times. (See particularly pages 21 and 126.)

Page 3: "This particular experiment has never been done in just this way." This statement was true at the date of publication (1965). The remarkable experimental progress since that date can be glimpsed through the following publications:

Claus Jönsson, "Elektroneninterferenzen an mehreren künstlich hergestellten Feinspalten," *Zeitschrift für Physik* **161** (1961) 454–474. Translated as "Electron diffraction at multiple slits," *American Journal of Physics* **42** (1974) 3–11. (Wave-like properties of electrons.)

A. Tonomura, J. Endo, T. Matsuda, T. Kawasaki, and H. Ezawa, "Demonstration of single-electron buildup of an interference pattern," *American Journal of Physics* **57** (1989) 117–120. (Simultaneous wave-like and particle-like properties of electrons.)

Movies of the above experiments are at
<http://www.hqrd.hitachi.co.jp/em/doubleslit.cfm>.

R. Gähler and A. Zeilinger, "Wave-optical experiments with very cold neutrons," *American Journal of Physics* **59** (1991) 316–324. (Wave-like properties of neutrons.)

Olaf Nairz, Markus Arndt, and Anton Zeilinger, "Quantum interference experiments with large molecules," *American Journal of Physics* **71** (2003) 319–325. (Wave-like properties of C_{60}.)

Michael S. Chapman, David E. Pritchard, *et al.*, "Photon scattering from atoms in an atom interferometer: Coherence lost and regained," *Physical Review Letters* **75** (1995) 3783–3787. (Observing atoms as they pass through the double slits, as discussed on page 7 of this book.)

E. Buks, R. Schuster, M. Heiblum, D. Mahalu, and V. Umansky, "Dephasing in electron interference by a 'which-path' detector," *Nature* **391** (1998) 871–874. (More on observing electrons, as or after they pass through the double slits.)

Paul R. Berman, editor, *Atom Interferometry* (Academic Press, San Diego, 1997).

Helmut Rauch and Samuel A. Werner, *Neutron Interferometry: Lessons in Experimental Quantum Mechanics* (Oxford University Press, New York, 2000).

Page 21: In the generalization to time, it helps to think of the holes in Fig. 1-9 as being covered by shutters that open only during specific time intervals. Then a path is specified by a prescription like "through hole E_2 at time slice t_{17}, then through hole D_3 at time slice t_{29}," etc. In the limit that the screens are drilled away to nothingness, the shutters are always open.

Page 22: Feynman's hunch was wrong: in fact other consistent interpretations *are* possible. One such alternative is the de Broglie-Bohm formulation, described in

David Bohm and B.J. Hiley, *The Undivided Universe: An Ontological Interpretation of Quantum Theory* (Routledge, London, 1993).

Page 23: The desired "statistical mechanics of [the] amplifying apparatus" is being worked out under the name of *decoherence*. The vast technical literature of this field is best approached through

W.H. Zurek, "Decoherence, einselection, and the quantum origins of the classical," *Reviews of Modern Physics* **75** (2003) 715–775.

Page 26: The entity called the kernel here is more often called the "propagator" or the "Green's function." See, for example,

L.S. Schulman, *Techniques and Applications of Path Integration* (Wiley, New York, 1981).

Hagen Kleinert, *Path Integrals in Quantum Mechanics, Statistics, Polymer Physics, and Financial Markets*, third edition (World Scientific, River Edge, New Jersey, 2004).

Page 28, problem 2-2: *Hint:* This problem can be solved directly, but it is easier if you first integrate by parts to prove the theorem that, for a harmonic oscillator,

$$S_{cl} = \frac{m}{2}\left[x(t)\dot{x}(t)\right]_{t_a}^{t_b}.$$

Page 28, answer to problem 2-3: If $T = t_b - t_a$, then

$$S_{cl} = \frac{m(x_b - x_a)^2}{2T} + \frac{fT(x_b + x_a)}{2} - \frac{f^2T^3}{24\,m}.$$

Page 33, equation (2.22): (1) This kernel has dimensions 1/[length]. More generally, in s-dimensional configuration space the kernel has dimensions $1/[\text{length}]^s$. (2) The factor A is a complex quantity with phase $\pi/4$ and the dimensions of length. (3) In contrast to the situation for the Riemann sum, the path integral normalizing factor A^{-N} goes to infinity as $\epsilon \to 0$ and the subset of paths becomes more representative. (4) We don't really sum over all paths, but over all paths moving forward in time.

Page 47: Sections 3-2 and 3-3 can be skipped on a first reading.

Page 56, equation (3.40): This probability density is unnormalized. (As reflected by the fact that it has the wrong dimensions!) The normalized version, which is used in Fig. 3-6, is

$$P(x') = \frac{1}{4b_1}\left([C(u_+) - C(u_-)]^2 + [S(u_+) - S(u_-)]^2\right).$$

Page 63, equation (3.60): This expression is correct in magnitude, but the phase (i.e. the branch of $i^{1/2}$) is ambiguous. The correct expression (see for example N.S. Thorber and E.F. Taylor, "Propagator for the simple harmonic oscillator," *American Journal of Physics* **66** (1998) 1022–1024) is

$$e^{-i\theta}\left(\frac{m\omega}{2\pi\hbar\,|\sin\omega T|}\right)^{1/2} \qquad \text{where} \qquad \theta = \frac{\pi}{4}[1 + 2\,\mathrm{trunc}(\omega T/\pi)].$$

Here "trunc" denotes the "truncation" function: $\mathrm{trunc}(x)$ is the largest integer less than or equal to x.

Page 64, problem 3-10: *Hint:* First prove that, for this system,

$$S_{cl} = \frac{m}{2}\left[x\dot{x} + y\dot{y} + \dot{z}^2 t\right]_{t_a}^{t_b}.$$

Page 98: The argument leading from the probability density (5.4) to the wave function (5.5) is suggestive, not definitive. Any argument of this sort cannot uncover the phase of the wave function. This phase might be a physically insignificant constant $e^{i\theta}$, in which case ignoring it would be perfectly permissible. But the phase factor might be a function of momentum $e^{i\theta(p)}$, which does not change the probability density for momentum, $P(p)$, but which can dramatically change the probability density for position, $P(x)$.

Page 107: The argument leading from the probability (5.23) to the amplitude (5.25) has the same defect remarked upon in the previous paragraph.

Page 130: We distinguish between the arbitrary field point $\mathbf{r}$ and the location of the particle $\mathbf{x}$ within that field. In chapter 6 this distinction is largely pedantic, but in chapter 9 (Quantum electrodynamics) it is essential.

Page 142, equation (6.62): This result holds in the limit of long times $\dfrac{mr_b^2}{2\hbar t_b} \to 0$. *Hint:* Use the substitution $x^2 = \dfrac{mR_{bc}^2}{2\hbar(t_b - t_c)}$, and then equation (A.3).

Page 204: This kernel has dimensions $1/[\sqrt{\text{mass}} \times \text{length}]^n$.

Page 217, problem 8-4: *Hint:* First show that (for N odd)

$$L = \frac{1}{4}(\dot{Q}_0^c)^2 + \frac{1}{2}\sum_{\alpha=1}^{(N-1)/2}\left[(\dot{Q}_\alpha^c)^2 - \omega_\alpha^2(Q_\alpha^c)^2 + (\dot{Q}_\alpha^s)^2 - \omega_\alpha^2(Q_\alpha^s)^2\right]$$

Page 217, problem 8-5: It is also worth showing that

$$\frac{\langle\Phi_0|(Q_\alpha^c)^n|\Phi_0\rangle}{\langle\Phi_0|1|\Phi_0\rangle} = \frac{\langle\Phi_0|(Q_\alpha^s)^n|\Phi_0\rangle}{\langle\Phi_0|1|\Phi_0\rangle} = \begin{cases} 0 & \text{for } n \text{ odd} \\ (n-1)!!(\hbar/2\omega_\alpha)^{n/2} & \text{for } n \text{ even} \end{cases}$$

Page 218: Explicitly, in sums over k (as in equations 8.89 and 8.93),

$$\sum_{k=1}^{N} f(k) \quad \text{means} \quad \sum_{\alpha=0}^{N-1} f\left(\frac{2\pi}{L}\alpha\right).$$

Page 233, equation (8.138): This equation is often written in the form (obtained through the use of equation A.12)

$$G_{00} = \exp\left\{-\frac{1}{4\hbar M\omega}\int_0^T\int_0^T f(t)f(s)e^{-i\omega|t-s|}\,ds\,dt\right\}$$

Page 236: Caution! Do not attempt this chapter without first reading chapter 8 (Harmonic oscillators) and working problems 8-3, 8-4, and 8-5. Do this even if you think you aren't interested in harmonic oscillators, and even if you think you already know all about them.

Page 247: The normalization, spelled out in more detail, is

$$\iint \Phi_0^*(\bar{a}_{1,\mathbf{k}'},\bar{a}_{2,\mathbf{k}'})\bar{a}_{1,\mathbf{k}}\bar{a}_{1,\mathbf{k}}^*\Phi_0(\bar{a}_{1,\mathbf{k}'},\bar{a}_{2,\mathbf{k}'})\prod_{\mathbf{k}'} d\bar{a}_{1,\mathbf{k}'}\,d\bar{a}_{2,\mathbf{k}'}.$$

Page 252: For polarization 1, this manipulation results in (using $t = t_c$, $s = t_d$, so that we have our usual sequence of a to d to c to b)

$$\begin{aligned}\lambda_{ML}{}^{1} = \frac{i}{\hbar}\sum_{\mathbf{k}} 2i\frac{\pi}{kc}\int_{t_a}^{t_b} dt_c \int_{t_a}^{t_c} dt_d\, e^{-ikc(t_c-t_d)} \sum_N \int_{-\infty}^{\infty} dx_c \int_{-\infty}^{\infty} dx_d \\ \times e^{-(i/\hbar)E_M(t_b-t_c)}\psi_M^*(\mathbf{x}_c)\bar{j}_{1,\mathbf{k}}(\mathbf{x}_c,t_c)\psi_N(\mathbf{x}_c) \\ \times e^{-(i/\hbar)E_N(t_c-t_d)}\psi_N^*(\mathbf{x}_d)\bar{j}_{1,\mathbf{k}}^*(\mathbf{x}_d,t_d)\psi_L(\mathbf{x}_d)e^{-(i/\hbar)E_L(t_d-t_a)}\end{aligned}$$

Page 309, equation (11.41): The numerator is the kernel from x_a to x_c at "time" $\breve{u}$, times $f(x_c)$, times the kernel from x_c to x_b at "time" β, integrated over all possible values of x_c:

$$\int_{-\infty}^{\infty} k(x_b,\beta;x_c,\breve{u})f(x_c)k(x_c,\breve{u};x_a,0)\,dx_c.$$

Then use expression (10.32) for the kernel.

Index